Application of Intelligent Techniques in Power System Stability, Control and Protection

Application of Intelligent Techniques in Power System Stability, Control and Protection

Veerapandiyan Veerasamy
Shailendra Singh
Sunil Kumar Singh

MDPI

Basel • Beijing • Wuhan • Barcelona • Belgrade • Novi Sad • Cluj • Manchester

Editors
Veerapandiyan Veerasamy
School of Electrical and Electronic Engineering
Nanyang Technological University, Singapore
Singapore

Shailendra Singh
Electrical Engineering Department
National Institute of Technology
Agartala
India

Sunil Kumar Singh
Electrical & Electronics Engineering Department
Shri Ramswaroop Memorial University
Barabanki
India

Editorial Office
MDPI AG
Grosspeteranlage 5
4052 Basel, Switzerland

This is a reprint of articles from the Special Issue published online in the open access journal *Energies* (ISSN 1996-1073) (available at: www.mdpi.com/journal/energies/special_issues/AIT_PSSCP).

For citation purposes, cite each article independently as indicated on the article page online and as indicated below:

Lastname, A.A.; Lastname, B.B. Article Title. *Journal Name* **Year**, *Volume Number*, Page Range.

ISBN 978-3-7258-2326-0 (Hbk)
ISBN 978-3-7258-2325-3 (PDF)
doi.org/10.3390/books978-3-7258-2325-3

Contents

About the Editors

Veerapandiyan Veerasamy

Veerapandiyan Veerasamy (Senior Member, IEEE) graduated with his bachelor's degree (Hons.) from Anna University, Chennai, India, in Electrical and Electronic Engineering in 2013, and his master's degree (Hons.) in Power Systems Engineering in 2015. He received his PhD in Electrical Power Engineering from University Putra Malaysia (UPM), Malaysia, 2021, and received the Graduation on Time Award from the University. He was an Assistant Professor in the Department of Electrical and Electronics Engineering, Rajalakshmi Engineering College, Chennai, India, from 2015 to 2018. He was a Research Fellow at the Clean Energy Research Lab, School of Electrical and Electronic Engineering, Nanyang Technological University (NTU), Singapore, from Dec 2021. He continued his research activities with the Energy Research Institute @ NTU (ERIAN) from Feb 2024 in Electrification Power Grid Center (EPGC), Jurong Island, Singapore. Since Aug 2024, he has been an Adjunct Lecturer at the School of EEE at NTU, Singapore. He has authored and co-authored 50 international journal articles and 20 book chapters at national and international conferences. He received a copyright in Mar 2021 for the code on "Load Flow Analysis using Generalized Linear Hopfield Neural Network via Moore-Penrose Pseudo-Inverse" from University Putra Malaysia, Malaysia.

His research areas include Power Flow Analysis, Optimal Power Flow, Fault and Power Quality Analysis, High Impedance Fault, Voltage and Frequency Control, P2P Energy Trading, DC to DC Converters, Distributed Controls, Design of AI-based Adaptive controllers, Federated Learning, and Recurrent Hopfield Neural Network. Further, he has a field expertise of operating 1-MW Microgrid at EPGC research centre, testing of 1-MW Lion Batteries (both 1st and 2nd Life) for Grid Ancillary Services, EV Chargers, and Power Management Solutions for EV Charging Stations.

Shailendra Singh

Shailendra Singh (S'15) received his Ph.D. degree in Power Systems from the Indian Institute of Technology (Banaras Hindu University), Varanasi, India, and M.Tech. Degree in Electrical Engineering with the specialization in Power Systems from the National Institute of Technology, Kurukshetra, India. He was a Visiting Research Scholar with the National Renewable Energy Laboratory (NREL), Golden, CO, USA, from April 2018 to October 2018. From March 2022 to September 2022, he worked as a Post Doctoral Research Fellow at the Centre for System Intelligence and Efficiency, School of Electrical and Electronics Engineering, Nanyang Technological University (NTU), Singapore. Since October 2022, Dr. Singh has been working as an Assistant Professor in Electrical Engineering Department, National Institute of Technology (NIT) Agartala.

Dr. Singh was the recipient of the Bhaskara Advance Solar Energy (BASE-2018) Fellowship supported by DST, Govt. of India and Indo-US Science and Technology Forum, New Delhi, India. He was awarded the POSOCO Power System Award (PPSA-2020) under Doctoral Category for his research work organized by Power System Operation Cooperation (Presently Grid Controller of India) Ltd in association with Fundamental for Innovation Technology Transfer (FITT), IIT Delhi. Recently, Dr. Singh has been awarded the prestigious Young Professional Research Award at the 3rd International Symposium on Sustainable energy and Technological Advancements during 23–24 Feb 2024 organized by NIT Meghalaya in recognition of his outstanding professional and research achievements in the field of energy technologies.

His research interests include Smart Energy Distribution Systems, Micro-Grid Control, Electric Vehicles Integration, and AI applications in Electrical Energy System.

Sunil Kumar Singh

Sunil Kumar Singh received his Ph.D. degree in Power Systems from the Indian Institute of Technology (Banaras Hindu University), Varanasi, India, and M.Tech. Degree in Electrical Engineering with specialization in Power Systems from the National Institute of Technology, Hamirpur, (H.P.) India.

Since August 2019, Dr. Singh has been working as an Assistant Professor in the Department of Electrical & Electronics Engineering, Shri Ramswaroop Memorial University, Lucknow-Deva Road, Uttar-Pradesh, India. He had served as Vice-chairman of IEEE student branch chapter at Indian Institute of Technology, (BHU). He had served as a Guest Editor for the Special Issues "Application of Intelligent Techniques in Power System Stability, Control and Protection" in *Energies*, MDPI Journal, 2023–2024. He had published more than 20 research articles and book chapters.

His current research interests includes application AI and ML in Power System, Renewable Energy Optimization, and Battery Charging Management in Electric Vehicles.

Preface

The modern power system is going through a significant transition due to global concerns on the use of conventional fossil fuels and meeting the emission limitation protocols. These additional regulations compel developing countries to place greater emphasis on renewable energy resources. Renewable resources are widely dispersed and often located over remote geographical areas. Although these resources are mostly small scale, they are rapidly growing and have not only enhanced the capacity of but also catalyzed the power system. The increasing penetration of renewable energies (RE) into the grid introduces numerous challenges for power utility engineers. The stochastic nature and strong dependencies on the atmospheric condition of renewable resources create stability issues and sometimes cascading failure in the grid system. The context of energy transition presents several challenges associated with grid stability, reliability, protection, and security, as well as energy production/consumption optimization regarding the inclusion of RE resources. Addressing these challenges requires the development of scalable advanced optimization techniques for energy consumption in order to facilitate the penetration (integration) of distributed/centralized renewable energy systems into electric grids, to reduce the peak load, to maintain frequency and voltage stability, and to reinforce grid protection. The advent of prosumers as participants in the grid restricts the application of conventional relaying methodologies in modern power systems. All these factors have forced power engineers to develop advanced and competent artificial intelligence-based control and protection systems to handle all network uncertainties that have evolved due to RE sources.

The central objective of this reprint is to present the vital factors associated with smart grid systems, reliability assessment, protection methodologies, and design advancements through recent research findings to utility engineers, engineering students, research scholars, and power managers.

This reprint includes ten chapters, covering all the prime critical aspects of modern power systems like uncertainty in electricity pricing due to solar or wind systems, distribution system coordination, frequency control challenges due to RE, and artificial intelligence-based protection methodologies. In addition, it also covers optimization issues and challenges of electric vehicles' integration in the modern grid network. Finally, we thank the Editorial team, authors, co-authors, and MDPI; without their valuable support, this reprint would not have been so successful.

Veerapandiyan Veerasamy, Shailendra Singh, and Sunil Kumar Singh
Editors

energies

MDPI

Article

Bidding Strategy for Wind and Thermal Power Joint Participation in the Electricity Spot Market Considering Uncertainty

Zhiwei Liao *, Wenjuan Tao, Bowen Wang and Ye Liu

School of Electric Power Engineering, South China University of Technology, Guangzhou 510640, China
* Correspondence: epliao@scut.edu.cn

Abstract: As the proportion of new energy sources, such as wind power, in the electricity system rapidly increases, their participation in spot market competition has become an inevitable trend. However, the uncertainty of clearing price and wind power output will lead to bidding deviation and bring revenue risks. In response to this, a bidding strategy is proposed for wind farms to participate in the spot market jointly with carbon capture power plants (CCPP) that have flexible regulation capabilities. First, a two-stage decision model is constructed in the day-ahead market and real-time balancing market. Under the joint bidding mode, CCPP can help alleviate wind power output deviations, thereby reducing real-time imbalanced power settlement. On this basis, a tiered carbon trading mechanism is introduced to optimize day-ahead bidding, aiming at maximizing revenue in both the electricity spot market and carbon trading market. Secondly, conditional value at risk (CVaR) is introduced to quantitatively assess the risks posed by uncertainties in the two-stage decision model, and the risk aversion coefficient is used to represent the decision-maker's risk preference, providing corresponding strategies. The model is transformed into a mixed-integer linear programming model using piecewise linearization and McCormick enveloping. Finally, the effectiveness of the proposed model and methods is verified through numerical examples.

Keywords: carbon capture; electricity spot market; bidding strategy; wind and thermal power; conditional value at risk

Citation: Liao, Z.; Tao, W.; Wang, B.; Liu, Y. Bidding Strategy for Wind and Thermal Power Joint Participation in the Electricity Spot Market Considering Uncertainty. *Energies* **2024**, *17*, 1714. https://doi.org/10.3390/en17071714

Academic Editors: Veerapandiyan Veerasamy, Shailendra Singh and Sunil Kumar Singh

Received: 11 March 2024
Revised: 24 March 2024
Accepted: 27 March 2024
Published: 3 April 2024

1. Introduction

In the context of developing a low-carbon economy, augmenting the integration of renewable energy into the power system and decarbonizing traditional fossil energy (e.g., coal and oil) are important means to achieve the goals of carbon emission reduction and carbon neutrality as scheduled [1]. With the accelerated construction and development of the electricity spot market, China's renewable energy sector has essentially shifted from the initial support phase involving benchmark electricity prices and preferential subsidies to a subsidy-free grid-connection phase at parity [2]. The generation capacity of renewable energy sources, including wind and photovoltaic, has experienced consistent growth. Nevertheless, they are no longer afforded the comprehensive grid-connection protection facilitated by the power grid; instead, they engage in competition within the electricity spot market alongside conventional power sources.

The strong randomness and volatility of wind power output make it difficult to accurately predict [3]. The participation of wind power in the spot market not only poses challenges to the secure operation of the power system but also results in penalties for wind power producers arising from the imbalances between the actual output and the winning bid [4]. This significantly weakens the profitability of wind power companies, which is not conducive to the enthusiasm for subsidy-free bidding and the healthy development of new energy. By optimizing resource allocation in the market [5], the combination of wind power with thermal power, energy storage systems, pumped storage power stations, and other types of flexible resources enables energy complementation. As a result, this method efficiently alleviates the anticipated revenue losses arising from deviation penalties, as referenced in [6–8]. For the electricity market, this approach is beneficial for solving the problem of insufficient grid reserves, ensuring a continuous and reliable power supply, thereby alleviating the pressure on grid dispatching. Ref. [9] discussed the participation of "new energy + storage" power plants as independent entities in the day-ahead market by controlling the charging and discharging behavior of the storage system to optimize the plant's power output. Ref. [10] considered the strong uncertainty of wind power output and formulated a bidding strategy for the wind-storage system in the energy market and the frequency regulation auxiliary service market, also taking into account the cycle life cost of storage. Ref. [11] designed a data-driven classification aggregation method for electric vehicles and proposed a day-ahead bidding strategy for the joint participation of electric vehicle fleets and wind power in the electricity and frequency regulation markets. Ref. [12] conducted joint bidding with demand response aggregators and wind power companies, introducing a balancing coefficient to balance the interests of multiple entities. In the above literature, wind power and new market entities such as energy storage collaborate to reduce upper and lower deviations. However, as the installed capacity of new energy continues to increase, there is still an issue with the insufficient scale of storage allocation due to high investment costs.

As traditional market entities, hydroelectric and thermal power possess excellent controllability, enabling them to meet the flexibility demands of integrating large-scale wind power into the grid within a short timeframe. Moreover, the collaborative participation in the market of these traditional market entities alongside wind farms also contributes to enhancing the stability of the electricity market. Ref. [13] investigated the bidding strategy of a coalition of multiple wind farms and pumped storage power stations, using the Shapley value and core solution to address revenue allocation among wind-pumped storage power stations and multiple wind farms. Hydropower is characterized by its sustainability and efficiency, but it entails significant construction costs and resource demands. Furthermore, the establishment of hydropower stations is constrained by geographical conditions. Currently, in China's "Three Norths" area, where wind power is developed intensively, the power source structure is still dominated by thermal power units and the wind–thermal bundling approach is applied to promote the grid connection of large-capacity wind farms. Therefore, encouraging thermal power companies and new energy companies to carry out substantive joint operations is an effective solution to promote the high-quality development of new energy. In [14–16], thermal power units were leveraged to offset stochastic wind power output, demonstrating that the joint bidding strategy can effectively increase the bidding profits for both parties. Ref. [14] proposed a dual-objective bidding strategy for wind–thermal–photovoltaic systems in the electricity and spinning reserve markets. Ref. [15] established a multi-objective, three-stage bidding and scheduling model for maximizing profits and minimizing carbon emissions of a wind–thermal–storage system based on mixed-integer programming. In [16], a wind–thermal joint bidding model was established utilizing the traditional information-gap decision theory method, with the cogeneration units supplying reserved capacity to minimize real-time imbalances. However, traditional thermal power units have a slow ramping rate, resulting in significant costs due to frequent starts and stops for offering adjustment services, and their high carbon emissions are contrary to China's current "dual carbon" targets.

In the context described above, this paper studies the joint bidding strategy for wind farms and CCPP in the spot market, taking into account imbalance costs. The transformation of traditional thermal power plants into CCPP not only results in lower carbon emission intensity but also provides a wider operating range and faster ramping rate [17], enhancing the environmental, efficiency, and flexibility levels of thermal power units. Consequently, it can serve as an ideal low-carbon power source to implement joint bidding with wind power. On this basis, a tiered carbon trading mechanism is introduced to reduce carbon emissions from thermal power units to achieve the maximum economic benefits in both electricity and carbon markets. The conditional value at risk (CVaR) is introduced to quantify various uncertainties, allowing decision-makers to set an appropriate risk aversion coefficient according to their risk tolerance, thus coordinating profit and risk optimization. Finally, the effectiveness of the proposed model is verified through numerical examples.

2. Theory and Methods

2.1. Mechanism Analysis of Flexible Operation of CCPP

Post-combustion carbon capture technology separates CO_2 directly from the flue gas produced by fuel combustion in boilers, capturing it through three stages: absorption, desorption, and compression. This technology has reached a mature state and is widely applied [18]. However, there is a coupling issue between the scrubber and the stripper in CCPP. Specifically, when the ratio of flue gas entering the scrubber from the venting channel to that directly discharged into the atmosphere remains constant, the CCPP cannot independently adjust the energy distribution of the carbon capture facilities. The sole available control option is the adjustment of power output in the CCPP. This results in conflicts between the loads of the generation system and the carbon capture system, posing a challenge for CCPP to harmonize the relationship between carbon capture energy consumption and net power output.

To address these challenges, this paper introduces an integrated, flexible operating mode for CCPP. The carbon flow and energy flow within a CCPP are shown in Figure 1.

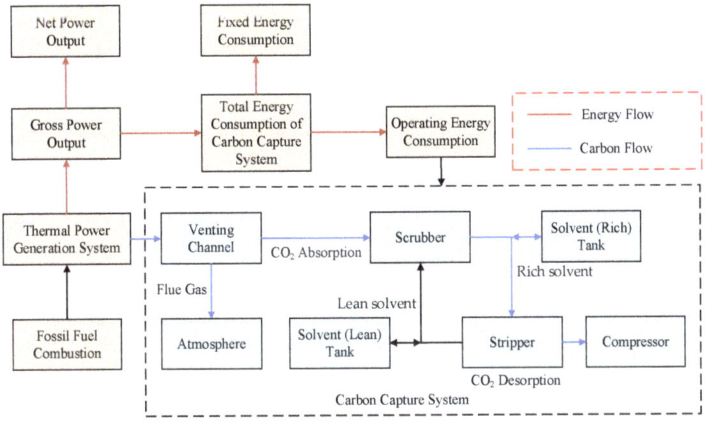

Figure 1. Carbon flow and energy flow of CCPP under the integrated, flexible operation mode.

This mode introduces two solvent storage tanks between the scrubber and the stripper, disrupting the traditional balance of solvent volumes within each. Firstly, the proportion of flue gas passing through the scrubber can be adjustable by the venting channel. The CO_2-containing flue gas has intimate contact with liquid absorbent within the scrubber. Meanwhile, CO_2 is absorbed by the MEA (monoethanolamine) solvent. Subsequently, the "rich" solvent, containing a higher concentration of CO_2, is directed to the stripper, where it desorbs the CO_2 and regenerates the "lean" solvent. A significant amount of thermal energy is required during the regeneration of the solvent during this process. The "lean" solvent is then recirculated back to the scrubber for the next absorption cycle, ensuring a strictly equal bidirectional flow of solvent between the scrubber and the stripper. During the above process, the "rich" solvent storage tank can store the "rich" solvent from the scrubber and pump it to the stripper for further processing when needed. By introducing these auxiliary facilities and decoupling the loads of power generation and carbon capture facilities, the CCPP is able to control the volume of CO_2 captured, thus giving the CCPP the opportunity to be operated flexibly [19].

The formula for calculating the mass of CO_2 captured by the carbon capture system is generally represented as follows:

$$\begin{cases} E_{Q,t} = \eta \gamma_t E_{G,t} + E_{H,t} \\ E_{G,t} = B_G P_{G,t} \end{cases}, \tag{1}$$

where $E_{G,t}$ represents the carbon emissions of the thermal power unit before being captured at hour t. η stands for the carbon capture efficiency. γ_t denotes the proportion of flue gas entering the scrubber through the venting channel at hour t. B_G is the carbon emission coefficient of the thermal power unit, representing the CO_2 emission intensity per unit of gross power output. $E_{H,t}$ is the mass of CO_2 released by the "rich" solvent storage tank at hour t, with CO_2 mass related to the solvent volume as follows:

$$d_{f,t} = \frac{E_{H,t} M_{MEA}}{\varepsilon \rho_f \sigma_f M_{CO_2}}, \tag{2}$$

where $d_{f,t}$ represents the volume of "rich" solvent pumped out from the "rich" solvent storage tank at hour t. M_{MEA} is the molar mass of the MEA solution. M_{CO_2} is the molar mass of CO_2. E is the quantity of desorption in the stripper. ρ_f and σ_f, respectively, denote MEA solvent concentration and density.

The gross power output of the generation system equals the sum of the net power output and the power consumed by the carbon capture system. When the gross power output is fixed, an increase in power consumption by the carbon capture system leads to a decrease in the net power output. This enhances the spinning reserve capacity of the thermal power unit, ensuring greater flexibility and reliability. Under the integrated, flexible operation mode, the solvent storage tank can provide additional CO_2. This allows the carbon capture facilities to capture more CO_2, resulting in increased capture energy consumption. As a result, the thermal power unit is capable of achieving a significant reduction in its net power output, enabling the CCPP to possess an extended range for net output adjustment.

Compared to conventional coal-fired power units, where the generation cycle is limited by their large thermal inertia and boiler state changes require significant time, the carbon capture facilities in post-combustion CCPP operate downstream of the generation system, providing greater independence. This independence allows for direct adjustment of the energy supply to the carbon capture system, enabling rapid adjustments to the net power output. Additionally, it can store CO_2, which is not to be treated in the stripper, so the thermal power unit can provide more power for the load during the peak load period. The carbon capture power plant can realize "energy time shifting" by adjusting the solvent storage, effectively resolving the contradiction between power generation and carbon capture.

2.2. *Joint Bidding Strategy for Wind Farms and CCPPs*

The generation alliance of wind and thermal power makes bidding decisions at two stages: the day-ahead market and the real-time balancing market. The day-ahead market bid is based on disclosed market information, generation cost functions, forecasted market clearing price, and wind power output. In the first stage, the day before the operation, the generation alliance submits its offering curve that aligns its economic interests to the electricity trading institution. In the MCP (market clearing price) mechanism, the settlement of all the winning generators is based on the unified clearing price λ. Since the capacity of the generation alliance only accounts for a small ratio in the whole market transaction, it is not a determinant of the market clearing price. Its bidding behavior exerts minimal impact on the system's marginal electricity price. Therefore, this paper assumes that the generation alliance acts as a price taker in the market. Thus, it only needs to submit power bid quantities instead of bidding curves to the market. In the second stage, the power generation plan formulated in the day-ahead market is put into action. Given the existence of forecasting errors, the CCPP performs rapid and precise control of real-time power output on an operational day, ensuring that the actual generation quantities align closely with the scheduled quantities in the day-ahead plan. When the actual electricity deviation of the wind farm surpasses the adjustment capability of the CCPP, the generation alliance enters the real-time balancing market. The imbalance price is used for the settlement of imbalanced electricity, which involves shortages or surpluses of electricity in real time.

Currently, the spot market mechanism in China is not yet mature. To prevent speculators from exploiting the arbitrage between day-ahead and real-time price differences, a dual-pricing mode is adopted in the real-time balancing market to incentivize participants to enhance their forecasting capabilities and provide reliable electricity supply [13]. The penalty coefficients meet the following conditions:

$$0 \leq \pi_{\text{down}} \leq 1 \leq \pi_{\text{up}}, \tag{3}$$

$$\begin{cases} \lambda_t^{\text{down}} = \pi_{\text{down}} \lambda_t^{\text{da}} \\ \lambda_t^{\text{up}} = \pi_{\text{up}} \lambda_t^{\text{da}} \end{cases}, \tag{4}$$

where π_{down} and π_{up}, respectively, represent the penalty coefficients for positive and negative imbalances. λ_t^{da} represents the day-ahead market clearing price. λ_t^{down} and λ_t^{up} represent the positive and negative imbalance prices, respectively.

The two-stage decision-making framework is illustrated in Figure 2; the uncertainties of clearing prices and wind power output are modeled via a set of scenarios. Assume that the forecast errors of random variables follow a normal distribution, and the scenarios are generated via random sampling from the probability distribution. Subsequently, a Kantorovich distance-based scenario reduction technique is utilized to establish a set of typical scenarios and their corresponding probabilities. The bidding strategy aims to maximize expected profits and minimize risk as multi-objectives. The risk aversion coefficient is determined based on the risk tolerance levels of generators. The optimal solution is calculated by combining all scenarios along with their respective probabilities. The wind power uncertainty is described through scenarios where the CCPP adjusts its actual net power output based on wind power variations within each scenario.

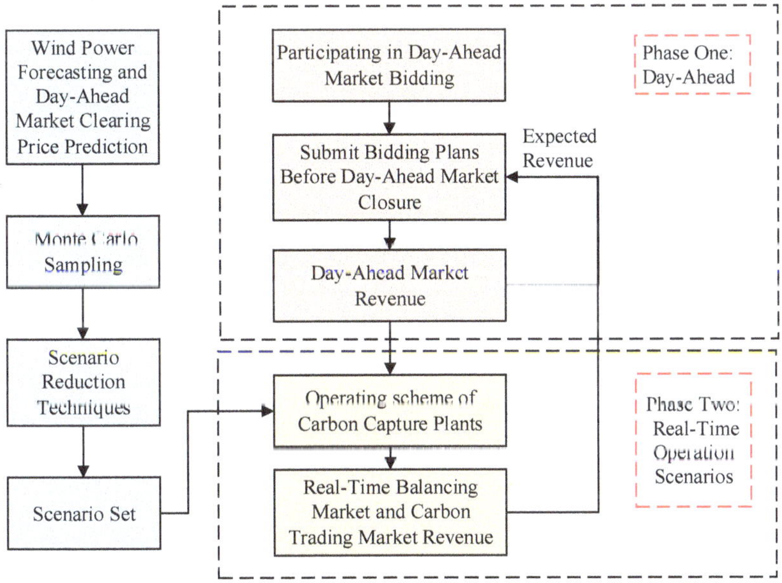

Figure 2. The decision-making framework.

2.3. Conditional Value at Risk

Should the day-ahead forecast results deviate significantly from the actual values, it can lead to a significant divergence between actual and expected profits. To mitigate the adverse impact of risks on the bidding model, CVaR is utilized as a risk measurement indicator to assess risk, thereby bolstering the robustness of the model. Unlike value at risk (VaR), which focuses on estimating the maximum potential loss within a confidence level, CVaR considers the expected value of losses that exceed the VaR threshold.

The loss function f (**x**,**y**) is related to the decision vector and uncertain vector. Here, **x** represents the decision variables, and **y** denotes the random variables associated with risk factors. When the joint probability density function $p(\mathbf{y})$ of **y** is known, the conditional value at risk function under continuous distribution can be formulated as follows:

$$F_{\beta}(\mathbf{x}) = \frac{1}{1-\beta} \int_{f(\mathbf{x},\mathbf{y}) \geq z_{\beta}(\mathbf{x})} f(\mathbf{x},\mathbf{y}) p(\mathbf{y}) dy, \tag{5}$$

$$z_{\beta}(\mathbf{x}) = \min\{z \in R : \psi(\mathbf{x}, z) \geq \beta\}, \tag{6}$$

where β is the confidence level. z_{β} represents the VaR value at confidence level β. $\psi(\mathbf{x},\alpha)$ denotes the probability that f (**x**,**y**) does not exceed the threshold z. $F_{\beta}(\mathbf{x})$ is the expected value of f (**x**,**y**) exceeding the VaR.

To address the integral problem, we can utilize a set of sampled scenarios to obtain an equivalent calculation formula for CvaR under discrete distribution as follows:

$$F_{\beta}(\mathbf{x}, \alpha) = z_{\beta} + \frac{1}{1-\beta} \sum_{s=1}^{N} \phi_s [f(\mathbf{x}, \mathbf{y}) - z_{\beta}]^{+}, \tag{7}$$

$$[f(\mathbf{x}, \mathbf{y}) - z_{\beta}]^{+} = \max[f(\mathbf{x}, \mathbf{y}) - z_{\beta}, 0], \tag{8}$$

where N is the total number of sampled scenarios. Φ_s is the probability of the scenario s.

CVaR is often used as a risk measure for cost-based loss functions. A smaller CVaR implies a lower level of tail cost, indicating a relatively better risk profile. For benefit-based functions, the loss function could be defined as the portion where actual benefits fall below expected benefits [20]. For ease of calculation, this paper defines CVaR as the average benefit when the outcome falls beyond a specified confidence level. CVaR provides a measure of the expected benefit in extreme adverse scenarios, focusing solely on the tail of the distribution beyond the chosen confidence level. The formula is articulated as follows:

$$X^{\mathrm{CV}}(\beta) = X^{\mathrm{V}}(\beta) - \frac{1}{1-\beta} \sum_{s=1}^{\mathrm{N}} \phi_s \left\{ X^{\mathrm{V}}(\beta) - E_s \right\}^{+}, \tag{9}$$

where X^{V} represents the VaR, which is the variable in the decision model. E_s is the benefit corresponding to scenario s. X^{CV} represents the average benefits that are less than the X^{V} threshold.

To facilitate calculation, auxiliary variables ε_s (s = 1, 2, . . ., N) are introduced to relax Equation (9), resulting in a linearized calculation formula for CVaR as follows:

$$X^{\mathrm{CV}}(\beta) = X^{\mathrm{V}}(\beta) - \frac{1}{1-\delta} \sum_{s=1}^{\mathrm{N}} \phi_s \varepsilon_s, \tag{10}$$

$$X^{\mathrm{V}}(\beta) - E_s \leq \varepsilon_s, \tag{11}$$

$$\varepsilon_s \geq 0. \tag{12}$$

2.4. Two-Stage Joing Bidding Model Based on CVaR

2.4.1. Objective Function

The objective function of the joint bidding decision model can be expressed as follows:

$$\max F = \sum_{s=1}^{N} \phi_s E_s + \delta X^{CV}(\beta), \tag{13}$$

$$E_s = R_s^{DA} + R_s^{PH} - C_s^{G} - C_s^{M} - C_s^{O} - C_s^{E}, \tag{14}$$

where E_s is the profit function, X^{CV} is the conditional value at risk. δ is the risk aversion coefficient, with $\delta > 0$ indicating an attitude towards risk aversion. It reflects the balance between expected profit and the degree of profit fluctuation, simplifying multi-objective optimization into a single objective through a weighted sum method. R_s^{DA} is the day-ahead market revenue under scenario s. R_s^{PH} is the real-time balancing market revenue under scenario s. C_s^{G}, C_s^{M}, C_s^{O}, and C_s^{E}, respectively, represent the fuel cost, solvent loss cost, start-up and shutdown cost, and carbon trading cost under scenario s.

$$R_s^{DA} = \sum_{t=1}^{T} \lambda_{s,t}^{da} P_{y,t}^{bid}, \tag{15}$$

$$R_s^{PH} = \sum_{t=1}^{T} (\lambda_{s,t}^{down} P_{y,s,t}^{down} \quad \lambda_{s,t}^{up} P_{y,s,t}^{up}), \tag{16}$$

where T is the number of hours. $P_{y,t}^{bid}$ represents the bidding quantity at hour t. $P_{y,s,t}^{down}$ and $P_{y,s,t}^{up}$, respectively, represent the positive and negative imbalance electricity under scenario s at hour t.

The operating costs of CCPP primarily consist of power generation costs, depreciation costs of carbon capture facilities, and MEA solvent loss costs. Generation costs mainly arise from the energy supplied to the carbon capture system and the net output. The depreciation costs of carbon capture facilities and storage tanks are constants and are unrelated to the operation state of the CCPP; thus, they are not considered in the model.

(1) Fuel Cost for Thermal Power Units

$$C_s^{G} = \sum_{t=1}^{T} [a(P_{G,s,t})^2 + bP_{G,s,t} + U_{s,t}c], \tag{17}$$

where a, b, and c are the quadratic, linear, and constant coefficients of the coal cost function for the thermal power generation unit. $U_{s,t}$ is a binary variable representing the on/off status of the thermal power unit at hour t, where 0 and 1 represent shutdown and start-up, respectively. $P_{G,s,t}$ is the gross output power at hour t under scenario s.

(2) Solvent Loss Cost

During the carbon capture process, the daily solvent loss of the MEA solution is calculated as follows:

$$C_s^{M} = \sum_{t=1}^{T} c_r g_r E_{Q,s,t}, \tag{18}$$

where c_r is the solvent cost coefficient. g_r is the operating loss coefficient. $E_{Q,s,t}$ is the mass of CO_2 being captured at hour t under scenario s.

(3) Carbon Trading Cost

The carbon trading market organizes the buying and selling of carbon emission rights among market members, thereby incentivizing power generators subject to carbon emissions assessments to manage power generation and carbon emissions, promoting carbon reduction on the generation side. If a power generator's actual carbon emissions exceed government-issued carbon allowances, additional carbon emission rights must be purchased. Conversely, if a power generator holds more carbon allowances than its actual emissions, it can sell the surplus carbon allowances. This paper breaks down carbon trading costs on a daily basis, recording actual carbon emissions and settling carbon trading costs at the end of each electricity spot market cycle. The quantity of carbon allowances involved in trading in the carbon market is as follows:

$$E_{\mathrm{act},s} - E_{\mathrm{ceq},s} = \sum_{t=1}^{\mathrm{T}} (E_{\mathrm{G},s,t} - E_{\mathrm{J},s,t} - P_{\mathrm{G},s,t} Z_{\mathrm{ceq}}), \tag{19}$$

where $E_{\mathrm{act},s}$ represents the actual carbon emissions under scenario s. $E_{\mathrm{ceq},s}$ represents the carbon allowances under scenario s. Z_{ceq} is the carbon emission baseline per unit of electricity for the thermal power unit. $E_{\mathrm{G},s,t}$ is the carbon emissions of the thermal power unit at hour t under scenario s. $E_{\mathrm{J},s,t}$ is the mass of CO_2 captured at hour t under scenario s. $\gamma_{s,t}$ is the proportion of flue gas entering the scrubber at hour t under scenario s.

A tiered carbon trading mechanism establishes distinct intervals for the trading of carbon emission rights. When the purchased quantity of carbon emission rights surpasses a certain interval, the carbon trading price increases accordingly. Compared to a uniform pricing mechanism, the tiered pricing mechanism can strengthen penalties for exceeding carbon emissions and incentives for carbon reduction. The cost calculation model for tiered carbon trading is as follows:

$$C_s^{\mathrm{E}} = \begin{cases} -\lambda(2+\sigma)d + \lambda(1+2\sigma)(E_{\mathrm{act},s} - E_{\mathrm{ceq},s} + 2d), \\ -E_{\mathrm{ceq,max}} \le E_{\mathrm{act},s} - E_{\mathrm{ceq},s} \le -2d \\ -\lambda d + \lambda(1+\sigma)(E_{\mathrm{act},s} - E_{\mathrm{ceq},s} + d), \\ -2d \le E_{\mathrm{act},s} - E_{\mathrm{ceq},s} \le -d \\ \lambda(E_{\mathrm{act},s} - E_{\mathrm{ceq},s}), -d \le E_{\mathrm{act},s} - E_{\mathrm{ceq},s} \le d \\ \lambda d + \lambda(1+\tau)(E_{\mathrm{act},s} - E_{\mathrm{ceq},s} - d), \\ d \le E_{\mathrm{act},s} - E_{\mathrm{ceq},s} \le 2d \\ \lambda(2+\tau)d + \lambda(1+2\tau)(E_{\mathrm{act},s} - E_{\mathrm{ceq},s} - 2d), \\ 2d \le E_{\mathrm{act},s} - E_{\mathrm{ceq},s} \le 3d \\ \lambda(3+3\tau)d + \lambda(1+3\tau)(E_{\mathrm{act},s} - E_{\mathrm{ceq},s} - 3d), \\ 3d \le E_{\mathrm{act},s} - E_{\mathrm{ceq},s} \le E_{\mathrm{G,max}} - E_{\mathrm{ceq,max}} \end{cases}, \tag{20}$$

where λ is the carbon trading benchmark price. τ is the penalty coefficient for tiered carbon trading. σ is the compensation coefficient for tiered carbon trading. d is the interval length of carbon emission rights.

(4) Start-up and Shutdown Costs for Thermal Power Units

$$C_s^{\mathrm{O}} = \sum_{t=1}^{\mathrm{T}} [U_{s,t}(1 - U_{s,t-1}) + U_{s,t-1}(1 - U_{s,t})] c_{\mathrm{o}}, \tag{21}$$

where c_{o} represents the cost of a single start-up or shut-down of the thermal powerunit.

(5) Operating Costs of Wind Farms

$$C_s^{\mathrm{W}} = \sum_{t=1}^{\mathrm{T}} c_{\mathrm{W}} P_{\mathrm{w},s,t}^{\mathrm{rt}}, \tag{22}$$

where c_{w} represents generation cost coefficients for wind farms.

2.4.2. Constraints

(1) Power Balance Constraint

$$P_{\mathrm{c},s,t}^{\mathrm{rt}} + P_{\mathrm{w},s,t}^{\mathrm{rt}} - P_{\mathrm{y},t}^{\mathrm{bid}} = P_{\mathrm{y},s,t}^{\mathrm{down}} - P_{\mathrm{y},s,t}^{\mathrm{up}}, \tag{23}$$

where $P^{rt}_{c,s,t}$ and $P^{rt}_{w,s,t}$, respectively, represent the net power output of the CCPP and the actual power output of the wind farm under scenario s at hour t.

The power generation status of the alliance under each scenario s and each hour t exists in one of three states: overgeneration, undergeneration, or the amount of output exactly equal to the bid quantities. Thus, the constraints for positive/negative imbalance power must meet the constraint $P^{up}_{y,s,t}P^{down}_{y,s,t} = 0$. To avoid introducing Boolean variables, the constraint conditions are equivalently transformed as follows.

$$P^{up}_{y,s,t} \geq P^{bid}_{y,t} - P^{rt}_{c,s,t} - P^{rt}_{w,s,t}, \tag{24}$$

$$P^{down}_{y,s,t} \geq P^{rt}_{c,s,t} + P^{rt}_{w,s,t} - P^{bid}_{y,t}, \tag{25}$$

$$P^{up}_{y,s,t} \geq 0, P^{down}_{y,s,t} \geq 0. \tag{26}$$

(2) Bidding Quantity Constraints

$$0 \leq P^{bid}_{y,t} \leq P_{c,\max} + P_{w,\max}, \tag{27}$$

where $P_{c,\max}$ and $P_{w,\max}$, respectively, refer to the maximum net power output of the thermal power unit and the capacity of the wind farm.

(3) Gross Power Output Constraints for Thermal Power Units

$$U_{s,t}P_{G,\min} \leq P_{G,s,t} \leq U_{s,t}P_{G,\max}, \tag{28}$$

where $P_{G,\max}$ and $P_{G,\min}$, respectively, denote the maximum and minimum technical power output of the thermal power unit.

(4) Ramp Rate Constraints of Thermal Power Units

$$\begin{cases} P_{G,s,t} - P_{G,s,t-1} \leq P_{UR} \\ P_{G,s,t-1} - P_{G,s,t} \leq P_{UR} \end{cases}, \tag{29}$$

where P_{UR} represents the maximum ramp rate of the thermal power unit.

(5) Start-up and Shutdown Constraints for Thermal Power Units

$$\begin{cases} \sum\limits_{k=t}^{t+T^{on}_{\min}-1} U_{s,k} \geq T^{on}_{\min}(U_{s,t} - U_{s,t-1}) \\ \sum\limits_{k=t}^{t+T^{off}_{\min}-1} (1 - U_{s,k}) \geq T^{off}_{\min}(U_{s,t-1} - U_{s,t}) \end{cases}, \tag{30}$$

where $T^{on}_{\min}$ and $T^{off}_{\min}$, respectively, represent the minimum continuous operating and shutdown durations.

(6) Operation Constraints for CCPPs

$$\begin{cases} P_{G,s,t} = P^{rt}_{c,s,t} + P_{ccs,s,t} \\ P_{ccs,s,t} = P_{B,t} + P_{I,s,t} \\ E_{J,s,t} = \eta\gamma_{s,t}E_{G,s,t} \\ F_{Q,c,t} = F_{J,c,t} + F_{H,c,t} \\ P_{I,s,t} = \varphi E_{Q,s,t} \end{cases}, \tag{31}$$

where $P_{ccs,s,t}$ represents the carbon capture system energy consumption under scenario s at hour t. $P_{B,t}$ is the fixed energy consumption of the carbon capture system, which is considered a constant value resulting from structural changes in the power plant. $P_{I,s,t}$ is the operating energy consumption of the carbon capture system under scenario s at hour t. φ is the power consumption per unit of CO_2 captured.

The constraints for the proportion of flue gas entering the scrubber and the captured CO_2 mass by the carbon capture system at each hour t are as follows:

$$0 \le \gamma_{s,t} \le 1, \tag{32}$$

$$0 \le E_{Q,s,t} \le \eta B_G \mu_{max} P_{G,max}, \tag{33}$$

where μ_{max} represents the maximum operation state coefficient of the stripper and compressor.

(7) Solvent Storage Tank Capacity Constraints

The formula for calculating the volume of the rich and lean solvent in the storage tanks is as follows:

$$\begin{cases} V_{F,s,t} = V_{F,s,t-1} - d_{f,s,t} \\ V_{P,s,t} = V_{P,s,t-1} + d_{f,s,t} \end{cases}, \tag{34}$$

where $V_{F,s,t}$ and $V_{P,s,t}$ represent the volume of the rich and lean solvent in the storage tanks, respectively, under scenario s at hour t.

These volumes satisfy the following constraints:

$$\begin{cases} 0 \le V_{P,s,t} \le V_{LSP,max} \\ 0 \le V_{F,s,t} \le V_{LSP,max} \\ V_{F,s,24} = V_{P,s,24} = V_{LSP,inv} \\ V_{F,s,0} = V_{P,s,0} = V_{LSP,inv} \end{cases}, \tag{35}$$

where $V_{LSP,max}$ represents the maximum capacity of the storage tanks. $V_{LSP,inv}$ refers to the initial volume of the rich and lean solvent in the storage tanks.

(8) CVaR constraints are as shown in Equations (11) and (12).

3. Model Solution Strategy

The solving process of the mixed-integer nonlinear programming (MINLP) model is relatively complex, resulting in higher computational costs and longer computation time. The complicated model with nonconvex and nonlinear features was transformed into a stochastic mixed-integer linear programming (MILP) model so that the previous bidding problem could be solved directly by a commercial solver.

(1) For Equation (17), a piecewise linearization approach is adopted, with the specific process as follows. Based on the required accuracy, the power output of thermal power units is discretely divided into L intervals on average:

$$\begin{cases} P_{d,min} = P_{d,1} < P_{d,2} < \ldots P_{d,L} = P_{d,max} \\ P_G = \sum\limits_{q=1}^{L+1} w_q P_{d,q} \end{cases}, \tag{36}$$

where $P_{d,q}$ represents the breakpoints of the q interval. $P_{d,min}$ and $P_{d,max}$ are, respectively, 0 and $P_{G,max}$. w_q are L+1 continuous auxiliary variables introduced. The additional constraint conditions to be incorporated are as follows:

$$\begin{cases} w_1 + w_2 + \ldots + w_{L+1} = 1 \\ z_1 + z_2 + \ldots + z_L = 1 \\ w_1 \ge 0, w_2 \ge 0, \ldots, w_{L+1} \ge 0 \\ w_1 \le z_1, w_2 \le z_1 + z_2, \ldots, \\ w_L \le z_{L-1} + z_L, w_{L+1} \le z_L \end{cases}, \tag{37}$$

where $[z_1, z_2, \ldots, z_L]$ are L binary auxiliary variables.

The nonlinear function in Equation (17) can be approximated by the following linearization function:

$$f(P_G) = \sum_{q=1}^{L+1} w_q f(P_{d,q}), \tag{38}$$

where f is the nonlinear function for the squared term in Equation (17).

(2) For the tiered carbon trading cost function in Equation (20), by introducing continuous variable $[g_1, g_2, \ldots, g_7]$ and binary variable $[a_1, a_2, \ldots, a_6]$, the linearized function for carbon trading costs can be derived as follows:

$$\begin{cases} E_{act} - E_{ceq} = \sum_{i=1}^{7} g_i \Delta E_i \\ C^{E,L} = \sum_{i=1}^{7} g_i C^E(\Delta E_i) \end{cases}, \tag{39}$$

where ΔE_i represents the breakpoints of $E_{act} - E_{ceq}$. The constraint conditions for the auxiliary variables g and a are as follows:

$$\begin{cases} g_1 + g_2 + \ldots + g_7 = 1 \\ a_1 + a_2 + \ldots + a_6 = 1 \\ g_1 \geq 0, g_2 \geq 0, \ldots, g_7 \geq 0 \\ g_1 \leq a_1, g_2 \leq a_1 + a_2, \ldots, \\ g_6 \leq a_5 + a_6, g_7 \leq a_6 \end{cases}. \tag{40}$$

(3) In Equation (21), for the binary product term of two binary variables $x_1 x_2$, a continuous auxiliary variable z is introduced. Then, the constraints equivalent to z and $x_1 x_2$ are as follows:

$$\begin{cases} z \geq x_1 + x_2 - 1 \\ z \geq 0 \\ z \leq x_i, i = 1, 2 \end{cases}. \tag{41}$$

(4) For the bilinear term γE_G in Equation (31), McCormick envelope linear relaxation is used to replace it, thereby reconstructing the original problem into an equivalent convex problem [21]. The specific expressions are as follows:

$$\begin{cases} h \geq \gamma^L E_G + \gamma E_G{}^L - \gamma^L E_G{}^L \\ h \geq \gamma^U E_G + \gamma E_G{}^U - \gamma^U E_G{}^U \\ h \leq \gamma^U E_G + \gamma E_G{}^L - \gamma^U E_G{}^L \\ h \leq \gamma E_G{}^U + \gamma^L E_G - \gamma^L E_G{}^U \end{cases}, \tag{42}$$

where the superscripts U and L, respectively, indicate the upper and lower limits of the variable.

4. Results and Discussion

The case study uses a 600 MW CCPP and a 370 MW wind farm. The detailed parameters for the thermal power unit of the carbon capture plant are shown in Appendix A, Table A1, and the parameters for the carbon capture facilities are referenced from [22]. The generation cost coefficient of the wind farm is CNY 100/MWh. The penalty coefficients for positive and negative imbalances are 0.4 and 1.6, respectively. The carbon emission baseline per unit of electricity is 0.7 t/(MW·h), with a carbon trading benchmark price of CNY 100/t, a compensation coefficient of 0.1, and a penalty coefficient of 0.2. The interval length of carbon emission rights is 2000 t. The forecast results for wind power output and day-ahead clearing price are shown in Figure 3, with standard deviations of the forecast errors set at 25% and 15%, respectively. The Monte Carlo method is used for random sampling to generate 1000 wind power out and price scenarios. Employing scenario reduction techniques, the scenarios are reduced to 10 typical ones, each with its corresponding probability. The risk aversion coefficient $\delta = 1$, and the CVaR confidence level $\beta = 0.95$. The model is solved using MATLAB software R2016b and the Gurobi optimizer 10.0.0.

4.1. Analysis of Bidding Results

Figure 3 illustrates the results of joint bidding under the risk-averse strategy. As depicted in Figure 3, from hours 5 to 9, wind power is abundant, and the predicted power output of the wind farm is substantial, yet the bidding quantities are relatively low. This is attributed to the lower electricity price during this period, which is close to the marginal cost of the thermal power unit, so the bidding quantities are relatively low. At the same time, in periods when the electricity price is high, the bidding quantities increase to maximize revenue from electricity sales. This enables the generation alliance to achieve profitability throughout the entire day. Additionally, limited by weather conditions such as wind speed, the inclusion of the thermal power unit also somewhat weakens the counter-peak characteristics of wind power, enabling the generation alliance to respond to clearing price signals and formulate an hourly generation plan. It is evident that the hourly bidding quantities align approximately with the trend of electricity price changes, thereby favoring a responsive approach to meet the market load demand.

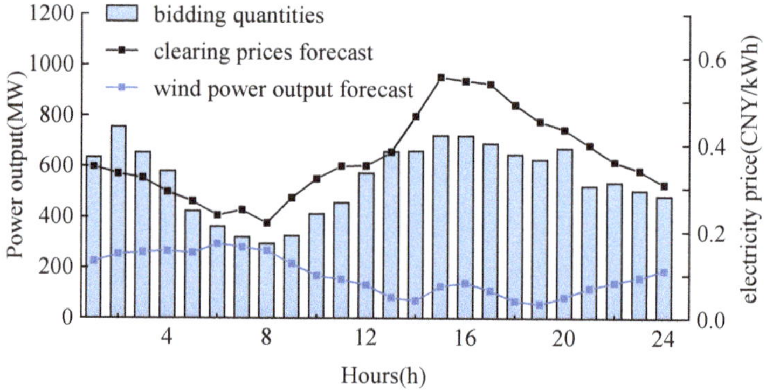

Figure 3. Optimized bidding results in the day-ahead market.

The bidding strategy takes into account revenues from both the day-ahead market and real-time balancing market settlements, providing a comprehensive decision-making perspective. The balancing market aims to eliminate imbalanced electricity. Given the strict penalties for generation shortage and wind curtailment, day-ahead bidding is conservative. Under an individual bidding strategy, the wind farm submits a generation plan based on the optimization results considering day-ahead revenues and real-time balancing market revenues for the scenario set. In the case of individual bidding by the CCPP, it assumes no power output deviation. Due to the higher costs associated with the thermal power unit compared to wind power, the CCPP facilitates the integration of wind power into the grid, thereby reducing overall operating costs. This results in joint bidding yielding higher profits than individual bidding, with lower risks associated with joint bidding compared to individual bidding.

According to Table 1, the total expected profits and total CVaR for individual bidding are CNY 1675.3 thousand and CNY 1547.6 thousand, respectively, while for joint bidding, they are CNY 1790.9 thousand and CNY 1647.9 thousand, with gains of CNY 115.6 thousand and CNY 100.3 thousand, respectively. Given the prioritization of wind power for grid integration under this strategy, thermal power inevitably stands to lose the profits it would have gained under individual bidding during the coordination process. Without adequate compensation, participating in joint bidding would be contrary to the interests of thermal power stakeholders. Using the Shapley value [23] to fairly allocate the total profits, Table 2 shows the profit for each member. The expected profit for wind power is CNY 875.3

thousand, an increase of 7.07%, while for thermal power, it is CNY 915.6 thousand, an increase of 6.74%. The analysis above demonstrates that the adoption of Shapley's value ensures mutually beneficial outcomes for participating members.

Table 1. Expected profit of individual bidding and joint bidding.

	Wind Power	Thermal Power	Total Expected Profits	Total CVaR
Unit	Thousand (CNY)	Thousand (CNY)	Thousand (CNY)	Thousand (CNY)
Individual bidding	817.5	857.8	1675.3	1547.6
Joint bidding	875.3	915.6	1790.9	1647.9

4.2. Comparison of Different Cases

This section considers the deployment of power generation alliance options. Three cases are considered: case 1 adopts the integrated flexible CCPP, equipped with solvent storage tanks. Case 2 employs a traditional CCPP, lacking solvent storage tanks. Case 3 refers to a conventional thermal power plant without carbon capture facilities.

Comparing the bidding results of each case with an individual wind farm, as shown in Table 2, it is evident that case 1 achieved the best results in both economic and low-carbon. The expected profit for case 1 increased by CNY 62.2 thousand and CNY 140.7 thousand compared to cases 2 and 3, respectively. Additionally, carbon trading costs decreased by CNY 622.5 thousand and CNY 1050.3 thousand, respectively, confirming that CCPPs have significantly lower carbon emission intensity compared to conventional thermal power plants.

Due to the fact that both case 1 and case 2 supply energy for the carbon capture system, the fuel costs are higher than those of case 3. However, the differences between the revenues from day-ahead market bidding and the total costs for case 1 and case 2 are, respectively, CNY 1810.8 thousand and CNY 1765.2 thousand, while for case 3, it is only CNY 1715.6 thousand. Thus, the economic feasibility of carbon capture power plants is ensured.

The carbon trading costs under conditions without and with solvent storage tanks are CNY 96.7 thousand and CNY −525.8 thousand, respectively, and the solvent loss costs are CNY 57.0 thousand and CNY 128.1 thousand, respectively. This implies that solvent storage tanks have increased the amount of CO_2 that can be captured. The amount of net carbon emissions for case 1 is lower than the amount of corresponding carbon allowances, allowing the CCPP to benefit in the carbon trading market and thus bringing additional revenue. Furthermore, case 3 submitted the least total bidding amount in the day-ahead market compared to case 2. The reason is that the conventional thermal power plant has a high carbon emission intensity, and to avoid incurring high carbon trading costs from the tiered carbon trading mechanism, they reduced the electricity generation amount. Case 1's total bidding amount is less than case 2, allowing it to provide greater adjustment capacity during the real-time stage.

The individual wind farm's positive and negative imbalances settlement in the balancing market were CNY 51.8 thousand and CNY −182.3 thousand, respectively, while for case 3, they were CNY 16.7 thousand and CNY −82.2 thousand, respectively. This indicates that the joint bidding mode can mitigate wind power output deviations. The imbalance settlements for the three cases decreased in sequence, with case 1's imbalance settlements being CNY 1.8 thousand and CNY −21.7 thousand, respectively. This demonstrates that case 1 has the smallest imbalance of electricity amount, thus promoting wind power grid integration to a greater extent. It also further illustrates that CCPPs consider solvent storage tanks to offer a more flexible operation, with a stronger capability to facilitate the integration of wind power compared to conventional thermal power plants and traditional CCPPs.

Table 2. Comparison of different cases.

	Unit	Case 1	Case 2	Case 3	Individual Wind Farm
Total Bidding Amount	MWh	13238	14947	14105	4012
Day-Ahead Market Revenue	Thousand (CNY)	5123.0	5663.4	5477.2	1358.4
Fuel Cost	Thousand (CNY)	3299.6	3334.3	2726.7	—
Solvent Loss Cost	Thousand (CNY)	128.1	57.0	—	—
Start-up and Shutdown Costs for Thermal Power Units	Thousand (CNY)	0	0	100.0	—
Carbon Trading Cost	Thousand (CNY)	−525.8	96.7	524.5	—
Total Cost	Thousand (CNY)	3312.2	3898.2	3761.6	410.3
Positive Imbalance Settlement	Thousand (CNY)	1.8	3.3	16.7	51.8
Negative Imbalance Settlement	Thousand (CNY)	−21.7	−39.7	−82.2	−182.3
Expected Profit	Thousand (CNY)	1790.9	1728.7	1650.2	817.5

4.3. Analysis of Risk Preference

This section explores the expected profits and CVaR values under different risk aversion coefficients, with δ values ranging from 0 to 10. Figure 4 illustrates the efficient frontier curve of CVaR, where a δ value of 0 represents a risk-neutral attitude, focusing solely on maximizing bidding profits. It is observed that the joint bidding strategy, when not considering risk, achieves a maximum expected profit of CNY 1799.4 thousand, which is CNY 23.5 thousand higher than the strong risk-averse joint bidding strategy (δ is 10). The purpose of risk aversion is to reduce the expected profit gap between different scenarios by sacrificing economic benefits. Hence, the resulting CVaR value when δ is 10 is CNY 1653.3 thousand, which is an increase of CNY 41.7 thousand compared to the situation when δ is 0.

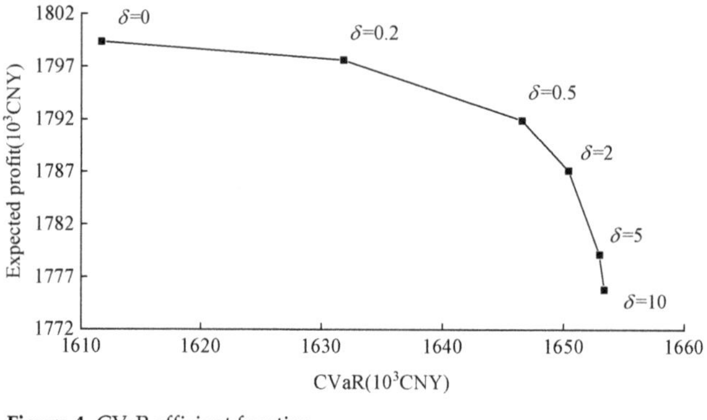

Figure 4. CVaR efficient frontier.

As the δ value increases, the expected profit exhibits a declining trend, while CVaR shows an ascending trend. This implies a positive correlation between risk and profit, wherein decision-makers aiming for higher profits also face greater risks. Within the δ range of 1 to 10, the decision-making objective shifts from balancing profits and risks ($\delta = 1$) to placing a stronger emphasis on risk ($\delta > 1$). As δ increases, the CVaR value grows, accompanied by a significant decrease in expected profits. The bidding strategy becomes more focused on the tail of the profit distribution under a given confidence level. Therefore, it is necessary to measure the risk attitude of decision-makers in bidding and find the optimal balance between risk and revenue, thereby formulating appropriate bidding strategies for market participants.

4.4. Analysis of the Flexible Operation Characteristics of CCPP

The power generation alliance needs to strategically determine its power output based on the bidding quantities and day-ahead clearing price. Figure 5 illustrates the actual power output of the wind farm and CCPP. Between hours 1 to 6, in cases where wind power output exceeds its forecast, the CCPP minimizes wind power curtailment by reducing its net power output. The carbon capture system can serve as a dispatchable load. From hours 5 to 10, the net power output of the thermal power unit is significantly below its minimum technical output. Between hours 18 and 24, the CCPP compensates for the shortage in wind power output, aiming to closely align the actual power generation quantities of the generation alliance with the day-ahead bidding quantities (day-ahead generation plan).

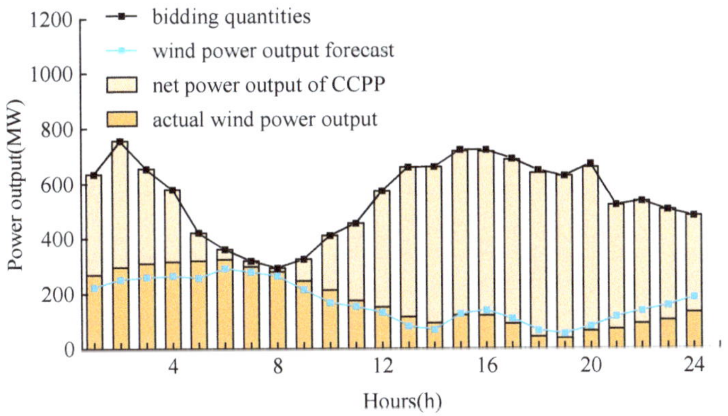

Figure 5. Power output of the wind farm and CCPP.

To further investigate the flexible operation mechanism of CCPP, an analysis of the power output and carbon flow is conducted. As shown in Figure 6, the net power output of a CCPP is the difference between its gross power output and the energy consumption of the carbon capture system. During low electricity price periods, it is observed that CCPP tend to increase the energy supply to carbon capture systems. This strategy is motivated by the opportunity to capture CO_2 and obtain revenue by selling the excess carbon allowances in the carbon trading market.

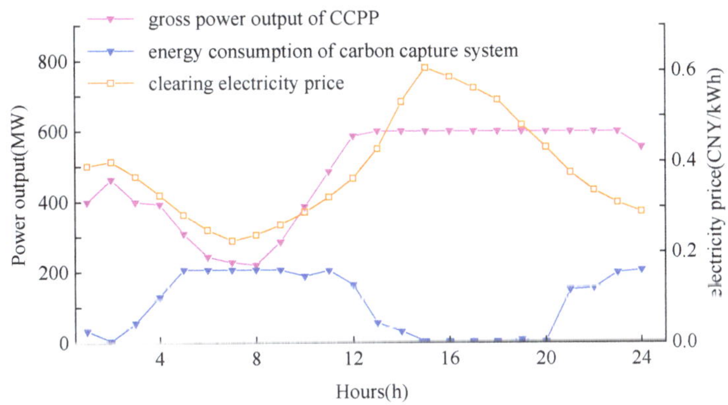

Figure 6. Power allocation of CCPP.

Conversely, during periods of high prices, the CCPP increases its net power output by raising its gross output and reducing the energy supply to the capture system. This strategy is motivated by the significantly higher revenue in the electricity spot market during these times, making it economically advantageous to sell electricity. In summary, the analysis reveals that the generation preferences of the CCPP depend on electricity prices. By adjusting the energy supply to the carbon capture system, the CCPP can flexibly modify net power output, enabling it to operate at an economically optimal point. The adoption of distinct strategies during peak and off-peak electricity prices is crucial for maximizing profits throughout the entire operational cycle.

As indicated by Figure 7, the scrubber and stripper of the integrated flexible CCPP no longer operate synchronously due to the installation of the solvent storage tank. During low electricity price periods, as the output of the thermal power unit decreases, the amount of CO_2 produced by the plant also decreases. Consequently, a carbon capture system, functioning as a dispatchable load to absorb excess electricity, may not have a sufficient supply of CO_2 for capture during such periods. From hours 4 to 11, as depicted in Figure 7, there is a reduction in the volume of the "rich" solvent in the tank, as solvents from the scrubber and "rich" solvent tank are transferred to the stripper. This leads to a substantial increase in carbon capture energy consumption, leading to a significant reduction in the net power output of the CCPP.

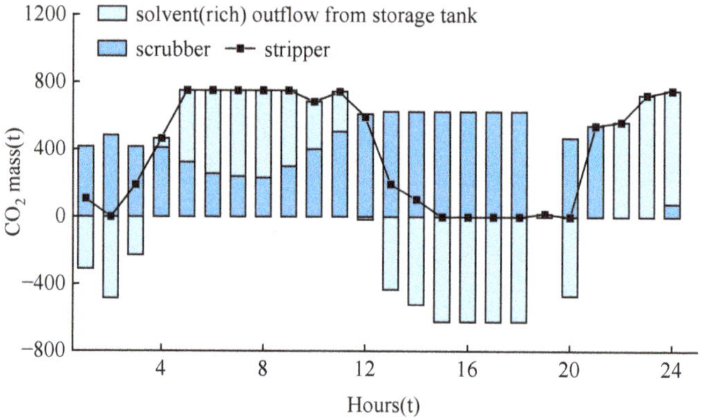

Figure 7. Distribution of carbon flow in CCPPs.

During peak electricity price periods, the CCPP requires a higher net power output to meet the day-ahead generation plan. Simultaneously, the plant will also produce a significant amount of CO_2. The CCPP increases the flue gas entering the scrubber and stores the absorbed CO_2 inside the "rich" solvent tank. This not only reduces emissions of CO_2 into the atmosphere but also reduces the energy supply to the carbon capture system by storing CO_2 without immediate capture. As a consequence, during these periods, the volume of solvent processed within the stripper decreases, resulting in an increase in the net power output of the CCPP. Finally, the volumes of both the lean and rich solvents in the storage tank are restored to their initial levels at the end of the cycle, ensuring the long-term operation of the carbon capture system.

4.5. Comparison of Results between MILP and MINLP

Table 3 compares the solution results of the MILP and MINLP models using the Gruobi optimizer. As shown in Table 3, compared to the MILP model, which was solved in 15.61 s, the MINLP solution was obtained in 1443.37 s. Regarding results accuracy, the MILP optimization results in an objective function value of 3438.9, while MINLP optimization yields a value of 3440.8, with a mere difference of 1.9 between the two. The difference between the two objective function values is rather small. As depicted in Figure 8, the optimal bidding electricity quantities in the day-ahead market obtained from MILP and MINLP are rather similar. In summary, based on the comparison of the exact optimal solution between the MILP and MINLP approaches, MILP can find acceptable exact solutions in our bidding problem much faster, demonstrating that the MILP approach holds advantages over the MINLP technique from the viewpoint of solution time.

Table 3. Comparison of solution results between MILP and MINLP.

	Solution Time/s	Objective Functions/ Thousand (CNY)	Expected Profit/ Thousand (CNY)
MILP	15.61	3438.9	1790.9
MINLP	1443.37	3440.8	1792.0

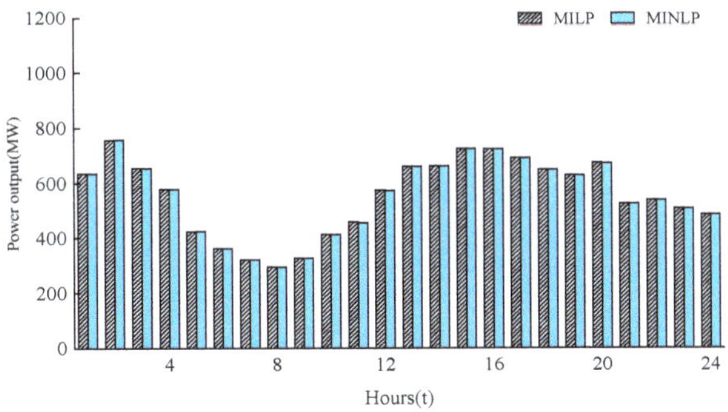

Figure 8. Comparison of the optimized bidding quantities in the day-ahead market.

5. Conclusions

Currently, China is in the early stages of constructing electricity spot markets and carbon trading markets. In anticipation of wind power participating in the electricity spot market in the future, a two-stage decision-making model is established, which combines CCPP and wind farm bidding in the spot market to mitigate the imbalance costs under the uncertainty of wind power output. The specific conclusions are as follows:

(1) Compared to individual bidding, the joint wind-thermal bidding strategy reduces risk while obtaining excess profits in the spot market.
(2) This paper compares three cases and verifies that the integrated flexible CCPP has clear advantages in terms of economy, low carbon emissions, and wind power integration. This is attributed to the installation of a solvent tank, which not only enhances the flexibility of the CCPP but also improves energy utilization efficiency. Thus, it deeply reduces carbon emissions while optimizing economic benefits.
(3) The decision model facilitates risk control, providing a comparative result for decision-makers to choose different risk coefficients. A risk-averse model has been established, demonstrating a proportional relationship between risk and profit. This can provide a better reference for market bidding.
(4) This study considers wind farms and CCPPs as price takers jointly participating in the spot market. As regional market trading rules continue to evolve and develop, and with the increasing share of renewable energy participation in the market, future work will focus on the price-making strategies of power generators.

Author Contributions: Conceptualization, Z.L.; Methodology, Z.L.; Software, Z.L.; Validation, Z.L.; Writing—original draft, W.T.; Writing—review and editing, B.W.; Supervision, Y.L. All authors have read and agreed to the published version of the manuscript.

Funding: This research was funded by the National Key Research and Development Program of China (2022YFB2403503).

Data Availability Statement: The data presented in this study are available on request from the corresponding author due to privacy.

Acknowledgments: Zhiwei Liao thanks Wenjuan Tao, Bowen Wang, and Ye Liu for their valuable discussions and their helpful advice with this paper.

Conflicts of Interest: The authors declare no conflicts of interest.

Appendix A

Table A1. Parameters of thermal power units.

Parameter	Unit	Value	Parameter	Unit	Value
Coal Cost Coefficient *a*	CNY/MW^2·h	0.012	Carbon Emission Intensity	t/MW·h	1.16
Coal Cost Coefficient *b*	CNY/MW·h	250	Ramp Rate	MW/h	100
Coal Cost Coefficient *c*	CNY	8700	Per Start-up and Shutdown Cost	Thousand CNY	50
Maximum Technical Output	MW	600	Minimum Start-up and Shutdown Time	h	4
Minimum Technical Output	MW	200			

References

1. Li, Z.; Chen, S.Y.; Dong, W.; Liu, P.; Du, E.; Ma, L.; He, J. Low Carbon Transition Pathway of Power Sector under Carbon Emission Constraints. *Proc. CSEE* **2021**, *41*, 3987–4000.
2. National Development and Reform Commission, National Energy Administration. Notice on Actively Promoting the Work of Wind Power and Photovoltaic Power Generation Grid Interconnection without Subsidies at Parity [EB/OL]. Available online: https://www.nea.gov.cn/2019-01/10/c_137731320.htm (accessed on 17 January 2024).
3. Ahmed, S.D.; Al-Ismail, F.S.M.; Shafiullah; Al-Sulaiman, F.A.; El-Amin, I.M. Grid integration challenges of wind energy: A review. *IEEE Access* **2020**, *8*, 10857–10878. [CrossRef]
4. Wu, Z.; Zhou, M.; Li, G.; Zhao, T.; Zhang, Y.; Liu, X. Interaction between balancing market design and market behaviour of wind power producers in China. *Renew. Sustain. Energy Rev.* **2020**, *132*, 110060. [CrossRef]
5. Xu, X.; Wang, H.; Yan, Z.; Lu, Z.; Kang, C.; Xie, K. Overview of Power System Uncertainty and Its Solutions under Energy Transition. *Autom. Electr. Power Syst.* **2021**, *45*, 2–13.
6. Dadashi, M.; Zare, K.; Seyedi, H.; Shafie-Khah, M. Coordination of wind power producers with an energy storage system for the optimal participation in wholesale electricity markets. *Int. J. Electr. Power Energy Syst.* **2022**, *136*, 107672. [CrossRef]
7. Cao, D.; Hu, W.; Xu, X.; Dragičević, T.; Huang, Q.; Liu, Z.; Chen, Z.; Blaabjerg, F. Bidding strategy for trading wind energy and purchasing reserve of wind power producer–A DRL based approach. *Int. J. Electr. Power Energy Syst.* **2020**, *117*, 105648. [CrossRef]
8. Alashery, M.K.; Xiao, D.; Qiao, W. Second-order stochastic dominance constraints for risk management of a wind power producer's optimal bidding strategy. *IEEE Trans. Sustain. Energy* **2019**, *11*, 1404–1413. [CrossRef]
9. Yang, B.; Tang, W.; Wu, F.; Wang, H.; Sun, W. Day-ahead market bidding strategy for "renewable energy + energy storage" power plantsconsidering conditional value-at-risk. *Power Syst. Prot. Control.* **2022**, *50*, 93–101.
10. Wang, H.; Chen, J.; Zhu, T.; Wu, M.; Chen, Q.; Zhu, J.; Liu, M. Joint Bidding Model and Algorithm of Wind-storage System Considering Energy Storage Life and Frequency Regulation Performance. *Power Syst. Technol.* **2021**, *45*, 208–215.
11. Zhang, Q.; Wu, X.; Deng, X.; Huang, Y.; Li, C.; Wu, J. Bidding strategy for wind power and Large-scale electric vehicles participating in Day-ahead energy and frequency regulation market. *Appl. Energy* **2023**, *341*, 121063. [CrossRef]
12. He, S.; Wang, L.; Zhao, S. Stochastic Cournot Game Based Distributed Optimization Strategy for Joint Bidding of Demand Response Aggregator and Multi-stakeholder Wind Power Producers. *Power Syst. Technol.* **2022**, *46*, 4789–4799.
13. Wu, Z.; Zhou, M.; Yao, S.; Li, G.; Zhang, Y.; Liu, X. Optimization Operation Strategy of Wind-storage Coalition in Spot Market Based on Cooperative Game Theory. *Power Syst. Technol.* **2019**, *43*, 2815–2824.
14. Khaloie, H.; Abdollahi, A.; Shafie-Khah, M.; Siano, P.; Nojavan, S.; Anvari-Moghaddam, A.; Catalão, J.P. Co-optimized bidding strategy of an integrated wind-thermal-photovoltaic system in deregulated electricity market under uncertainties. *J. Clean. Prod.* **2020**, *242*, 118434. [CrossRef]
15. Khaloie, H.; Abdollahi, A.; Shafie-Khah, M.; Anvari-Moghaddam, A.; Nojavan, S.; Siano, P.; Catalão, J.F.S. Coordinated wind-thermal-energy storage offering strategy in energy and spinning reserve markets using a multi-stage model. *Appl. Energy* **2020**, *259*, 114168. [CrossRef]
16. Peng, F.; Sui, X.; Hu, S.; Zhou, W.; Sun, H.; Chen, X. Joint bidding Strategy for Wind and Thermal Power Based on Information Gap Decision Theory. *Power Syst. Technol.* **2021**, *45*, 3379–3388.
17. Xi, H.; Zhu, M.; Lee, K.Y.; Wu, X. Multi-timescale and control-perceptive scheduling approach for flexible operation of power plant-carbon capture system. *Fuel* **2023**, *331*, 125695. [CrossRef]
18. Zhu, M.; Liu, Y.; Wu, X.; Shen, J. Dynamic modeling and comprehensive analysis of direct air-cooling coal-fired power plant integrated with carbon capture for reliable, economic and flexible operation. *Energy* **2023**, *263*, 125490. [CrossRef]
19. Kou, Y.; Wu, J.; Zhang, H.; Yang, J.; Jiang, H. Low carbon economic dispatch for a power system considering carbon capture and CVaR. *Power Syst. Prot. Control.* **2023**, *51*, 131–140.
20. Wang, X.; Gao, C. Two-stage Decision-making Model of Power Generation and Coal Purchase Arrangement for Power Generation Companies in Medium and Long-term Market. *Power Syst. Technol.* **2021**, *45*, 3992–3999.
21. Khajavirad, A. On the strength of recursive McCormick relaxations for binary polynomial optimization. *Oper. Res. Lett.* **2023**, *51*, 146–152. [CrossRef]

22. Cui, Y.; Deng, G.; Zhao, Y.; Zhong, W.; Tang, Y.; Liu, X. Economic Dispatch of Power System with Wind Power Considering the Complementarity of Low-carbon Characteristics of Source Side and Load Side. *Proc. CSEE* **2021**, *41*, 4799–4815.
23. Wang, X.; Zhang, H.; Zhang, S.; Wu, L. Impacts of joint operation of wind power with electric vehicles and demand response in electricity market. *Electr. Power Syst. Res.* **2021**, *201*, 107513. [CrossRef]

energies

MDPI

Review

High Impedance Fault Models for Overhead Distribution Networks: A Review and Comparison with MV Lab Experiments

Juan Carlos Huaquisaca Paye [1,2,*], João Paulo A. Vieira [1], Jonathan Muñoz Tabora [1,3], André P. Leão [1], Murillo Augusto M. Cordeiro [1], Ghendy C. Junior [4], Adriano P. de Morais [4] and Patrick E. Farias [5]

[1] Institute of Technology, Electrical Engineering Faculty, Federal University of Pará, Belém 66075-110, Brazil; jpavieira@ufpa.br (J.P.A.V.); jonathan.munoz@unah.edu.hn (J.M.T.); leaoap2017@gmail.com (A.P.L.); murilloaugustocordeiro@gmail.com (M.A.M.C.)

[2] Electrical Engineering Department, The Public University of El Alto (UPEA), La Paz 00591, Bolivia

[3] Electrical Engineering Department, National Autonomous University of Honduras (UNAH), Tegucigalpa 04001, Honduras

[4] Department of Electrical Engineering, Federal University of Santa Maria, Santa Maria 97105-900, Brazil; ghendy@ufsm.br (G.C.J.); adriano@ctism.ufsm.br (A.P.d.M.)

[5] Federal Institute of Education, Science and Technology, Federal Institute of Rio Grande do Sul, Farroupilhas 95174-274, Brazil; patrick.farias@farroupilha.ifrs.edu.br

* Correspondence: juancarloshuaquisaca@gmail.com

Abstract: Detecting and locating high impedance faults (HIF) in overhead distribution networks (ODN) remains one of the biggest challenges for manufacturers and researchers due to the complexity of this phenomenon, where the electrical current magnitude is similar to that of the loads. To simulate HIF, the selection of the HIF model is important, because it has to correctly reproduce the characteristics of this phenomenon, so that it does not negatively influence the simulations results. Therefore, HIF models play a fundamental role in proposing solutions and validating the effectiveness of the proposed methods to detect and localize HIF in ODN. This paper presents a systematic review of HIF models. It is intended to facilitate the selection of the HIF model to be considered. The models are validated based on experimental data from medium voltage (MV) laboratories, specifically, recorded waveforms from two HIF tests conducted in an MV lab were analyzed and compared with three established HIF models. The efficacy of these models was assessed against MV lab test data to ensure a precise representation of both transient and steady-state conditions for fault conductance and current waveforms. The findings show that the two nonlinear resistor models better approximate the waveforms obtained in the experimental tests performed in this study.

Keywords: high impedance faults (HIFs); HIF models; overhead distribution networks (ODN); lab experiment

Citation: Huaquisaca Paye, J.C.; Vieira, J.P.A.; Tabora, J.M.; Leão, A.P.; Cordeiro, M.A.M.; Junior, G.C.; Morais, A.P.d.; Farias, P.E. High Impedance Fault Models for Overhead Distribution Networks: A Review and Comparison with MV Lab Experiments. *Energies* **2024**, *17*, 1125. https://doi.org/10.3390/en17051125

Academic Editors: Veerapandiyan Veerasamy, Shailendra Singh and Sunil Kumar Singh

Received: 31 December 2023
Revised: 2 February 2024
Accepted: 6 February 2024
Published: 27 February 2024

1. Introduction

1.1. General Considerations

A high impedance fault (HIF) is defined as a disturbance caused by a conductor that touches a high-resistance nonmetallic surface connected to the ground, producing an arcing fault current lower than the residual relay pickup setting. According to [1], HIFs involving downed conductors in contact with a surface, correspond to 10% of all faults occurring in overhead distribution networks (ODN) [1]. The percentage of HIFs in ODN is even higher since not all HIF events are reported [2]. In addition, HIFs can occur without rupture or downfall of conductors on a surface, therefore, the HIF current is so low that it cannot be detected by conventional overcurrent protection functions [3]. This protection issue has been a challenge for distribution utility engineers for a long time [4].

To further complicate matters, HIF current waveforms have diverse characteristics, such as asymmetry, nonlinearity, build-up, shoulder, intermittence, and randomness [5].

High impedance faults in ODN may cause serious and harmful effects, such as (1) risk of electric shock; (2) risk of fire in material assets and forests; (3) electricity supply interruptions; and (4) long service restoration times. These effects continuously motivate the development of new ideas and innovative solutions for HIF detection and location, since they are challenging issues remaining unsolved or partially solved by the power distribution industry. In that sense, more reliable algorithms for HIF studies should be developed based on the above-mentioned HIF current waveform characteristics, which can be represented by mathematical models that can reproduce them.

Test data from many early staged HIF studies have provided valuable information toward understanding and characterizing HIFs [6,7]. To test the effectiveness and robustness of these algorithms, time-domain computational simulations must be performed considering the asymmetric, nonlinear, intermittent, and random nature of the HIF phenomenon. It is worth highlighting that HIFs are a random and dynamic phenomenon since they produce arcing and flashing at the point of contact, and the conductor may move around due to electromagnetic force [8].

The use of accurate HIF models for purposes of designing and testing HIF detection and location algorithms is essential. HIF computational simulations can provide, at the very least, initial data for preliminary research. On the other hand, field tests on an actual medium voltage (MV) distribution network are not a recommended practice due to the inherent danger involving this type of arcing fault in an uncontrolled environment.

Modeling of HIFs in ODN can be classified into three main categories: (1) model based on active and passive circuit elements; (2) model based on passive circuit elements; and (3) arcing model. Modeling based on active and passive circuit elements was proposed by [9] and analyzed in [10]. The model is based on antiparallel diodes and its parameters are considered to remain constant during the analyzed period. However, this model cannot reproduce transient-state conditions of the HIF current waveform. A model based on passive circuit elements was introduced by [11]. Some proposed modifications by [12] were included in the model, which allows to simulate HIF on different types of ground surfaces. Yet, this model depends on the V $\times$ I curve at the HIF point for each type of ground surface. The arcing model was initially introduced by [13,14] in the form of a first-order differential equation. This model is a simple physical representation that describes the behavior of an electric arcing. In the last years, arcing models have been investigated in [15,16]. However, representing HIFs by arc models that can reproduce varied characteristics of the phenomenon is not always possible, which motivates the research in HIF modeling.

1.2. Motivation and Contribution

In the literature there are works related to the review of HIF detection and localization techniques [5,17–19], but the main goal is not to analyze HIF models. And the works that include HIF models [10,20], do not perform a validation of the analyzed models, on the other hand, some authors only validate the selected model separately [12,16,21,22], without comparing the waveform reproduction, using the different HIF models that exist in the literature. In that sense, the main contributions of this work are:

- A systematic review of existing works on HIF models for ODN with a critical analysis on reproducing HIF characteristics. The paper shows the evolution, popularity, limitations, advantages, and disadvantages of HIF models. As a result, researchers and specialists can save time in the selection of the appropriate HIF model for their studies.
- The primary contribution of this work is the comparison of three well-known HIF models for ODN with actual experiments conducted in a high-voltage laboratory at the Federal University of Para. This analysis highlights both the strengths and limitations of these models across various applications.

The motivation and contribution of this study arise from the limitations that the authors may experience at the beginning of the research on issues developed in HIF, as occurred in previous studies presented by the authors in [21,23–28]. Based on the above, this study contributes directly to the proposal and validation of the most appropriate detection or localization method in HIF.

2. High Impedance Fault Characteristics

The HIF characteristics contained in the current waveform, as defined follow:

- Asymmetry: Peak values of current are different in the positive and negative half cycle. The asymmetric nature of HIF current is influenced by the porosity and moisture of the surface contact. The presence of silica in the contact surfaces causes asymmetry, according to [9]. The heated silica forms a type of cathode spot that absorbs electrons, causing voltage drops when the cable is subjected to a positive voltage.
- Nonlinearity: Voltage x current characteristic curve is highly nonlinear. This characteristic is caused by the electric arc associated with the nonlinearity of high-impedance objects [9,29,30].
- Build-up: HIF current magnitude gradually increases up to its maximum value. This is due to: (a) the physical accommodation of the cable in the soil, since the cable can move or settle into the soil [12–31]; and (b) the arc penetrates the soil surface and causes soil ionization, increasing the effective area of the equivalent electrode [30].
- Shoulder: HIF current magnitude maintains constant right after the Build-up end.
- Intermittence: HIF electric arc is extinct during a time due to the loss of moisture in the surface and physical accommodation of the cable.
- Randomness: Peak values of current randomly oscillate at each half cycle within a relatively small range due to the random behavior of the electric arc.

Describe these characteristics related to [7], published in 1982. The works in [12] and in general, works addressing HIFs in ODN describe the first five characteristics of HIF mentioned above, since they are consolidated in the literature [32,33]. The randomness characteristic of a HIF, mentioned in [15,34–38], is manifested by a random oscillation in the HIF current waveform. In the technical literature, the characteristic of randomness and some others can be observed in actual HIF current waveforms presented in [11,12,15,34,37–43], in which the randomness issue is not addressed [11], among others, consider that the HIF current waveform can be divided into two distinct conditions: transient and steady-state conditions. Some HIF models only consider the steady-state condition, in which the HIF current waveform of a cycle is repeated over time. The transient state condition lasts only the first few cycles, during the build-up. The steady-state condition starts after the build-up end. The nonlinearity and asymmetry characteristics are contained in the steady-state current waveforms. The presence of randomness in the HIF current waveform leads the steady-state condition to a quasi-steady-state condition, as shown in Figure 1. It is observed that, after the current peaks build up, a random oscillation appears. HIF current peak values are within a standard deviation. This characteristic of randomness is contained practically in any HIF current waveform.

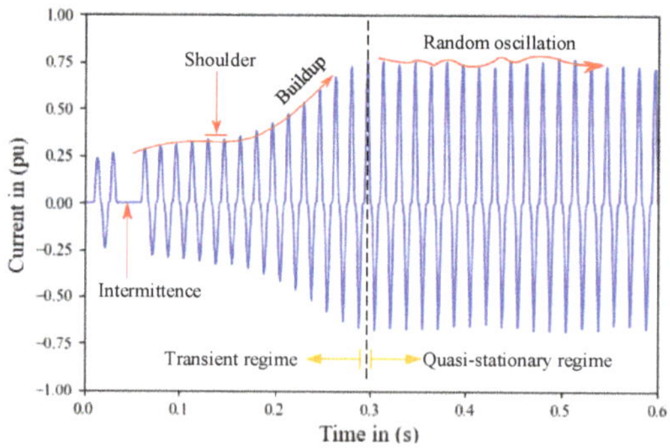

Figure 1. Typical HIF current waveform.

References [5,34,38,44] mention other characteristics contained in the HIF current waveform. However, this paper does not focus on other characteristics, since they are dependent on the main characteristics (e.g., the harmonic content depending on the nonlinearity characteristic). According to [30], HIFs are considered single-phase faults because an incident involving two or more phases is typically a high current fault event or can be treated as two isolated single-phase HIF events. Also, it should be borne in mind that the HIF current is exactly in phase with the voltage at the fault point and the resulting harmonic currents tend to reach their peaks at the same time as the HIF current fundamental component, according to [30,45,46].

3. Categories of HIF Models for ODN

3.1. Models Based on Active and Passive Circuit Elements

This section shows the evolution of the models of active and passive circuit elements used in the studies of the last 30 years for each HIF model in the ODN, classified in terms of their elements, representation and year of publication. The base model (model "a") and the rest of the models in this category can be constructed using Figure 2 and Table 1. Figure 2 shows the different active and passive elements that can be connected in series within elements 1, 2, 3 and 4. For example, in model "a", element 1 consists of a DC voltage source (V_P) and a diode (D_P) connected in series. Element 2 consists of a DC voltage source (V_N) and a diode (D_N) connected in series, and element 3 consists of a resistor (R) and an inductor (X) connected in series.

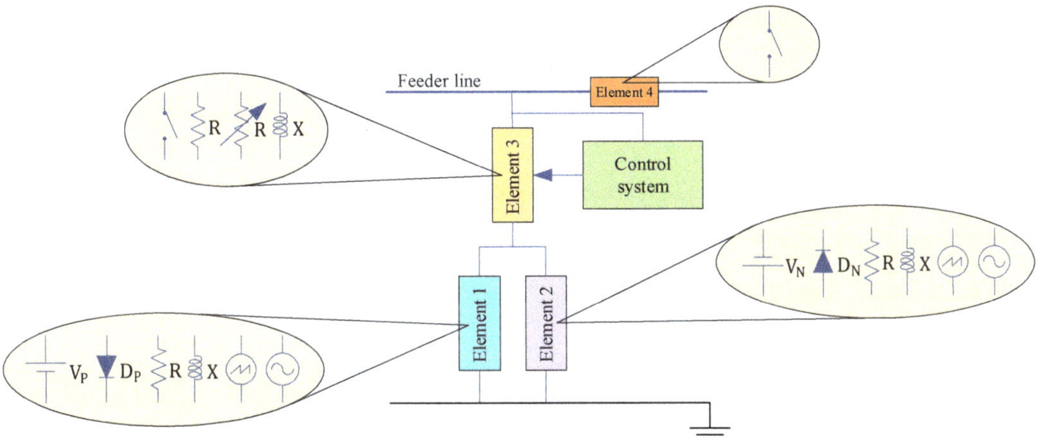

Figure 2. Representation of HIF models based on active and passive circuit elements according to Table 1.

The electrical circuit elements that compose a HIF model consist of a constant impedance (resistance R and reactance X) in series with two antiparallel diodes (D_P and D_N). Each diode is in series with a DC voltage source (V_P or V_N). This antiparallel-diode configuration allows the current during positive half cycles to circulate only through one branch and the current during negative half cycles flows through another branch. The fault current magnitude is controlled by the impedance value, but the source voltage values also influence the current magnitude control, as observed in Figure 3, which presents simulation results to understand the behavior of the model proposed in [9].

Table 1. Timeline of HID models based on active and passive circuit elements.

Model	Year	Ref.	Element 1				Element 2				Element 3			Element 4	Control System
a	1990	[9]	D_P	V_P	-	-	D_N	V_N	-	-	X	R	-	-	-
b	1993	[47]	D_P	R	V_P	-	D_N	R	V_N	-	(switch)	-	-	-	-
c	1998	[48]	D_P	(source)	-	-	D_N	(source)	-	-	(switch)	(switch)	R (variable)	(switch)	TACS
d	2004	[49]	D_P	V_P	-	-	D_N	V_N	-	-	(switch)	R (variable)	-	-	-
e	2005	[50]	D_P	V_P	(AC source)	-	D_N	V_N	(AC source)	-	R	R (variable)	-	-	MODELS
f	2006	[51]	D_P	X	R	V_P	D_N	X	R	V_N	-	-	-	-	-
g	2010	[52]	D_P	R	V_P	-	D_N	R	V_N	-	(switch)	-	-	-	-
h	2011	[53]	D_P	R	(source)	-	D_N	R	(source)	-	R (variable)	-	-	-	-
i	2015	[54]	D_P	R	V_P	-	D_N	R	V_N	-	(switch)	R (variable)	-	-	-
j	2016	[55]	D_P	V_P	-	-	D_N	V_N	-	-	X	R (variable)	-	-	-
k	2017	[31]	D_P	V_P	-	-	D_N	V_N	-	-	(switch)	R (variable)	-	(switch)	MODELS TACS

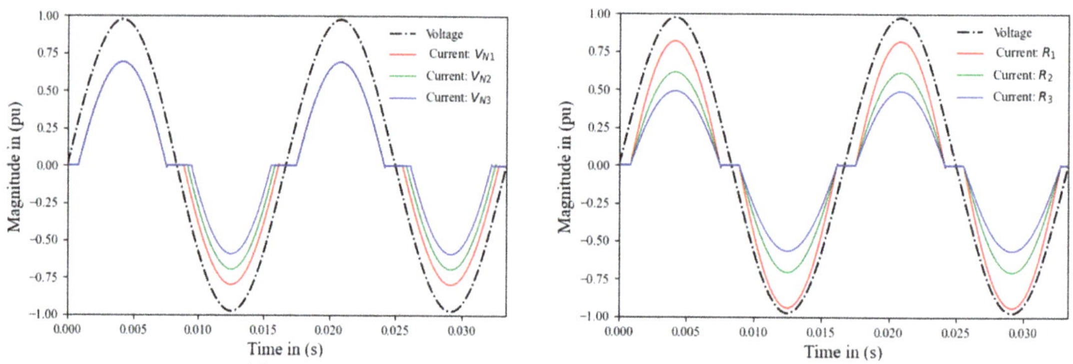

(**a**) (**b**)

Figure 3. High impedance fault current waveforms: (**a**) when source voltage varies; (**b**) when resistance varies.

In Figure 3a, the variation of the current for different values of voltage V_N are presented, where: $V_{N1} < V_{N2} < V_{N3}$, maintaining both constant impedance and constant voltage V_P. It can be concluded that the variable parameter (V_N) simultaneously modifies the current magnitude and the current conduction instants during negative semi-cycles (the three currents are superimposed during positive semi-cycles). By analogy, this behavior will be repeated during the positive semi-cycle, but varying V_P. Therefore, the variations of V_P and V_N only affect their respective semi-cycles. Figure 3b shows the current variation for different resistance values, where $R_1 < R_2 < R_3$, keeping constant the voltage values of the DC source (V_P and V_N). It is clearly observed that the variable parameter influences on current during two half-cycles and controls the current magnitude. It is important to note that the phase angle difference between the voltage at the fault point and the fault current is negligible, as indicated in the previous section. This fact must be considered to assign a minimum value of inductive reactance to models that consider this element.

As described in Section 2, the HIF current changes over time. However, in the model based on active and passive circuit elements, when it considers constant parameters from the beginning to the end of the simulation, produces a current waveform in which all cycles are equal over time, not allowing full reproduction of this characteristic. Regarding the non-linearity characteristic of HIF, the model does not reproduce this characteristic in the complete cycle, the linearity is maintained within the HIF current conduction intervals. Therefore, this model does not fully reproduce this feature, a fact that can be seen in Figure 4, where the current waveform of a typical HIF is shown, as well as the waveform reproduced from simulations (with the model "a" of Table 1) for a given voltage.

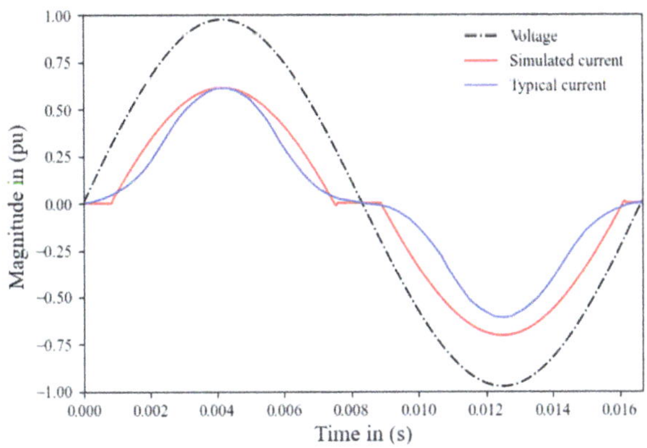

Figure 4. High impedance fault current and voltage waveforms.

In 1993, the model proposed in [9], derived from works [47,56], is modified as shown in the model "b", of Table 1. The reactance is neglected, and a resistance is considered in each branch (R_P and R_N), allowing independent control of the current magnitude at each half cycle, but without solving other model limitations.

Reference [57] clearly shows how to reproduce different fault current magnitudes, asymmetries, arc extinctions and arc restarts by modifying V_P, V_N, R_P and R_N. A model proposed later in [51,58] does not neglect the reactance, unlike [9], but it considers a resistance for each branch, as observed in the model "f" of Table 1, represented in Figure 2. The work in [59] proposes two non-linear resistances R_P and R_N, as observed in the model "b" of Table 1 in Figure 2, but it does not provide knowledge of this characteristic. With this proposal, the non-linear characteristic can be reproduced in the conducting and non-conducting intervals as presented in Figure 4. Since HIF is a random and nonlinear phenomenon, authors in [60–64] propose that the parameters V_P, V_N, R_P and R_N vary randomly between certain limits at each time interval (ms), following a uniform and/or Gaussian distribution.

The use of random values for elements of this model has been extended to several studies, such as [65–68]. By proposing the time-varying parameters, the fault current waveform is no longer the same as that of the other cycles. However, the *build-up* and *shoulder* characteristics are not guaranteed due to random behavior. Modified models of [9], configured in parallel, are proposed in [52,69] (see model "g" of Table 1 represented in Figure 2). The energization of each of the models in parallel is done by employing switches. According to the authors, this model can reproduce the characteristics of randomness and *build-up* but adding a greater number of parameters and a switching control system.

In [48,53], DC sources are replaced by sawtooth waveform generators (see models "c" and "h" of Table 1 represented in Figure 2), to control the phase difference between the voltage at the fault point and the fault current. But as described in Section 2, it is not necessary to control this difference, because it is negligible. In addition, a variable resistance is added to the model. In the model in Figure 2, switch 1, normally open, isolates the downstream load from the fault point, and switch 2, normally closed, connects the HIF model to the electrical system. Switch 3 controlled by TACS controls the re-ignition and extinction of the electric arc, to reproduce the intermittency, which is one of the main characteristics of the HIF.

Afterwards, in [54], it is proposed to randomly modify the value of the resistance, in each cycle, and the model is according to the model "i" of Table 1. The references described ([48,53,54]) are not intended to reproduce buildup and other features. In [49,70], the reactance of the model proposed in [9] is neglected and the fixed resistance becomes a non-linear resistance, as can be observed in the model "d" of Table 1, represented in Figure 2. In [49], the voltage values of the DC sources vary randomly at each half-cycle. In [70], the resistance value varies randomly under a uniform distribution. However, in [49] there is the problem of dependence between the beginning of the current conduction and the magnitude of the current waveform, as can be observed in Figure 3a. The HIF models proposed in [49,70] reproduce the current waveform nonlinearity and the current magnitude random behavior.

In [50], a model modification is proposed based on a combination of the model proposed in [9] with the arc model based on a differential equation, according to the model "e" of Table 1. This model is based on works [9,48,71]. Some new features have been added to the model, such as the impedance from [9] which has been replaced by a linear resistor and a time-varying non-linear resistor, representing the earth resistance and the dynamic arc, respectively. AC sources have also been added to the model guaranteeing the variable point of arc ignition and cooling. This model is not simple, since it is practically the union of two models classified into different categories. Therefore, the model complexity is increased due to the greater number of elements composing the model.

In [55], the constant resistance of the model from [9] is replaced by a time-varying resistance to correctly simulate the build-up and shoulder. The model remained simple, as can be observed in the model "j" of Table 1 in Figure 2. The time-varying resistance is modeled by a polynomial function and a constant for transient and quasi-steady states, respectively.

The modeling proposed in [55] can be associated with [11]. In reference [39], regarding the circuit configuration, a resistance is added to each branch that represents the arc resistance. In reference [72], the model proposed in [9] is adapted to be used in the frequency domain, demonstrating the importance of this modeling in the literature. In short, this HIF model can reproduce the characteristics of asymmetry, nonlinearity, build-up, and shoulder considering transient and quasi-steady states.

The model "k" in Table 1, represented by Figure 2 shows the last model based on the combining of electric circuit elements, which was proposed in [31]. It is basically the model proposed in [9] with some features of other models. The main difference is that the model proposed in [31] has a complex control system of resistance and a sectionalizing switch that controls the current circulation. The resistance consists of three portions that add up to the total resistance value. The first portion is established to reproduce the build-up. The second portion is responsible for simulating the asymmetry between half cycles of the same cycle. Finally, the last part reproduces the arc resistance variation, similarly to [11].

It is important to highlight that the evolution of this category of HIF models has not been independent of other categories. Some features present in models of other categories have helped in the evolution of models in this category and vice-versa. A clear example is the way described in [11] to model the build-up and shoulder.

An important observation is that, when this category of HIF models is used, the fault current waveform does not faithfully reproduce waveforms of an actual HIF, especially in the vicinity of zero crossings. In Figure 4, it can be observed the difference between the simulated current (I_{sim}) using the model from [9] and the typical current (I_{typ}) for a type of surface (e.g., gravel). It is worth noting that the nonlinear resistance can reproduce nonlinearity in the current conduction interval, but not in the non-conduction interval. Last, when DC sources are considered in the model, the reproduced current has limitations in the zero-crossing.

3.2. Models Based on Passive Circuit Elements

Figure 5 and Table 2 illustrates a representation of the different HIF models based on passive elements such as resistance and/or reactance. It can be observed that the evolution of this category of models is not based on a single model, unlike the evolution of models presented in Section 3.1. First, in 1981, authors in [73] proposed a HIF model based on two series nonlinear resistance as a function of the current as shown in the model "a" of Table 2 and represented in Figure 5.

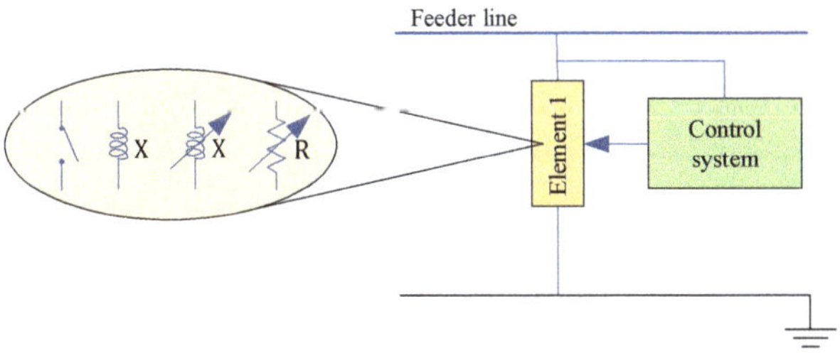

Figure 5. Representation of high impedance fault models based on passive circuit elements according to Table 2.

Table 2. High impedance fault models based on passive circuit elements.

Model	Year	Ref.	Element 1		Control System
a	1981	[73]	R	R	-
b	1996	[74]	X	R	-
c	1998	[46]		R	MODELS
d	2000	[75]	X	R	-
e	2001	[11]	R	R	TACS

One resistance represents the arc resistance, and the other represents the earth resistance. Subsequently, in [74], HIF is modeled by a resistance and an inductance, both non-linear as a function of the current, subject to a second-order polynomial, as can be observed in Figure 5 and the model "b" of Table 2. Similarly, in [75], HIF is simulated in two ways: (a) constant resistance and reactance; and (b) non-linear resistance as a function of current and constant reactance, as can be observed in Figure 5 and model "d" of Table 2. These models have a common ability to reproduce the characteristic of nonlinearity.

In 1985, the work in [76] used the simplest model consisting of a constant resistance connected at the fault point. This model reproduces only the low HIF magnitude. In recent years, some authors used this model, such as [77], published in 2015. Afterward, the authors in [78] proposed a model based on a combination of 16 nonlinear impedances with different degrees of nonlinearity and with an arc start for different voltage values. HIF simulations can be performed by applying any combination of 16 impedances, that is, one impedance is used for each half-cycle of the voltage waveform. However, this model reproduces with limitations in the shoulder and build-up characteristics through a combination sequence that increases the model complexity. In reference [45], the random nature of the arc has stochastically been simulated using the MODELS language in ATP software (Model "c" of Table 2 and Figure 5). This model consists of a normally open switch connected in series with a variable fault resistance to produce a variety of fault current waveforms each half cycle. After the fault starts, periodic switching operations are determined by the voltage values of the corresponding arc, chosen randomly for each half cycle. This model has the ability to reproduce the intermittency characteristic, but the reproduction of the build-up and shoulder characteristics is not guaranteed due to the stochastic simulation.

The work in [11], published in 2001, presents the most important HIF model this category of models based on passive elements, illustrated in Figure 5 and Table 2, model "e", since it is capable of reproducing several characteristics of HIF. This model consists of two nonlinear resistances in series, one resistance is responsible for simulating the asymmetry and nonlinearity characteristics, and the other resistance is controlled by TACS to simulate the build-up and shoulder characteristics. The two resistances are determined from curves of voltage at the fault point versus fault current.

In reference [12], published in 2013, the model from [11] is combined with the configuration of switches proposed by [48] (see model "c" of Table 1 in Figure 2), and switches S_1 and S_2 are used for disconnecting downstream load from the fault point and connecting the HIF model at the fault point, respectively. In reference [12], experimental tests were carried out to establish HIFs of six surface types. The test results were used to model each HIF associated with a surface type. Subsequently, ref. [79] has proposed a control in the switch S_3 to simulate the intermittency characteristic. Switch operations are the same adopted in [48], as seen in model "c" of Table 1 in Figure 2. The evolution of this model allows us to reproduce five basic HIF characteristics described in Section 2. Reference [80] proposes the use of two series connected resistors, one dielectric resistance and the other constant bad contact resistance between the conductor and dielectric. The dielectric resistance behavior is based on the solid dielectric electrical breakdown theory. The author shows that the proposed nonlinear resistance model is more consistent in reproducing current waveform at zero-crossing points than the traditional arc model based on differential equations (Mayr model). However, the work does not show the reproduction of the build-up and shoulder characteristics.

3.3. Arc Model

This subsection presents the third category of HIF modeling, which is expressed in the form of a first-order differential equation and represented by the Figure 6 and the Table 3. This HIF model is well known as the arc equation and basically consists of: (1) a single resistance subject to the arc model, as illustrated in Figure 6 and model "a" of Table 3; and (2) two resistances, one that represents the arc resistance (modeled by a differential equation) and the other that can be a constant resistance, shown in Figure 6 and model "b" of Table 3, or a variable over time, shown in Figure 6 and model "c" of Table 3, which represents the resistance of a surface. According to [22], the thermal model has been used since the first descriptions of the arc electrical conductivity shown by Cassie [13] and Mayr [14], in the form of a first-order differential equation. This model has subsequently been improved and modified according to needs. For example, ref. [81] details the implementation of the arc model described in [71] using EMTP. Known as the arc digital model and derived from the description of Hochrainer's arc, the model from [71] is derived from the works of Mayr and Cassie.

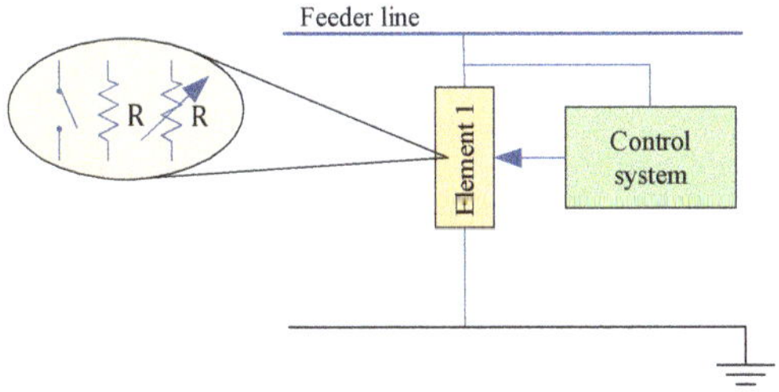

Figure 6. Representation of high-impedance fault models based on the physical process involved in the electric arc, according to Table 3.

Table 3. High impedance fault models based on the physical process involved in the electric arc.

Model	Year	Ref.	Element 1			Control System
a	2006	[82]			-	MODELS
b	2007	[21]				MODELS
c	2008	[83]			-	MODELS

In [82], HIF is modeled by a single resistance based on the first-order differential equation in (1), as originally proposed in [71]. To simulate the random behavior of the phenomenon, the arc voltage per unit length parameter of the differential equation assumes random values within pre-defined limits (G_0 is a function of the arc voltage per unit length parameter). In reference [22], HIF is also modeled using the detailed arc description from [71], and to implement the model, the implementation method from [81] is used. Reference [22] models a HIF caused by the cable contact with the tree. This model is represented by two resistances in series, one constant that represents the tree resistance and the other variable that represents the arc resistance.

$$\frac{dg_{(t)}}{dt} = \frac{G_0 - g_{(t)}}{\tau} \quad (1)$$

where G_0 is stationary arc conductance, τ is arc time constant, and $g(t)$ is time-varying arc conductance [22,71].

In works [83,84], two variable resistances in series are used, one that represents the surface resistance varying between two limits and the other that represents the arc resistance modeled by Equation (1). To reproduce the randomness characteristic, parameter τ takes random values with uniform distribution within predefined limit values.

References [15,85,86] model the HIF based on the Equation (1), whose parameters are obtained from the voltage waveform at the fault point and fault current waveform using the least squares method and the sum of squared deviations from theoretical and experimental values. In these works, the reproduction of nonlinearity and build-up characteristics can be observed. Among works addressing HIF modeling based on the arc model, ref. [15] is recommended because the model is described in a detailed way, showing the calculation of differential equation parameters. Reading [22] is also recommended. Finally, it is worth mentioning that, according to [80], the model based on the arc equation is not consistent in reproducing the HIF waveform at zero-crossing points (see Figure 3 in [80]) and therefore it has a similar limitation to models of the first category.

4. High Impedance Fault Models for ODN

The interaction between any HIF model with an ODN can be illustrated in Figure 7. The configurations of switches S_1, S_2, and S_3 are based on [48]. The switches S_1, S_2, and S_3 allow to simulate different HIF classes. Under normal system conditions, S_1 and S_2 are normally closed and S_3 is normally open. In the case of a load-side HIF, S_1 is normally open and S_2 and S_3 are normally closed, while if the fault is from the source side, S_1 and S_3 are closed, and S_2 is open, in both scenarios there is a similarity between the waveform, both from the source and from the load, as analyzed in [87]. It is also possible to simulate the cable downfall time, which is represented by the time S_3 takes to close. For faults involving the unbroken conductor in contact with a tree branch, the three switches are normally closed. It is worth highlighting that S_3 has two functions, one that is to connect the HIF model to the distribution system model and the other that is to simulate the intermittence characteristic.

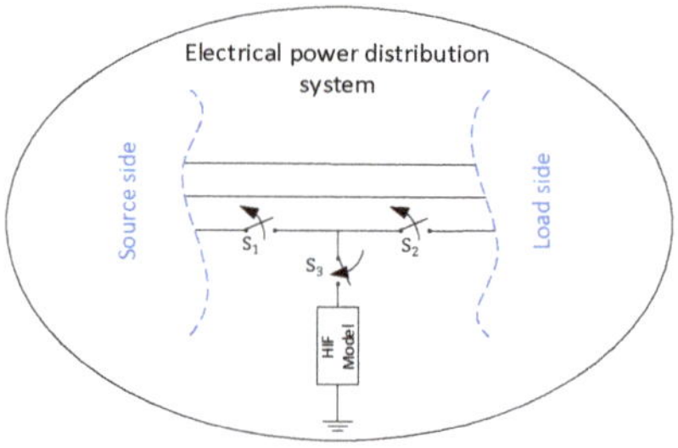

Figure 7. Interaction between the HIF model and distribution network model.

Figure 8a,b show the circuits that represent categories of HIF models based on a combination of electric circuit elements and passive elements, respectively. In the category of models based on a combination of electric circuit elements, the parameter values can be modified to reproduce other models in the same category.

Figure 8. High impedance fault models: (**a**) based on active and passive circuit elements; (**b**) based on passive circuit elements.

For example, making $R_P = R_N$, the circuit model of Figure 8a reproduces the same behavior as the circuit model "d" of Table 1 represented in Figure 2. In this model category, the inductive reactance X_L and other elements proposed to control the phase difference between the voltage at the fault point and fault current are neglected. Equation (2) relates the voltage at the fault point and the fault current using the sign function expressed in Equation (3) to represent the diodes behavior, similar to [88].

$$v_{(t)} = \left(R_P \cdot i_{(t)} + V_P\right) \cdot sgn\left(i_{(t)}\right) + \left(-R_N \cdot i_{(t)} + V_N\right) \cdot sgn\left(i_{(t)}\right) \quad (2)$$

where $sgn\left(i_{(t)}\right)$ is the current sign function, as follows.

$$sgn\left(i_{(t)}\right) = \begin{cases} 1 \; if \; i_{(t)} < 0, \\ 0 \; if \; i_{(t)} = 0, \\ -1 \; if \; i_{(t)} > 0. \end{cases} \tag{3}$$

Note that HIF models based on a combination of electrical circuit elements are easy to implement, especially when model parameters are constant. However, such models can become complex when the dynamic behavior of model parameters is considered due to the need to employ a control system. To use HIF models of this category, the values of resistances and DC sources must be known. To simulate the random behavior as suggested by many authors, it is necessary to randomly modify the interest parameter values during the simulation. In the case of nonlinear behavior of the resistances, the modification of parameter values over time can be done following a certain function (e.g., a polynomial function). It is worth noting that R_P and R_N can be composed of several resistances, as proposed in [31]. Figure 8b shows the electrical circuit that represents categories of HIF models based on passive elements and order first differential equations. It is possible to obtain different HIF models that compose these two categories, defining the constant or dynamic behavior of the resistances.

The relationship between the voltage at the fault point and fault current is given by:

$$v_{(t)} = \left(R_{1(t)} + R_{2(t)}\right) \cdot i_{(t)} \tag{4}$$

Equation (4) is apparently very simple; however, the model complexity is proportional to the dynamic behavior of the resistances, similar to the first category of HIF models.

The HIF models based on passive elements, which reproduce the greatest number of HIF characteristics, represent the dynamic behaviors of the two resistances. According to [11], dynamic behaviors of $R_1(t)$ and $R_2(t)$ are expressed by Equations (5) and (6), respectively. It is worth mentioning it is possible to reproduce the randomness characteristic adding to the model a third resistance that varies randomly.

$$R_1(t) = \frac{v(t)}{i_n + \frac{i_{n+1} - i_n}{v_{n+1} - v_n} \cdot (v(t) - v_n)} \tag{5}$$

$$R_2(t) = a_0 + a_1 \cdot t^1 + a_2 \cdot t^2 + \ldots + a_{n-1} \cdot t^{n-1} + a_n \cdot t^n, \tag{6}$$

For the HIF model based on the order first differential equation, $R_1(t)$ is determined by the arc equation in (1) and $R_2(t)$ is constant or time-varying. However, ref. [15] shows that is possible to model a HIF making $R_2(t) = 0$.

The solution from Equation (1) is expressed by Equation (7), in which the calculation of parameters G_0 and τ are described in detail by [15]. $R_1(t)$ is expressed by Equation (8).

$$g(t) = G_0 \cdot \left(1 - e^{-t/\tau}\right) \tag{7}$$

$$R_1(t) = \frac{1}{g(t)} \tag{8}$$

From the mathematic point of view, the three categories of HIF models do not present a clear difference in terms of model implementation complexity. To use HIF models based on a combination of electric circuit elements, four parameters (V_P, V_N, R_P, and R_N) of Equation (2) need to be determined. Simulation of a HIF becomes simple if these four parameters are constant. However, it is necessary to determine the four parameters if the resistances are time-varying, increasing the model implementation complexity. To use HIF models based on passive elements, it is necessary to determine the behaviors of the two resistances, which usually are time-varying. This HIF model becomes the simplest of all HIF models if the two resistances are constant. Lastly, to use HIF models based on the order first differential equation two parameters need to be determined. These two parameters usually are constant during simulation, which implies in limitations to reproduce all HIF characteristics. On the other hand, when these parameters are time-varying is necessary to determine a time-varying function.

5. Comparison of HIF Models for ODN with MV Lab Test Measured Data

5.1. Materials and Methods

Besides the systematic literature review, this study presents an experimental evaluation to verify the three predominant HIF models with real data obtained from oscillographs in laboratory facilities. Figure 9 shows the methodology used in this study. The computational evaluation was carried out using the ATP/EMTP software (https://atp-emtp.org/, accessed on 25 January 2020) for the simulation of all the HIF models. The following types of HIF models for ODN are compared in this section with lab test results:

- Model 1—model based on active and passive circuit elements;
- Model 2—model based on passive elements;
- Model 3—arcing model.

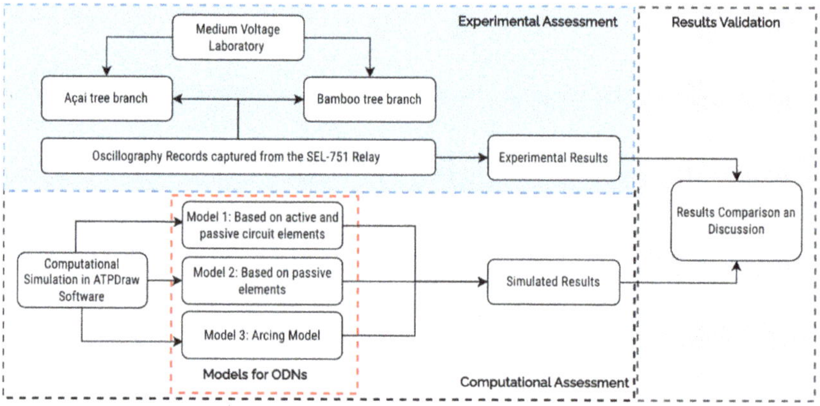

Figure 9. Flowchart of the methodology proposed for the model validation.

For the experimental assessment, the HIF experiments have been carried out at the MV distribution system laboratory or MV lab, Federal University of Para, Brazil. The MV lab was designed and developed for HIF testing purposes. Figure 10 shows the facilities and equipment where the experiments were carried out, and Figure 11 the three-line diagram of the MV laboratory. The MV lab comprises a 225 kVA, 0.22 V/13.8 kV transformer of the WEG manufacturer (Wetherill Park, Australia), SEL-751 feeder protection relay of the SEL manufacturer (Pullman, WA, USA), an alternating current (AC) contactor of the WEG manufacturer, breaker circuits, fuses, potential transformers (PTs), current transformers (CTs), conductors, 75 kVA, 13.8 kV/0.22 kV service transformer of the WEG manufacturer and load.

The 225 kVA transformer is delta connected on the low-voltage (LV) side (0.22 kV) and grounded wye connected on the medium voltage (MV) side (13.8 kV) to emulate a typical distribution substation. The MV system is a three-phase three-wire circuit, and the neutral conductor is grounded on the transformer. In this paper, oscillography records captured by the SEL-751 relay are used to compare the three HIF models. The oscillography records were obtained from two HIF experimental tests, one using a Bamboo tree branch and the other using an Açai tree branch. The SEL-751 relay's residual overcurrent element was used for tripping purposes with a sensitive setting.

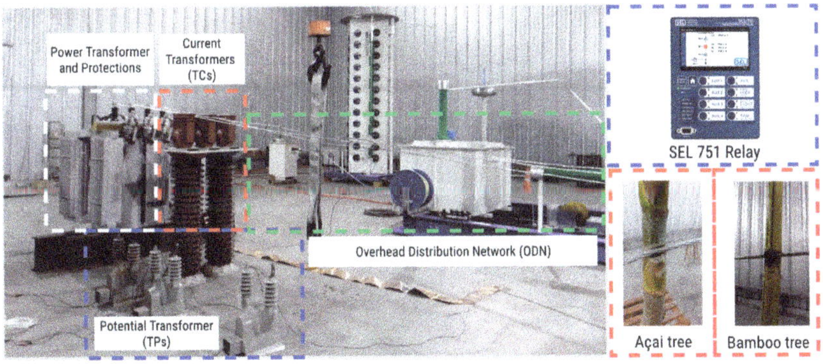

Figure 10. Medium voltage laboratory for the HI testing.

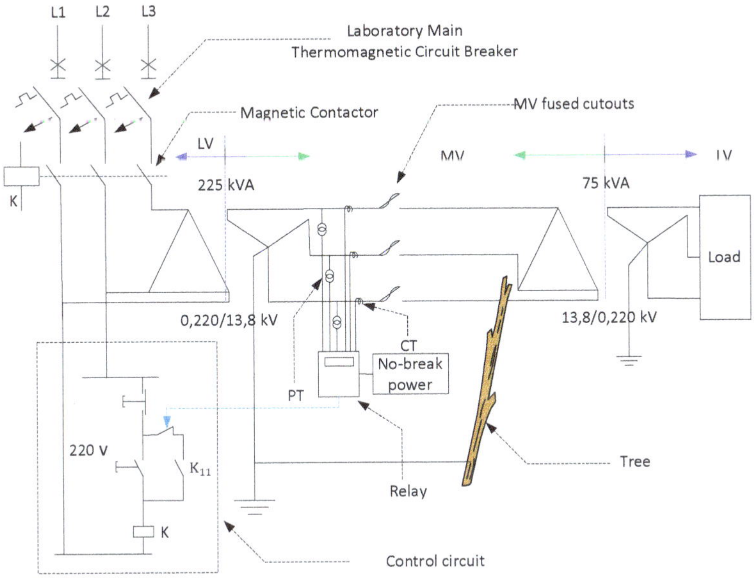

Figure 11. Three-line diagram of the MV lab for HIF testing.

5.2. Results and Discussion

From the measurements in the laboratory presented in Figure 10, data were obtained from the relay for further processing and analysis. Figures 12 and 13 show HIF current waveforms obtained through experimental laboratory tests using a bamboo tree branch and an açai tree branch (common trees in the Amazon region, site of the study), respectively, both subjected to a $13.8/\sqrt{3}$ kV phase voltage. The two waveforms were selected because they present the typical characteristics of a HIF, such as nonlinearity, asymmetry, shoulder and build up, especially because they have a marked envelope, one with an increasing envelope, where the magnitude of the instantaneous current exceeds 10 A, and the other with an oscillating envelope, where the maximum current does not exceed 3 A. To evaluate the limitations or capabilities of the studied models, the two HIF current waveforms are reproduced, using the three selected models. The models are implemented in the ATPDraw software version 7.0, and the results of the simulations are presented graphically, where the differences in the reproduced currents in relation to the measured current can be visually appreciated.

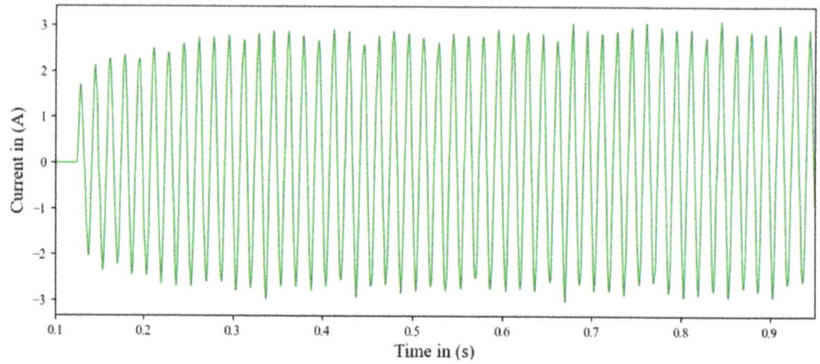

Figure 12. Measured current waveform obtained from a HIF experimental test with a Bamboo tree branch.

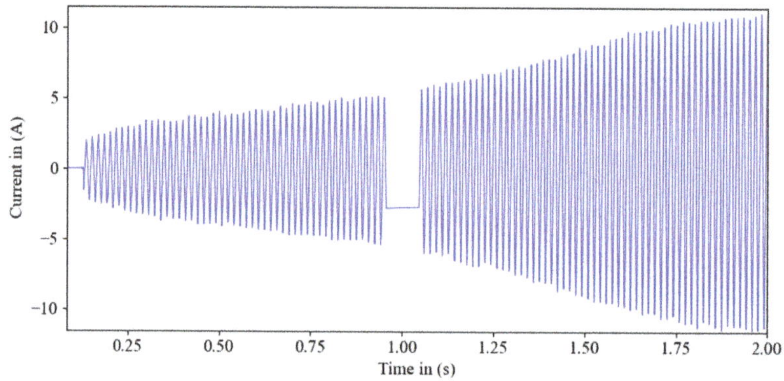

Figure 13. Measured current waveform obtained from a HIF experimental test with an Açai tree branch.

Figure 14 shows the comparison of the conductance reproduced by each type of HIF model with the measured conductance data (red color curve) from an experimental HIF test with a bamboo tree branch. The conductance is determined by the relationship between simulated or measured current and voltage waveforms. When Model 1 is used, the average conductance calculated for positive and negative cycles are 0.00023084 S and 0.00023094 S, respectively.

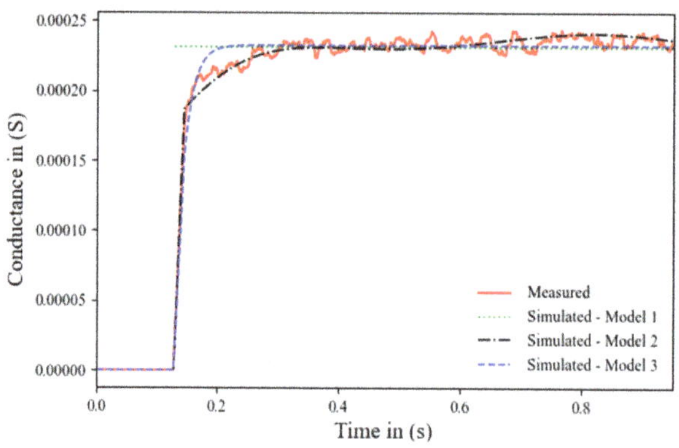

Figure 14. Comparison of the conductance reproduced by each type of HIF model with the measured conductance data from a HIF experimental test with a bamboo tree branch.

In this case, they are practically the same, indicating that the asymmetry is negligible. It is noteworthy that the parameters of Model 1 are constant and therefore the conductance is represented by a straight (cyan color line). The black color curve is the conductance calculated using Model 2, represented by a fifth-order polynomial expressed by Equation (6), where its coefficients were determined by fitting the curve. The blue color curve is the conductance calculated using Model 3, whose parameters $G_0 = 0.000232432$ S and $\tau = 0.016045$ s were estimated by the least-squares method. Note that the conductance calculated using the model 2 has better matching with the measured conductance (red color curve), both in the transient and steady-state conditions. The steady-state condition is reached when t = 0.35 s.

Figures 15 and 16 show comparisons of the current waveforms calculated by HIF models with the measured current waveforms resulting from a HIF experimentally generated by a bamboo tree branch, for transient and steady-state conditions, respectively. The steady-state condition is reached when t = 0.3 s. It can be observed that the Model 2 more accurately replicates the measured transient waveform when compared with Models 1 and 3. As for the steady-state waveform, all HIF models have similar and sufficiently accurate results.

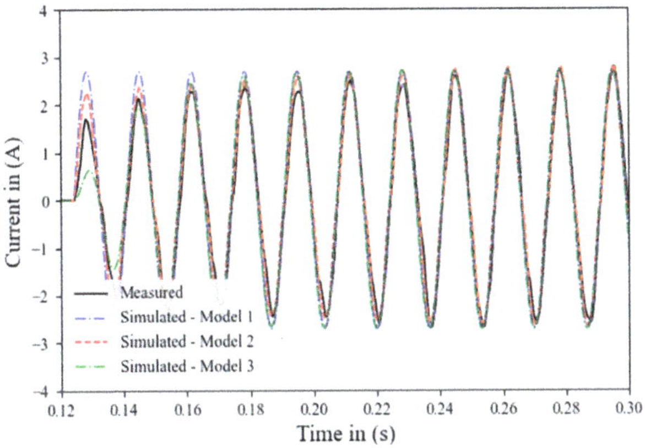

Figure 15. Comparison of the simulated current waveforms (transient state) with the measured current waveform resulting from an experimental test with a bamboo tree branch.

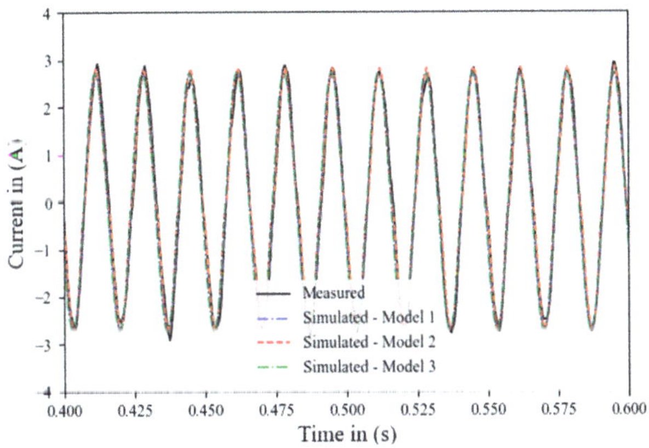

Figure 16. Comparison of the simulated current waveforms (steady-state condition) with the measured current waveform resulting from an experimental test with a bamboo tree branch.

Figure 17 shows the comparison of the conductance reproduced by each type of HIF model with the measured conductance data (red color curve) from an experimental HIF testing with açai tree branch. The average conductance calculated using Model 1 for positive and negative cycles are 0.00059475 S and 0.00059823 S, respectively. The black color curve is the conductance calculated using Model 2 represented by a sixth-order polynomial. The blue color curve is the conductance calculated using Model 3, whose parameters $G_0 = 0.001699$ S and $\tau = 2.345621$ S were estimated by the least-squares method. It can be observed that model 2 accurately reproduces the measured conductance in the transient and steady-state conditions.

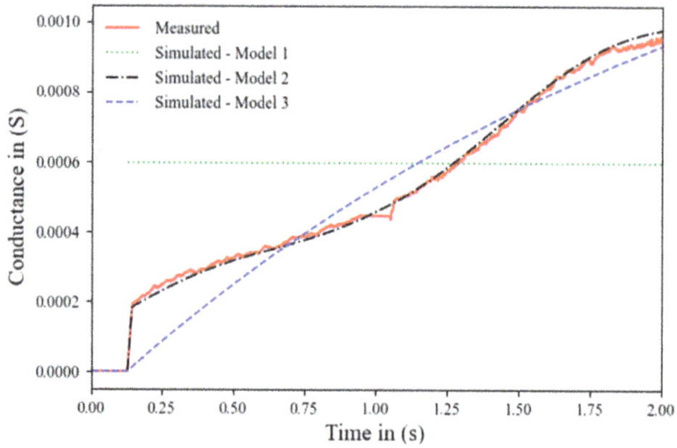

Figure 17. Comparison of the conductance reproduced by each type of HIF model with the measured conductance data from an HIF experimental test with the Açai tree branch.

Figures 18 and 19 show comparisons of the current waveforms calculated by HIF models with the measured current waveforms resulting from a HIF experimentally generated by an Açai tree branch, for transient and steady-state conditions, respectively. The steady-state condition is reached when t = 0.35 s. It can be observed that the Model 2 accurately replicates the measured current waveform when compared with Models 1 and 3, for both transient and steady-state conditions.

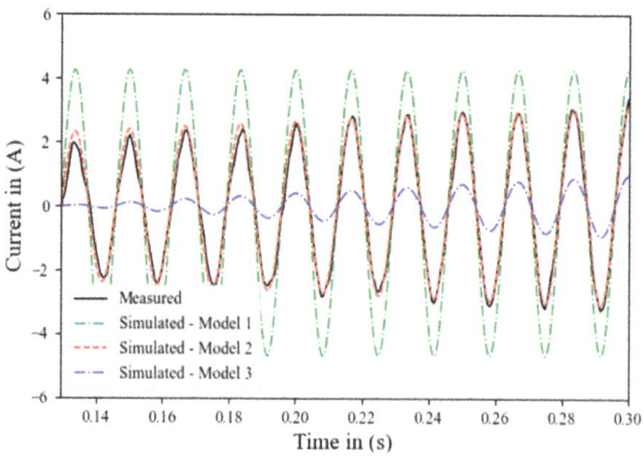

Figure 18. Comparison of the simulated current waveforms (transient state) with the measured current waveform resulting from an experimental test with an Açai tree branch.

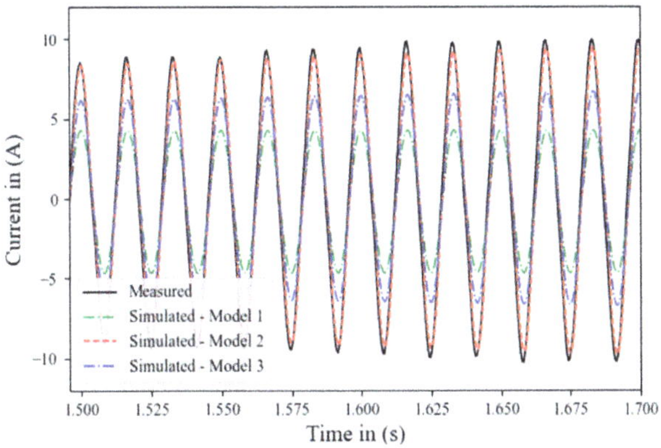

Figure 19. Comparison of the simulated current waveforms (steady-state condition) with the measured current waveform resulting from an experimental test with an Açai tree branch.

6. Conclusions

Currently, models based on active and passive circuit elements are the most widely used category in investigations related to the detection and location of HIF. This paper systematically reviewed three broad categories of HIF models for overhead distribution network studies performed by EMT-type programs and compared three well-known HIF models, one of each category, with MV laboratory experimental data. The comparison presented in this work showed that the model based on two non-linear resistances is more accurately matched with measured waveforms obtained from two experimental HIF tests using tree branches when compared with other models. The presented review, models, studies, and comparison will be useful to researchers and practicing distribution protection engineers who are actively using EMT simulation tools for modeling and analysis of overhead distribution networks. The lab experimental results reported in this paper encourage further research to develop new high-impedance fault models using a large database repository of lab experimental tests or actual field events.

Author Contributions: Conceptualization, J.C.H.P. and J.P.A.V.; methodology, A.P.L. and J.C.H.P.; software, M.A.M.C. and J.C.H.P.; validation, J.M.T. and J.C.H.P.; formal analysis, G.C.J., A.P.d.M. and P.E.F.; investigation, J.C.H.P., J.M.T. and A.P.L.; resources, J.P.A.V., G.C.J., A.P.d.M. and P.E.F.; data curation, J.C.H.P., A.P.L. and M.A.M.C.; writing—original draft preparation, J.C.H.P., J.M.T. and J.P.A.V.; writing—review and editing, J.C.H.P., J.M.T., A.P.L. and M.A.M.C.; supervision, J.P.A.V., G.C.J., A.P.d.M. and P.E.F.; funding acquisition, J.P.A.V. All authors have read and agreed to the published version of the manuscript.

Funding: This work was supported by the Energy Equatorial Group at the Research and Development Project Number PD-00371-0039/2019 and the publication taxes by the Pro-Rectory of Research and Post-Graduate Studies-PROPESP/UFPA with call PI O10ACO9403N-PAPQ.

Conflicts of Interest: The authors declare no conflict of interest.

References

1. Adamiak, M.; Wester, C.; Thakur, M.; Jensen, C. High Impedance Fault Detection on Distribution Feeders. 12p. Available online: https://www.aer.gov.au/system/files/Mr%20Marcus%20Steel%20-%20Attachment%20B%20to%20submission%20on%20Ergon%20Energy%20application%20for%20ring%20fencing%20waiver%20-%20December%202015.pdf (accessed on 3 September 2022).
2. Ghaderi, A.; Mohammadpour, H.A.; Ginn, H.L.; Shin, Y.-J. High-Impedance Fault Detection in the Distribution Network Using the Time-Frequency-Based Algorithm. *IEEE Trans. Power Deliv.* **2015**, *30*, 1260–1268. [CrossRef]
3. Theron, J.C.J.; Pal, A.; Varghese, A. Tutorial on high impedance fault detection. In Proceedings of the 2018 71st Annual Conference for Protective Relay Engineers (CPRE), College Station, TX, USA, 26–29 March 2018; pp. 1–23.

4. Ojanguren, I.; Steynberg, G.; Channelière, J.-L.; Karsenti, L.; Das, R. *High Impedance Faults*; CIGRE BROCHURE by WG B5.94; CIGRE Publication: Paris, France, 2009.
5. Ghaderi, A.; Ginn, H.L.; Mohammadpour, H.A. High impedance fault detection: A review. *Electr. Power Syst. Res.* **2017**, *143*, 376–388. [CrossRef]
6. Balser, S.J.; Clements, K.A.; Kallaur, E. Detection of High Impedance Faults. Available online: http://royalcommission.vic.gov.au/Commission-Reports/Final-Report/Summary.html (accessed on 6 May 2020).
7. Russell, B.D. *Detection of Arcing Faults on Distribution Feeders. Final Report*; Research Foundation, Texas A and M University: College Station, TX, US, 1982.
8. Hou, D. Detection of high-impedance faults in power distribution systems. In Proceedings of the 2007 Power Systems Conference: Advanced Metering, Protection, Control, Communication, and Distributed Resources, Clemson, SC, USA, 13–16 March 2007; pp. 85–95.
9. Emanuel, A.E.; Cyganski, D.; Orr, J.A.; Shiller, S.; Gulachenski, E.M. High impedance fault arcing on sandy soil in 15 kV distribution feeders: Contributions to the evaluation of the low frequency spectrum. *IEEE Trans. Power Deliv.* **1990**, *5*, 676–686. [CrossRef]
10. Zamanan, N.; Sykulski, J. The evolution of high impedance fault modeling. In Proceedings of the 2014 16th International Conference on Harmonics and Quality of Power (ICHQP), Bucharest, Romania, 25–28 May 2014; pp. 77–81.
11. Nam, S.R.; Park, J.K.; Kang, Y.C.; Kim, T.H. A modeling method of a high impedance fault in a distribution system using two series time-varying resistances in EMTP. In Proceedings of the 2001 Power Engineering Society Summer Meeting, Vancouver, BC, Canada, 15–19 July 2001; Conference Proceedings (Cat. No.01CH37262). Volume 2, pp. 1175–1180.
12. dos Santos, W.C.; de Souza, B.A.; Brito, N.S.D.; Costa, F.B.; Paes, M.R.C. High Impedance Faults: From Field Tests to Modeling. *J. Control Autom. Electr. Syst.* **2013**, *24*, 885–896. [CrossRef]
13. Cassie, A.M. *Theorie Nouvelle des Arcs de Rupture et de la Rigidité des Circuits*; Cigre: Paris, France, 1939; pp. 588–608.
14. Mayr, O. Beiträge zur Theorie des statischen und des dynamischen Lichtbogens. *Arch. F. Elektrotechnik* **1943**, *37*, 588–608. [CrossRef]
15. Torres, V.; Guardado, J.L.; Ruiz, H.F.; Maximov, S. Modeling and detection of high impedance faults. *Int. J. Electr. Power Energy Syst.* **2014**, *61*, 163–172. [CrossRef]
16. Torres-Garcia, V.; Guillen, D.; Olveres, J.; Escalante-Ramirez, B.; Rodriguez-Rodriguez, J.R. Modelling of high impedance faults in distribution systems and validation based on multiresolution techniques. *Comput. Electr. Eng.* **2020**, *83*, 106576. [CrossRef]
17. Lavanya, S.; Prabakaran, S.; Kumar, N.A. Literature Review: High Impedance Fault in Power System and Detection Techniques. *Math. Stat. Eng. Appl.* **2022**, *71*, 944–958. [CrossRef]
18. Li, L.; Redfern, M.A. A review of techniques to detect downed conductors in overhead distribution systems. In Proceedings of the 2001 Seventh International Conference on Developments in Power System Protection (IEE), Amsterdam, The Netherland, 9–12 April 2001; pp. 169–172.
19. Aljohani, A.; Habiballah, I. High-Impedance Fault Diagnosis: A Review. *Energies* **2020**, *13*, 6447. [CrossRef]
20. Varghese, P.R.; Subathra, M.S.P.; Mathew, C.; George, S.T.; Sairamya, N.J. High Impedance Fault Arc Modeling—A Review. In *Intelligent Solutions for Smart Grids and Smart Cities*; Siano, P., Williamson, S., Beevi, S., Eds.; Springer Nature: Singapore, 2023; pp. 163–181.
21. Cordeiro, M.A.M.; Heringer, W.R.; Paye, J.C.H.; Sousa, A.L.; Leão, A.P.; Vieira, J.P.A.; Farias, P.E.; Wontroba, A.; Gallas, M.; Rossini, J.P.; et al. Validation of a High Impedance Fault Model for Overhead Distribution Networks Using Real Oscillography Data. In Proceedings of the 13th Latin-American Congress On Electricity Generation and Transmission, Santiago de, Chile, Chile, 20–23 October 2019; FDCT: Irvine, CA, USA, 2019.
22. Elkalashy, N.I.; Lehtonen, M.; Darwish, H.A.; Izzularab, M.A.; Taalab, A.I. Modeling and experimental verification of high impedance arcing fault in medium voltage networks. *IEEE Trans. Dielectr. Electr. Insul.* **2007**, *14*, 375–383. [CrossRef]
23. Leão, A.P.; Tostes, M.E.L.; Vieira, J.P.A.; Bezerra, U.H.; Heringer, W.R.; Sousa, A.L.; Cordeiro, M.A.M.; Paye, J.C.H.; Santos, M.C. Characteristics of High Impedance Faults in Overhead Distribution Networks in Bamboo Branches. In Proceedings of the 2020 IEEE PES Transmission & Distribution Conference and Exhibition—Latin America (T&D LA), Montevideo, Uruguay, 28 September–2 October 2020; pp. 1–6.
24. Wontroba, A.; Morais, A.P.; Cardoso, G.J.; Vieira, J.P.A.; Farias, P.E.; Gallas, M.; Rossini, J.P. High-impedance fault detection on downed conductor in overhead distribution networks. *Electr. Power Syst. Res.* **2022**, *211*, 108216. [CrossRef]
25. Wontroba, A.; de Morais, A.P.; Rossini, J.P.; Gallas, M.; Cardoso, G.; Vieira, J.P.A.; Farias, P.E.; Santos, M.C. Modeling and Real-Time Simulation of High Impedance Faults for Protection Relay Testing and Methods Validation. In Proceedings of the 2019 IEEE PES Innovative Smart Grid Technologies Conference—Latin America (ISGT Latin America), Gramado, Brazil, 15–18 September 2019; pp. 1–6.
26. Farias, P.E.; de Morais, A.P.; Rossini, J.P.; Cardoso, G. Non-linear high impedance fault distance estimation in power distribution systems: A continually online-trained neural network approach. *Electr. Power Syst. Res.* **2018**, *157*, 20–28. [CrossRef]
27. de Sousa, Á.L.; Vieira, J.P.A.; de Melo Cordeiro, M.A.; Leão, A.P.; Paye, J.C.H.; de Morais, A.P.; Junior, G.C.; Santos, M.C. Harmonic Analysis of High-Impedance Fault Experimental Current Waveforms Using Fast Fourier Transform. *Simpósio Bras. Sist. Elétricos SBSE* **2022**, *2*, 1958–1965. [CrossRef]

28. Heringer, W.R.; Cordeiro, M.A.M.; Paye, J.C.H.; Sousa, A.L.; Leão, A.P.; Vieira, J.P.A.; Santos, M.C.; Cardoso, G.; de Morais, A.P.; Wontroba, A.; et al. Reproduction of a High Impedance Double line-to-ground Fault Using Real Oscillography Data. In Proceedings of the 2020 IEEE PES Transmission & Distribution Conference and Exhibition—Latin America (T&D LA), Montevideo, Uruguay, 28 September–2 October 2020; pp. 1–6.
29. Aucoin, B.M.; Russell, B.D. Distribution High Impedance Fault Detection Utilizing High Frequency Current Components. *IEEE Power Eng. Rev.* **1982**, *PER-2*, 46–47. [CrossRef]
30. Jeerings, D.I.; Linders, J.R. Unique aspects of distribution system harmonics due to high impedance ground faults. *IEEE Trans. Power Deliv.* **1990**, *5*, 1086–1094. [CrossRef]
31. Trondoli, L.H.P.D.C. Modelo Estocástico Parametrizável Para o Estudo de Faltas de Alta Impedância em Sistemas de Distribuição de Energia Elétrica. Masters' Thesis, Universidade de São Paulo, São Carlos, Brazil, 2018.
32. Lopes, F.V.; Santos, W.C.; Fernandes, D.; Neves, W.L.A.; Brito, N.S.D.; Souza, B.A. A transient based approach to diagnose high impedance faults on smart distribution networks. In Proceedings of the 2013 IEEE PES Conference on Innovative Smart Grid Technologies (ISGT Latin America), Sao Paulo, Brazil, 15–17 April 2013; pp. 1–8.
33. Santos, W.C.; Costa, F.B.; Silva, J.A.C.B.; Lira, G.R.S.; Souza, B.A.; Brito, N.S.D.; Paes Junior, M.R.C. Automatic building of a simulated high impedance fault database. In Proceedings of the 2010 IEEE/PES Transmission and Distribution Conference and Exposition: Latin America (T D-LA), Sao Paulo, Brazil, 8–10 November 2010; pp. 550–554.
34. Masa, A.V.; Werben, S.; Maun, J.C. Incorporation of data-mining in protection technology for high impedance fault detection. In Proceedings of the 2012 IEEE Power and Energy Society General Meeting, San Diego, CA, USA, 22–26 July 2012; pp. 1–8.
35. Benner, C.L.; Russell, B.D. Practical high-impedance fault detection on distribution feeders. *IEEE Trans. Ind. Appl.* **1997**, *33*, 635–640. [CrossRef]
36. Aucoin, B.M.; Jones, R.H. High impedance fault detection implementation issues. *IEEE Trans. Power Deliv.* **1996**, *11*, 139–148. [CrossRef]
37. Sultan, A.F.; Swift, G.W.; Fedirchuk, D.J. Detecting arcing downed-wires using fault current flicker and half-cycle asymmetry. *IEEE Trans. Power Deliv.* **1994**, *9*, 461–470. [CrossRef]
38. Costa, F.B.; Souza, B.A.; Brito, N.S.D.; Silva, J.A.C.B.; Santos, W.C. Real-Time Detection of Transients Induced by High-Impedance Faults Based on the Boundary Wavelet Transform. *IEEE Trans. Ind. Appl.* **2015**, *51*, 5312–5323. [CrossRef]
39. Ramos, M.; Bretas, A.; Bernardon, D.; Pfitscher, L. Distribution networks HIF location: A frequency domain system model and WLS parameter estimation approach. *Electr. Power Syst. Res.* **2017**, *146*, 170–176. [CrossRef]
40. Macedo, J.R.; Resende, J.W.; Bissochi, C.A.; Carvalho, D.; Castro, F.C. Proposition of an interharmonic-based methodology for high-impedance fault detection in distribution systems. *Transm. Distrib. IET Gener.* **2015**, *9*, 2593–2601. [CrossRef]
41. Russell, B.D.; IEEE Power Engineering Society; Power Engineering Education Committee; Power Systems Relaying Committee; Transmission and Distribution Committee; Substations Committee. *IEEE Tutorial Course, Detection of Downed Conductors on Utility Distribution Systems*; Publications Sales Department, Institute of Electrical and Electronics Engineers: New York, NY, USA; IEEE Service Center: Piscataway, NJ, USA, 1989.
42. Mangueira Lima, É.; dos Santos Junqueira, C.M.; Silva Dantas Brito, N.; de Souza, B.A.; de Almeida Coelho, R.; Gayoso Meira Suassuna de Medeiros, H. High impedance fault detection method based on the short-time Fourier transform. *Transm. Distrib. IET Gener.* **2018**, *12*, 2577–2584. [CrossRef]
43. Sedighi, A.-R.; Haghifam, M.-R.; Malik, O.P.; Ghassemian, M.-H. High impedance fault detection based on wavelet transform and statistical pattern recognition. *IEEE Trans. Power Deliv.* **2005**, *20*, 2414–2421. [CrossRef]
44. Sarwar, M.; Mehmood, F.; Abid, M.; Khan, A.Q.; Gul, S.T.; Khan, A.S. High impedance fault detection and isolation in power distribution networks using support vector machines. *J. King Saud Univ. Eng. Sci.* **2019**, *32*, 524–535. [CrossRef]
45. Snider, L.A.; Yuen, Y.S. The artificial neural-networks-based relay algorithm for the detection of stochastic high impedance faults. *Neurocomputing* **1998**, *23*, 243–254. [CrossRef]
46. Dong, X.; Shi, S. Single Phase to Ground Fault Processing. In *Fault Location and Service Restoration for Electrical Distribution Systems*; John Wiley & Sons, Ltd.: Hoboken, NJ, USA, 2016; pp. 163–203. ISBN 978-1-118-95028-9.
47. Sharaf, A.M.; Snider, L.A.; Debnath, K. A neural network based relaying scheme for distribution system high impedance fault detection. In Proceedings of the Proceedings 1993 The First New Zealand International Two-Stream Conference on Artificial Neural Networks and Expert Systems, Dunedin, New Zealand, 24–26 November 1993; pp. 321–324.
48. Wai, D.C.T.; Yibin, X. A novel technique for high impedance fault identification. *IEEE Trans. Power Deliv.* **1998**, *13*, 738–744. [CrossRef]
49. Sheng, Y.; Rovnyak, S.M. Decision tree-based methodology for high impedance fault detection. *IEEE Trans. Power Deliv.* **2004**, *19*, 533–536. [CrossRef]
50. Michalik, M.; Rebizant, W.; Lukowicz, M.; Lee, S. J.; Kang, S. H. Wavelet transform approach to high impedance fault detection in MV networks. In Proceedings of the 2005 IEEE Russia Power Tech, St. Petersburg, Russia, 27–30 June 2005; pp. 1–7.
51. Zamanan, N.; Sykulski, J.K. Modelling arcing high impedances faults in relation to the physical processes in the electric arc. *WSEAS Trans. Power Syst.* **2006**, *1*, 1507–1512.
52. Sedighi, A.R.; Haghifam, M.R. Simulation of high impedance ground fault in electrical power distribution systems. In Proceedings of the 2010 International Conference on Power System Technology, Hangzhou, China, 24–28 October 2010; pp. 1–7.

53. Baqui, I.; Zamora, I.; Mazón, J.; Buigues, G. High impedance fault detection methodology using wavelet transform and artificial neural networks. *Electr. Power Syst. Res.* **2011**, *81*, 1325–1333. [CrossRef]
54. Moravej, Z.; Mortazavi, S.H.; Shahrtash, S.M. DT-CWT based event feature extraction for high impedance faults detection in distribution system. *Int. Trans. Electr. Energy Syst.* **2015**, *25*, 3288–3303. [CrossRef]
55. Ferraz, R.G.; Iurinic, L.U.; Filomena, A.D.; Gazzana, D.S.; Bretas, A.S. Arc fault location: A nonlinear time varying fault model and frequency domain parameter estimation approach. *Int. J. Electr. Power Energy Syst.* **2016**, *80*, 347–355. [CrossRef]
56. Sharat, A.M.; Snider, L.A.; Debnath, K. A neural network based back error propagation relay algorithm for distribution system high impedance fault detection. In Proceedings of the 1993 2nd International Conference on Advances in Power System Control, Operation and Management, APSCOM-93, Hong Kong, China, 7–10 December 1993; Volume 2, pp. 613–620.
57. Lai, T.M.; Snider, L.A.; Lo, E.; Sutanto, D. High-impedance fault detection using discrete wavelet transform and frequency range and RMS conversion. *IEEE Trans. Power Deliv.* **2005**, *20*, 397–407. [CrossRef]
58. Shannon, R.E. Introduction to the art and science of simulation. In Proceedings of the 1998 Winter Simulation Conference, Washington, DC, USA, 13–16 December 1998; Proceedings (Cat. No.98CH36274). Volume 1, pp. 7–14.
59. Samantaray, S.R.; Panigrahi, B.K.; Dash, P.K. High impedance fault detection in power distribution networks using time-frequency transform and probabilistic neural network. *Transm. Distrib. IET Gener.* **2008**, *2*, 261–270. [CrossRef]
60. Soheili, A.; Sadeh, J.; Bakhshi, R. Modified FFT based high impedance fault detection technique considering distribution non-linear loads: Simulation and experimental data analysis. *Int. J. Electr. Power Energy Syst.* **2018**, *94*, 124–140. [CrossRef]
61. Gautam, S.; Brahma, S.M. Detection of High Impedance Fault in Power Distribution Systems Using Mathematical Morphology. *IEEE Trans. Power Syst.* **2013**, *28*, 1226–1234. [CrossRef]
62. Ghaderi, A.; Mohammadpour, H.A.; Ginn, H. High impedance fault detection method efficiency: Simulation vs. real-world data acquisition. In Proceedings of the 2015 IEEE Power and Energy Conference at Illinois (PECI), Champaign, IL, USA, 20–21 February 2015; pp. 1–5.
63. Ghaderi, A.; Mohammadpour, H.A.; Ginn, H. Active fault location in distribution network using time-frequency reflectometry. In Proceedings of the 2015 IEEE Power and Energy Conference at Illinois (PECI), Champaign, IL, USA, 20–21 February 2015; pp. 1–7.
64. Nayak, P.K.; Sarwagya, K.; Biswal, T. A novel high impedance fault detection technique in distribution systems with distributed generators. In Proceedings of the 2016 National Power Systems Conference (NPSC), Bhubaneswar, India, 19–21 December 2016; pp. 1–6.
65. Soheili, A.; Sadeh, J.; Lomei, H.; Muttaqi, K. A new high impedance fault detection scheme: Fourier based approach. In Proceedings of the 2016 IEEE International Conference on Power System Technology (POWERCON), Wollongong, NSW, Australia, 28 September-1 October 2016; pp. 1–6.
66. Kavi, M.; Mishra, Y.; Vilathgamuwa, D.M. Detection and identification of high impedance faults in single wire earth return distribution networks. In Proceedings of the 2016 Australasian Universities Power Engineering Conference (AUPEC), Brisbane, Australia, 25–28 September 2016; pp. 1–6.
67. Kavi, M.; Mishra, Y.; Vilathgamuwa, M.D. High-impedance fault detection and classification in power system distribution networks using morphological fault detector algorithm. *Transm. Distrib. IET Gener.* **2018**, *12*, 3699–3710. [CrossRef]
68. Soheili, A.; Sadeh, J. Evidential reasoning based approach to high impedance fault detection in power distribution systems. *Transm. Distrib. IET Gener.* **2017**, *11*, 1325–1336. [CrossRef]
69. Sedighi, A.-R. A New Model for High Impedance Fault in Electrical Distribution Systems. Available online: http://www.isroset.org/ (accessed on 9 June 2020).
70. Hong, Y.-Y.; Huang, W.-S.; Chang, Y.-R.; Lee, Y.-D.; Ouyang, D.-C. Locating high-impedance fault in a smart distribution system using wavelet entropy and hybrid self-organizing mapping network. In Proceedings of the 2017 IEEE PES Innovative Smart Grid Technologies Conference Europe (ISGT-Europe), Torino, Italy, 26–29 September 2017; pp. 1–6.
71. Kizilcay, M.; Pniok, T. Digital simulation of fault arcs in power systems. *Eur. Trans. Electr. Power* **1991**, *1*, 55–60. [CrossRef]
72. Nunes, J.U.N.; Bretas, A.S.; Bretas, N.G.; Herrera-Orozco, A.R.; Iurinic, L.U. Distribution systems high impedance fault location: A spectral domain model considering parametric error processing. *Int. J. Electr. Power Energy Syst.* **2019**, *109*, 227–241. [CrossRef]
73. Carr, J. Detection of High Impedance Faults on Multi-Grounded Primary Distribution Systems. *IEEE Trans. Power Appar. Syst.* **1981**, *PAS-100*, 2008–2016. [CrossRef]
74. Sharaf, A.M.; El-Sharkawy, R.M.; Al-Fatih, R.; Al-Ketbi, M. High impedance fault detection on radial distribution and utilization systems. In Proceedings of the 1996 Canadian Conference on Electrical and Computer Engineering, Calgary, AB, Canada, 26–29 May 1996; Volume 2, pp. 1012–1015.
75. Sharaf, A.M.; Abu-Azab, S.I. A smart relaying scheme for high impedance faults in distribution and utilization networks. In Proceedings of the 2000 Canadian Conference on Electrical and Computer Engineering, Quebec City, QC, Canada, 13–16 May 2000; Conference Proceedings. Navigating to a New Era (Cat. No.00TH8492). Volume 2, pp. 740–744.
76. Lee, R.E.; Bishop, M.T. A Comparison of Measured High Impedance Fault Data to Digital Computer Modeling Results. *IEEE Trans. Power Appar. Syst.* **1985**, *PAS-104*, 2754–2758. [CrossRef]
77. Dasco, A.; Marguet, R.; Raison, B. Fault distance estimation in distribution network for high impedance faults. In Proceedings of the 2015 IEEE Eindhoven PowerTech, Eindhoven, The Netherlands, 29 June-2 July 2015; pp. 1–6.
78. Yu, D.C.; Khan, S.H. An adaptive high and low impedance fault detection method. *IEEE Trans. Power Deliv.* **1994**, *9*, 1812–1821. [CrossRef]

79. dos Santos, W.C. Identificação de Faltas de Alta Impedância em Sistemas de Distribuição. Ph.D. Thesis, Universidade Federal De Campina Grande, Campina Grande, Brazil. Available online: http://dspace.sti.ufcg.edu.br:8080/xmlui/handle/riufcg/602 (accessed on 14 January 2023).
80. Wang, B.; Geng, J.; Dong, X. High-Impedance Fault Detection Based on Nonlinear Voltage–Current Characteristic Profile Identification. *IEEE Trans. Smart Grid* **2018**, *9*, 3783–3791. [CrossRef]
81. Darwish, H.A.; Elkalashy, N.I. Universal arc representation using EMTP. *IEEE Trans. Power Deliv.* **2005**, *20*, 772–779. [CrossRef]
82. Michalik, M.; Rebizant, W.; Lukowicz, M.; Lee, S.-J.; Kang, S.-H. High-impedance fault detection in distribution networks with use of wavelet based algorithm. *IEEE Trans. Power Deliv.* **2006**, *21*, 1793–1802. [CrossRef]
83. Cui, T.; Dong, X.; Bo, Z.; Richards, S. Integrated scheme for high impedance fault detection in MV distribution system. In Proceedings of the 2008 IEEE/PES Transmission and Distribution Conference and Exposition: Latin America, Bogota, Columbia, 13–15 August 2008; pp. 1–6.
84. Cui, T.; Dong, X.; Bo, Z.; Klimek, A.; Edwards, A. Modeling study for high impedance fault detection in MV distribution system. In Proceedings of the 2008 43rd International Universities Power Engineering Conference, Padova, Italy, 1–4 September 2008; pp. 1–5.
85. Torres, V.; Maximov, S.; Ruiz, H.F.; Guardado, J.L. Distributed Parameters Model for High-impedance Fault Detection and Localization in Transmission Lines. *Electr. Power Compon. Syst.* **2013**, *41*, 1311–1333. [CrossRef]
86. Torres, V.; Ruiz, H.F.; Maximov, S.; Ramírez, S. Modeling of high impedance faults in electric distribution systems. In Proceedings of the 2014 IEEE International Autumn Meeting on Power, Electronics and Computing (ROPEC), Ixtapa, Mexico, 5–7 November 2014; pp. 1–6.
87. Thomas, M.S.; Bhaskar, N.; Prakash, A. Voltage Based Detection Method for High Impedance Fault in a Distribution System. *J. Inst. Eng. India Ser. B* **2016**, *97*, 413–423. [CrossRef]
88. Iurinic, L.U.; Herrera-Orozco, A.R.; Ferraz, R.G.; Bretas, A.S. Distribution Systems High-Impedance Fault Location: A Parameter Estimation Approach. *IEEE Trans. Power Deliv.* **2016**, *31*, 1806–1814. [CrossRef]

energies

MDPI

Article

An Autonomous Distributed Coordination Strategy for Sustainable Consumption in a Microgrid Based on a Bio-Inspired Approach

Marcel García [1,2], Jose Aguilar [2,3,4,*] and María D. R-Moreno [5]

1 Laboratorio de Prototipos, Experimental National Universidad de Táchira, San Cristóbal 5001, Venezuela; mgarciam@unet.edu.ve
2 CEMISID, Universidad de Los Andes, Mérida 5101, Venezuela
3 GIDITIC, Universidad EAFIT, Medellín 050022, Colombia
4 IMDEA Networks Institute, 28918 Madrid, Spain
5 Universidad de Alcalá, Escuela Politécnica Superior, ISG, 28805 Alcalá de Henares, Spain; malola.rmoreno@uah.es
* Correspondence: aguilar@ula.ve or aguilarjos@gmail.com or jose.aguilar@imdea.org or jlaguilarc@eafit.edu.co

Abstract: Distributed energy resources have demonstrated their potential to mitigate the limitations of large, centralized generation systems. This is achieved through the geographical distribution of generation sources that capitalize on the potential of their respective environments to satisfy local demand. In a microgrid, the control problem is inherently distributed, rendering traditional control techniques inefficient due to the impracticality of central governance. Instead, coordination among its components is essential. The challenge involves enabling these components to operate under optimal conditions, such as charging batteries with surplus solar energy or deactivating controllable loads when market prices rise. Consequently, there is a pressing need for innovative distributed strategies like emergent control. Inspired by phenomena such as the environmentally responsive behavior of ants, emergent control involves decentralized coordination schemes. This paper introduces an emergent control strategy for microgrids, grounded in the response threshold model, to establish an autonomous distributed control approach. The results, utilizing our methodology, demonstrate seamless coordination among the diverse components of a microgrid. For instance, system resilience is evident in scenarios where, upon the failure of certain components, others commence operation. Moreover, in dynamic conditions, such as varying weather and economic factors, the microgrid adeptly adapts to meet demand fluctuations. Our emergent control scheme enhances response times, performance, and on/off delay times. In various test scenarios, Integrated Absolute Error (IAE) metrics of approximately 0.01% were achieved, indicating a negligible difference between supplied and demanded energy. Furthermore, our approach prioritizes the utilization of renewable sources, increasing their usage from 59.7% to 86.1%. This shift not only reduces reliance on the public grid but also leads to significant energy cost savings.

Keywords: emergent control; energy management systems; distributed control systems; response threshold model; smart grid

Citation: García, M.; Aguilar, J.; R-Moreno, M.D. An Autonomous Distributed Coordination Strategy for Sustainable Consumption in a Microgrid Based on a Bio-Inspired Approach. *Energies* **2024**, *17*, 757. https://doi.org/10.3390/en17030757

Academic Editors: Veerapandiyan Veerasamy, Shailendra Singh and Sunil Kumar Singh

Received: 3 December 2023
Revised: 26 January 2024
Accepted: 1 February 2024
Published: 5 February 2024

1. Introduction

Distributed energy systems can solve the problems associated with centralized generation, such as [1]: high energy losses, hazardous waste generation, and expensive infrastructure, among others. Specifically, RES with energy storage devices can be integrated to formulate HES for a reliable power supply (saving and releasing energy at the adequate time) to overcome the limitations in the wind and solar resources since their performance characteristics depend on the environment, which reduces the reliability of the power supply and imposes additional costs for providers and/or consumers [2]. On the other hand, it is necessary to consider the effects of energy management on the consumer, especially the controllable loads, to ensure an efficient energy system [3].

An MG is a distributed energy network with diverse small energy sources that work autonomously with respect to the main network [4]. An MG is classically made up of a local generation system, an energy storage system, consumers, and a connection point to the grid. The main goal of an MG is to reach a more efficient and reliable energy supply, to provide a higher quality of energy service, and to make it safer and more sustainable. The MG control problem is a distributed control problem, which requires coordination and control mechanisms for its agents without a central authority to manage them [1].

One of the incentives of this paper is to effectively coordinate and integrate the various agents in an energy system, both on the production and consumption side, to meet energy needs, maximize the use of renewable energy, and reduce energy from the main network [5]. In this way, it is expected to minimize energy costs and the need to manage the natural uncertainty of RES, along with the limited storage capacity of batteries [6]. In this sense, optimal management of these energy resources is essential to have a balance between demand and supply.

In general, an autonomous management approach of an MG for efficient demand management must be able, among other things, to turn off controllable loads when the cost of energy is high, and charge storage devices using only renewable energy, among other things. For this, new distributed control strategies, such as those based on emergent control, are necessary.

This paper aims to make use of the RTM [7] to define a self-control mechanism, so that each component takes part in the energy exchange when the circumstances are favorable. To create an emerging control proposal that enables effective energy management, it is necessary to identify the local variables and the rules for decision-making, as well as the feedback mechanisms between them. The use of an emergent control strategy based on bio-inspired techniques that allows micro-level coordination distinguishes this paper from earlier ones in the context of MG. The advantage of using the RTM over other control methods is based on the potential for creating a distributed control strategy. Each agent is prompted to act or not by its two main variables, the stimulus and threshold. The threshold is linked to the accomplishment of the goals, and the stimulus takes context information into account. Both employ RL as a self-control mechanism to control the agent's response, enabling autonomous coordination actions. The main contributions of our paper are:

- Specification of an autonomous MG management approach and self-regulating demand and load generation through emergent control mechanisms, enabling MG self-organization.
- Definition of an emergent control approach based on the RTM, applicable to both the generation and demand sides. Traditionally used for task assignment problems, this is the first application of the RTM in coordination and distributed control contexts.
- Development of a scheme to balance generated and demanded power, combining load from the main network and renewable sources, and controlling controllable loads and excess energy storage.
- Schematization of a methodology to apply our emergent control approach in MGs in any context. Following this methodology, an MG autonomy can be specified.

In summary, this paper's scientific novelty lies in its autonomous MG management approach, adapting the RTM—commonly used for distributed task assignment problems—as a solution for distributed control challenges in the energy sector. Our approach ensures autonomous component coordination to meet demand based on environmental conditions.

The paper is organized as follows: Section 2 discusses related work. Section 3 outlines the theoretical framework of our proposal and the process under study. Section 4 introduces our emergent control system design based on the RTM. Section 5 details the experiments conducted, quality assessment metrics, and comparisons with other works. The paper concludes with a summary and future work directions.

2. Related Work

Shahgholian [8] provides a comprehensive literature review on the operations, applications, and control of MGs, with a particular focus on local control and coordination among its components. The author also highlights various research areas, including control strategies, optimization methods, stability, and reactive power compensation techniques. This paper notably introduces control architectures like decentralized systems and variations based on MAS.

Liu et al. [9] introduce a consensus MAS DRL algorithm where each DER is an agent. These agents learn control strategies based on local states and messages received from neighboring agents.

Zhou et al. [10] discuss distributed communication network characteristics for MGs, distributed control strategies, and communication reliability issues. However, their work lacks experimental and case study analyses. Maomao et al. [11] analyze recent classifications, architectures, technologies, and methods for coordinating residential MGs, including game theory and MAS.

Zhou et al. [12] propose an MG cluster structure with an autonomous coordination control strategy that combines normalized droop-based control with adaptive control. The MG consists of an EP, a PEU, and several small MGs. Areekkara et al. [13] utilize the MAS paradigm for its autonomy, scalability, flexibility, and adaptability in MG energy management systems. Harmouch et al. [14] describe the implementation of a decentralized MAS-EMS for controlling an MG cluster. Each MG within the cluster employs a dedicated MAS-EMS for power dispatch aimed at minimizing grid power exchange. Another dedicated EMS manages power distribution among various MGs. Kofinas et al. [15] propose a cooperative MA system for managing a standalone MG's energy. They use a distributed, cooperative RL method in a continuous action-state space, enabling MG components to operate as independent learners while coordinating through shared state variables. Fuzzy Q-learning techniques were recommended for this purpose.

Hu et al. [11] analyze recent developments in residential MG architectures and technologies, reviewing key coordination and negotiation methods. Chouikhi et al. [16] present two EM mechanisms in the context of smart buildings. To begin with, they use negotiation and multi-leader–follower game techniques to simulate the interactions between energy suppliers and building occupants. Then, in order to reduce consumer interactions and maximize energy demand cost, they suggest a distributed game theory-based scheduling method for energy demand. In this paper, the system entities are modeled using MAS.

Xu et al. [17] define a home energy management framework based on RL, using a finite Markov decision process for hour-ahead energy consumption scheduling and an extreme learning machine for predicting electricity prices and solar generation.

Arnone et al. [18] suggest a Building Management System for controlling energy flows in smart buildings integrated with various energy networks and RES, using heuristic optimization for 24 h energy generation and consumption scheduling. Aguilar et al. [19] describe a scheduling system that automatically defines hours of utilization of the controllable load devices in a home, in such a way that the use of renewable energy is maximized. To do this, they construct an autonomous cycle of data analysis tasks made up of three tasks: a scheduling task to create appliance usage schedules, and two tasks to estimate the amount of renewable energy generated and the amount of demanded load. Garcia et al. [20] conducted a systematic literature review on emergent control systems, focusing on conceptual bases, methodologies, principles, architectures, and their application in Cyber-Physical Systems, particularly in Industry 4.0.

Du et al. [21] propose a control strategy for coordinating neighboring autonomous MGs using a smart switch, focusing on power sharing without considering internal component coordination. Chai and Peng [22] propose a decentralized coordinated power control approach for islanding MGs, combining bus voltage signals with fuzzy logic control, without inter-MG interconnections. Rahman et al. [6] design a cooperative MAS for energy sharing in MGs, using graph theory to establish communication links between controllable batteries and clients. Arwa and Folly [23] explore the use of RL algorithms for MG power management, highlighting challenges and solutions.

Zheng et al. [24] analyze the energy dispatch optimization of connected MGs, proposing a distributed predictive control model with local predictive control for each MG component. Kyriakou and Kanellos [25] propose a method for coordinated scheduling of the optimal operation of active distribution networks containing complex MGs (prosumers and electric vehicles), through a strategy of centralized control [26].

Our literature review indicates that while there are numerous studies on MG control and optimization strategies, as well as architectures and methodologies, few focus on distributed control architecture addressing DER coordination. Most employ the MAS paradigm, RL, or predictive or fuzzy control models. However, case studies are limited, lacking benchmarks for performance analysis and metrics for proposal quality evaluation. Therefore, our study makes qualitative comparisons with existing works. To our knowledge, this paper is the first to propose an autonomous management approach for MGs using emerging approaches with the RTM. MAS-based control schemes allow for MG adaptation, but our emergent control approach uniquely adjusts component roles based on contextual uncertainty.

3. Theoretical Framework

The RTM used in this paper is presented in this section, as well as the MG components that will be analyzed.

3.1. Response Threshold Model (RTM)

Drawing inspiration from the behavior of labor division in ant colonies, the RTM [27] mimics ants' sensitivity to external and internal stimuli, including the intensity of pheromone signals. This model allows the colony to adapt to varying environmental conditions [28], enabling ants to modify their behavior according to the needs and occurrences within the colony [29]. In general, the sensitivity to external/internal stimuli is quantified using a parameter called the response threshold (θ). An agent (ant) with a lower response threshold is more likely to become an active worker, while one with a higher threshold is less likely to be activated. In a conventional RTM for the division of labor [27], an agent j has the probability q_j of work in a given moment according to the Equation (1):

$$q_j = \frac{s_j(t)^2}{s_j(t)^2 + \theta_{ij}(t)^2} \tag{1}$$

where s_j is the external or internal stimulus and θ_{ij} is the ant's sensitivity (response threshold) to perform task j at time t. The task assignment problem, which assumes that each worker i responds to a stimulus for a specific task j when the intensity of the stimulus exceeds the worker's threshold for that stimulus, can be solved with the help of this model. Additionally, the ants change the intensity of the cumulative stimulus (s_j) in accordance with Equation (2) in order to exert control over the system through a process of individual or group learning by associating rewards with stimuli [27]:

$$s_j(t+1) = s_j(t) + \delta - \frac{\alpha \cdot N_{act}}{N} \tag{2}$$

In the conventional model, the variations in the stimulus's intensity are caused by the way the task is carried out; here, N_{act} denotes the individuals who are actively engaged in the colony, N are the potential participants, α denotes the task's efficiency, and δ denotes the stimulus's increase in intensity per unit of time. In contrast, the response threshold is fixed from a simple reinforcement process, such that the threshold decreases when the corresponding task is performed and increases when the corresponding task is not performed. These combined reinforcement processes [30] allow the appearance of specialized workers; that is, some workers are more sensitive to the stimuli with a particular task j, starting from a group of initially identical individuals. The response threshold incorporating RL is expressed as follows [7]:

$$\theta_{ij} = \theta_{ij} - y_{ij}\beta\Delta t + (1 - y_{ij})\gamma\Delta t \quad (3)$$

where y_{ij} is the proportion of entities of type i performing task j, and β and γ are the learning and forgetting rates, respectively. This means that according to Equation (3), in the next Δt time units, y_{ij} fraction of entities of type I do task j, and $(1 - y_{ij})$ fraction of entities do nothing or something else. In this way, the RTM refers to the potential for responding to stimuli connected to tasks to address the issue of labor division in a colony.

3.2. *System Modeling*

An MG is a distributed electrical network with various small energy sources running parallel to the main grid [4]. The aim of an MG is to provide a higher quality of energy service, catering to the growing demand for reliable, safe, sustainable, and efficient electricity supply. Our focus is on an MG composed of a PV system, an energy storage system, an external source (the main grid), and controllable loads. Below, we describe each component's model.

3.2.1. Photovoltaic System

A PV system generates voltage and current from solar cells arranged in series and parallel. Solar radiation affects the PV (P_{PV}) array's maximum power output in a variety of ways. Therefore, a control strategy is needed to efficiently utilize solar radiation to produce the most power. The output power is defined in the work of Antonanzas–Torres et al. [31] as:

$$P_{PV} = \eta S\phi[1 - 0.005(T_a + 25)] \quad (4)$$

where η is the conversion efficiency of the PV array, S is the area of the PV array (m^2), ϕ is the solar radiation (kw/m^2), and T_a is the ambient temperature [°C].

3.2.2. Energy Storage System (ESS)

One of the most popular tools to conserve electrical energy in various applications is the battery, but in addition, supercapacitors can be used to store electricity. In PV or wind energy conversion systems, batteries act as a backup. In addition to storing excess solar or wind energy during sunny or windy days, they are used to release the energy during stationary periods or at night. Various models have been put forth in the area of battery modeling. The modified simple battery model 3 is employed for this paper, but it is based on the power [W] rather than the voltage [V] [32].

$$V_t = V_{OC} - \left(R_{int} + \frac{K}{SOC}\right)I \quad (5)$$

where V_t is the battery's terminal voltage, V_{oc} is the voltage in the circuit without a load, R_{int} is the internal resistance of the battery, K is a polarization constant, I is the effective value of the discharge current, and SOC is the battery's state of charge at time t (in percent).

3.2.3. Controllable Loads

Controllable loads refer to devices with manageable energy consumption. They are categorized as follows [33]:

- Type I: this type consists of various residential loads, like refrigerators, air conditioners, water heating, etc. These loads can be interrupted or controlled (e.g., for reducing demand).
- Type II: this type contains battery storage, Vehicle-to-Grid, etc., and it can be charged from or discharged to the grid (it can inject power into the grid). Furthermore, this type of load can be controlled to accommodate its grid needs.

- Type III: this type includes the rest of the appliances schedulable, which can be deferred for a suitable moment. There are diverse appliances (phone chargers, microwaves, washing machines, tumble dryers, dishwashers, vacuum cleaners, etc.).

Our research primarily focuses on controlling the energy demand of Type III controllable load appliances.

4. Design of an Emergent Control System Based on the Response Threshold Model

This section describes the components of the MG considered in this study, and our emerging control approach based on the RTM.

4.1. Distributed Control Architecture of a Microgrid

The MG analyzed in this paper, depicted in Figure 1, is designed to meet the energy demands of a residential building. Surplus self-generated energy is stored in a battery, and the power balance in the MG is maintained as per the energy conservation principle, expressed by Equation (6):

$$P_T - P_L = \Delta P \tag{6}$$

where P_T is the total potential delivered or supplied, P_L is the total energy consumption of the building, made up mainly of non-controllable loads (LNC) and controllable loads (LC), and ΔP is the error in the supply–demand.

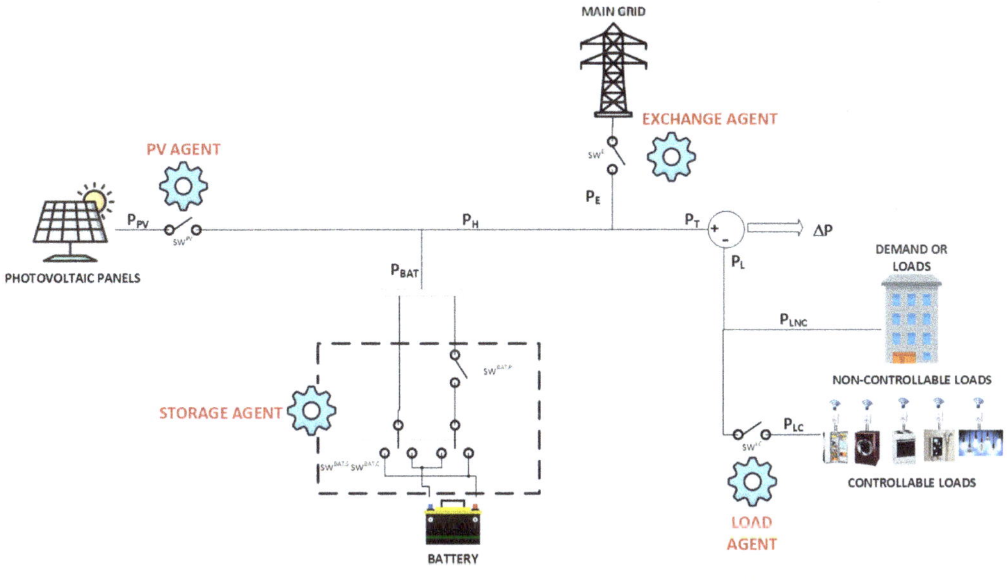

Figure 1. Distributed control architecture for an MG.

Figure 1 introduces four agents: the PV agent denotes the control mechanism to deactivate or activate solar energy; the exchange agent manages energy supply from non-renewable sources; the storage agent handles battery charging and discharging according to specific conditions; and the load agent deactivates controllable loads when there is no renewable energy (from solar energy or battery), or the demand is greater than the power of renewable energy delivered to the energy network, and the energy market cost exceeds the price that consumers are willing to pay.

Thus, the energy system (ES) includes a PV panel, a battery, a load demand side, and a connection to the main grid. The batteries store energy during surplus generation and release it during peak load demand. The main grid is taken as a standby producer that begins to compensate when there is a deficit for the demanded energy; that is, when the storage and/or renewable generation systems fail in the supply of demanded energy. Our hybrid system's overall power output (P_T) is indicated as follows (Equation (7)):

$$P_T = P_H + P_E \quad (7)$$

where P_H is the produced power by the renewable systems (e.g., wind turbines, PV panels (P_{PV}), etc.) and the power in the energy storage system (P_{BAT}) (see Equation (8)), and P_E is the main grid power.

$$P_H = P_{PV} + P_{BAT} \quad (8)$$

As well, the demand will be divided into controllable (P_{LC}) and non-controllable (P_{LNC}) (see Equation (9)):

$$P_L = P_{LC} + P_{LNC} \quad (9)$$

4.2. Specification of the MAS

Utilizing the MultiAgent Systems for INtegrated Automation (MASINA) methodology [34,35], we design the emergent control approach based on the RTM. MASINA requires two steps: (i) specifying the general architecture of the system to derive the MAS, and (ii) designing each agent.

The first step involves representing the distributed system as a MAS, identifying agents, relevant variables, parameters, and behaviors, among other aspects. Table 1 summarizes the agent characterization for the MG in Figure 1.

Table 1. Characterization of the agents.

Agent	Behavior	Environment	External Variables	Internal Variables or Parameters
Photovoltaic system	- It is activated when there is potential to generate solar energy	Solar radiation Weather conditions	Demand	Conversion efficiency Solar panel area
ESS	- It is a producer when the demand exceeds the solar potential and the energy is stored. - When there is an excess of renewable energy and not enough energy is stored, it becomes a consumer. - When none of the preceding suppositions are true, it does nothing.		Renewable sources Demand	SOC
Utility grid	- When batteries and renewable energy sources are unable to meet the demand, it is turned on.		Renewable sources SOC Demand	
Controllable load	- It is activated when there are conditions to consume energy (low prices, energy from renewable sources, etc.).		Renewable sources Demand Market cost	Comfort cost

Now it is necessary to identify the system variables which have the most influence on the agents' behavior, in our case:

- Solar radiation.
- Demand.
- Weather conditions.
- SOC.

- Energy costs.

The second step involves designing agents based on their behavioral models using the RTM (see Figure 2). With the RTM variables (probability, stimulus and response threshold) and the internal and external variables of each agent, it is established when an agent can be active (its ON/OFF states).

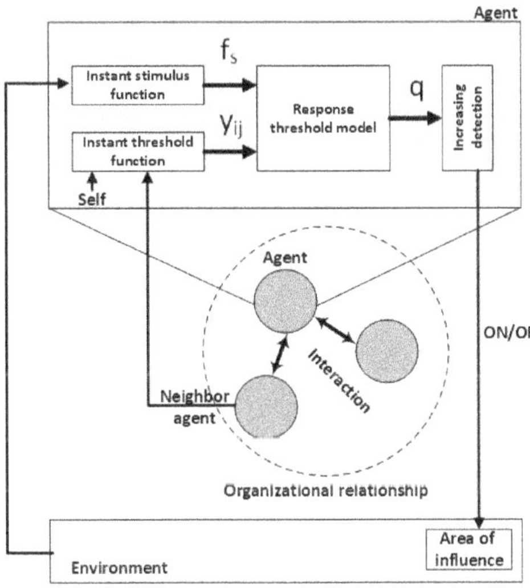

Figure 2. RTM-based agent architecture.

Figure 2 illustrates the RTM model instantiated in each agent of our MAS. The stimulus s_j of each agent (Equation (10)) is characterized by a cumulative function, in which f_s is the instantaneous stimulus based on the behaviors defined in Table 1 and w_j is the attenuation factor of the instantaneous stimulus function.

$$s_j(\mathrm{t}+1) = s_j(\mathrm{t}) + w_j f_s(t) \tag{10}$$

In turn, the expression of the RTM response threshold (Equation (3)), which is related to the sensitivity to the stimulus, represents the current needs in the denominator, and, in the numerator, the contribution of the agent to satisfy it. In the context of our MAS, it is specific to the role that each agent has, so it must consider aspects such as the requested demand and energy generation capacity, among other things. The following section explains these aspects in detail for each agent. Finally, the activation probability of an agent (Equation (1)) can be directly used by our agents to define their states.

In summary, the initial step in designing an emergent control system based on the RTM involves identifying the agents to define the MAS. This process includes determining their local and external variables. Once the agents are identified, we then proceed to model their behavior. This involves defining the stimulus and response threshold for each agent, dictating whether an agent will act or not. The stimulus is designed to assess context information, while the response threshold is aligned with the objectives. These components facilitate the implementation of an emergent control approach using the RTM, enabling autonomous energy management.

Equations (1), (3) and (10) are crucial for determining the behavior of each agent. They incorporate an implicit RL process as a self-control mechanism to regulate the agent's response, along with a feedback process for sharing experiences and information. This allows for autonomous actions by the agents (see the macro-algorithm of each agent in Algorithm 1).

Algorithm 1. Behavior of each agent based on the RTM.
Input: local/external variables
Procedure: 1. Definition of the stimulus using Equation (10) 2. Definition of the response threshold using Equation (3) 3. Determination of the state of the agent using Equation (1)
Output: state of the agent (OFF/ON)

4.3. *Emergent Control System for a Microgrid*

In this section, a control model for each agent is designed based on the RTM, identifying contextual variables and components to formulate mathematical expressions of stimuli and thresholds, whose integration will form the final control system.

4.3.1. Photovoltaic Agent

Each PV agent will be activated when the stimulus increases, which occurs when energy can be produced due to weather conditions (there is radiation (P_{PVmax}), see Equation (4)), and there is demand (P_L) or capability to store energy ($(1-Soc)Q_{CAP}$). In addition, it will be prone to not activate when its response threshold rises, i.e., when the power delivered to the electrical network is being covered by the PV ($\frac{P_{PV}}{P_L} > 0$ and $\left(1-\frac{P_{PV}}{P_L}\right) \cong 0$); causing the controller to turn off. Thus, the RTM for one PV agent, in an emergent control scheme, is as follows, where q_{PV} is its probability to generate solar energy:

$$s_{PV}(\mathrm{t}+1) = s_{PV}(\mathrm{t}) + w_{PV}(P_{PVmax}(P_L + (1-Soc)Q_{CAP})) \tag{11}$$

$$\theta_{PV}(\mathrm{t}+1) = \theta_{PV}(\mathrm{t}) - \beta_{PV}\left(1-\frac{P_{PV}}{P_L}\right)\Delta\mathrm{t} + \gamma_{PV}\frac{P_{PV}}{P_L}\Delta\mathrm{t} \tag{12}$$

$$q_{PV}(\mathrm{t}) = \frac{s_{\mathrm{PV}}(\mathrm{t})^2}{s_{\mathrm{PV}}(\mathrm{t})^2 + \theta_{PV}(\mathrm{t})^2} \tag{13}$$

where P_{PV} is the maximum power output of the PV agent that varies according to solar radiation (see Equation (4)) and w_{PV} is an attenuation factor to make the differences of the perceived signals less sensitive.

4.3.2. Energy Storage Agent

This section presents the RTM used with the energy storage agent. This emergent control system must manage three states: consumer, producer, or passive. The following are the equations for this emergent control system:

<u>Producer</u>:

Equation (14) defines that the stimulus for the production is when there are demand needs ($\mathrm{P_L}$) that are not covered by $\mathrm{P_{PV}}$ and there is stored energy (SOC). Thus, the threshold (Equation (15)) is minimized when the power supplied by the solar panels does not contribute to meeting the demand $\left(\left(1-\frac{P_{PV}}{P_L}\right) > 1\right)$ and there is also stored energy (Soc).

$$s_{BAT,S}(\mathrm{t}+1) = s_{BAT,S}(\mathrm{t}) + w_{BAT,S}Soc(P_L - P_{PV}) \tag{14}$$

$$\theta_{BAT,S}(\mathrm{t}+1) = \theta_{BAT,S}(\mathrm{t}) - \beta_{BAT,S}Soc\left(1-\frac{P_{PV}}{P_L}\right)\Delta\mathrm{t} + \gamma_{BAT,S}\frac{P_{PV}}{P_L}\Delta\mathrm{t} \tag{15}$$

$$q_{BAT,S}(\mathrm{t}) = \frac{s_{\mathrm{BAT,S}}(\mathrm{t})^2}{s_{\mathrm{BAT,S}}(\mathrm{t})^2 + \theta_{BAT,S}(\mathrm{t})^2} \tag{16}$$

Consumer:

Equation (17) defines that the stimulus happens when the battery is discharged (1 − Soc) and there is excess energy from the solar panels. The threshold (Equation (18)) is adjusted to active the agent when there is storage availability $((1 - Soc) \cong 1)$ and the solar panels cover the energy needs and there is an excess ($\frac{P_{PV}}{P_L} > 1$).

$$s_{BAT,C}(t+1) = s_{BAT,C}(t) + w_{BAT,C}(1 - Soc)(P_{PV} - P_L) \tag{17}$$

$$\theta_{BAT,C}(t+1) = \theta_{BAT,C}(t) - \beta_{BAT,C}(1 - Soc)\frac{P_{PV}}{P_L}\Delta t + \gamma_{BAT,C}\left(1 - \frac{P_{PV}}{P_L}\right)\Delta t \tag{18}$$

$$q_{BAT,C}(t) = \frac{s_{BAT,C}(t)^2}{s_{BAT,C}(t)^2 + \theta_{BAT,C}(t)^2} \tag{19}$$

Passive:

Thus, $q_{BAT,S}$ is the probability of supplying energy and $q_{BAT,C}$ is the probability of consuming energy. Finally, state 3 is a passive state (see Equation (20)).

$$q_{BAT,P}(t) = 1 - (q_{BAT,S}(t) + q_{BAT,C}(t)) \tag{20}$$

4.3.3. Main Grid Agent

The RTM applied for the exchange agent is defined in Equations (21)–(23). The agent is stimulated when the demand (P_L) is greater than the power produced by the renewable systems in the energy storage systems ($P_L - P_H > 1$); see Equation (21). The threshold is adjusted to activate the agent when the ratio between the available power and demand is not sufficient $\left(\frac{P_H}{P_L} \cong 0\right)$; see Equation (22). Lastly, $q_E(t)$ is the probability of activating, or not, the exchange agent:

$$s_E(t+1) = s_E(t) + w_E(P_L - P_H) \tag{21}$$

$$\theta_E(t+1) = \theta_E(t) - \beta_E\left(1 - \frac{P_H}{P_L}\right)\Delta t + \gamma_E\frac{P_H}{P_L}\Delta t \tag{22}$$

$$q_E(t) = \frac{s_E(t)^2}{s_E(t)^2 + \theta_E(t)^2} \tag{23}$$

4.3.4. Load Agent

This part of the study focuses on the RTM for the load agent, representing the controllable type III loads of customers. The load agent enables consumers to connect to the distribution network based on their needs, scheduling the use of controllable loads to utilize renewable energy or during periods of low energy prices. We assume that the uncontrollable and controllable loads of other types such as security systems, external lighting, refrigerators, and heating, are already covered.

Specifically, controllable loads of type III can be switched off when the price of electricity is high or there is no renewable energy or in the battery, among other reasons. In particular, in our context, the stimulus to turn on the controllable loads is when the energy comes from renewable sources ($P_H - P_L > 1$) or when the minimum acceptable price of the energy ($cost_{comfort}$) is not exceeded $\left(cost_{comfort} - cost_{real} > 1\right)$. $cost_{comfort}$ can be defined by the user or can be defined as the minimum or average cost of the previous day's invoice, or something similar. The threshold is adjusted to activate the agent when there is surplus energy that comes from renewable sources ($\frac{P_H}{P_L} > 1$) or the price to pay for energy is interesting ($\frac{cost_{comfort}}{cost_{real}} > 1$). When $cost_{comfort}$ is exceeded and there is no surplus energy, then this agent must be turned off to avoid overpayments. Finally, $q_{LC}(\mathrm{t})$ is the probability of activating, or not, one of the load agents (controllable loads of type III).

$$s_{LC}(\mathrm{t}+1) = s_{LC}(\mathrm{t}) + w_{LC}\mathrm{f}_{LC}(\mathrm{t}),\ \ \mathrm{f}_{LC}(\mathrm{t}) = \begin{cases} (P_H - P_L), & P_H \geq P_L \\ \left(cost_{comfort} - cost_{real}\right), & P_H < P_L \end{cases} \tag{24}$$

$$\theta_{LC}(\mathrm{t}+1) = \theta_{LC}(\mathrm{t}) - \beta_{LC}Y_{LC}(\mathrm{t})\Delta\mathrm{t} + \gamma_{LC}(1 - Y_{LC}(\mathrm{t}))\Delta\mathrm{t} \tag{25}$$

$$Y_{LC}(t) = \begin{cases} \frac{P_H}{P_L}, & P_H \geq P_L \\ \frac{cost_{comfort}}{cost_{real}}, & P_H < P_L \end{cases} \tag{26}$$

$$q_{LC}(\mathrm{t}) = \frac{s_{LC}(\mathrm{t})^2}{s_{LC}(\mathrm{t})^2 + \theta_{LC}(\mathrm{t})^2} \tag{27}$$

Table 2 summarizes the different stimulus and response threshold expressions for each agent of our MAS to model the behavior of each MG component. This allows an autonomous distributed coordination using our RTM-based emergent control approach.

Table 2. Stimulus and response threshold expressions of each agent.

Agent	Instant Stimulus $f_s(t)$	Instantaneous Threshold (y_{ij})
Photovoltaic system	$P_{PVmax}(P_L + (1 - Soc)Q_{CAP}$	$\frac{P_{PV}}{P_L}$
Battery	Producer: $Soc(P_L - P_{PV})$	Producer: $\frac{P_{PV}}{P_L}$
	Consumer: $(1 - Soc)(P_{PV} - P_L)$	Consumer: $1 - \frac{P_{PV}}{P_L}$
Utility grid	$P_L - P_H$	$\frac{P_H}{P_L}$
Controllable load	$\mathrm{f}_{LC}(t)$	$1 - Y_{LC}(t)$

5. Experiments

This section presents the experimental protocol and scenarios for analyzing the behavior of our emergent control system based on the RTM.

5.1. Experimental Protocol

To validate the proposed controller described in Section 4, we explore a series of case studies, each with its unique characteristics. The experiments simulate a range of operational situations that controllers might encounter, such as variations in renewable generation dynamics due to weather conditions, fluctuations in uncontrolled loads, battery charging and discharging processes, and compensation from the main grid in case of deficiencies in other MG supply components. Power units are expressed in Watts [W]. We use the following performance criteria [36] to evaluate the quality of the results:

- Integral square error (ISE): This metric penalizes errors with higher values more severely than those with lower values. It is particularly useful for indicating overshoots and aggressive control, which are common following a disturbance.

$$ISE = \int_{t_1}^{t_2} \varepsilon(t)^2 dt \tag{28}$$

- Integral Absolute Error (IAE): Unlike ISE, IAE does not differentiate between positive and negative errors. It is frequently used for online controller tuning and is suitable for typical operations and non-monotonic step responses.

$$IAE = \int_{t^1}^{t^2} |\varepsilon(t)| dt \tag{29}$$

We also define an objective function J for hyperparameter optimization of our models, which was defined by Equation (30), which reaches an error close to zero. Mostly, it was used for the search grid algorithm [37,38] to alter the parameters of our model.

$$Ji = \min\{ISE(\Delta P)\} \tag{30}$$

These metrics help compare different case studies in the absence of similar works, demonstrating the feasibility and potential of our proposal. Real scenarios are modeled to conduct proof-of-concept tests. The baseline for our paper is maintaining energy balance (Equation (6)), with on/off delay time inferred through the ISE and IAE metrics. Hyperparameter optimization is an ongoing process, evolving as new agents are incorporated into the ecosystem.

5.2. Energy Scenarios

In order to demonstrate the adaptability and effectiveness of our emergent control strategy, we conducted several case studies, each designed to test the system under different conditions. Following is an overview of each one:

5.2.1. Case Study 1: Solar Energy in Remote Rural Locations

This case study analyzes the calibration and performance of the PV agent in an environment without an ESS. Conducted over 10 sunny days (measured in hours) with a constant demand of 250 W, the study aims to understand the model's dynamics by examining the interplay between the stimulus variable and the threshold. The calibration process involved is as follows:

1. Setting w_{PV} to modulate the instantaneous stimulus signal such that it allows variations of the accumulated stimulus in the order of tenths, to prevent overflow.
2. Estimating preliminary values for β_{PV} and γ_{PV} with simulations indicating the temporal trends of the stimulus and the threshold as shown in Figure 3B, with the dotted gray line representing the orientation of the signals. The adjustment is made using Equation (30) as the objective function.
3. The space around the previously obtained initial values of β_{PV} and γ_{PV} is explored. For that, a sweep is made of β_{PV} between $1 \times 10^{-5}/100 \times 10^{-5}$ and γ_{PV} between $1 \times 10^{-5}/10 \times 10^{-5}$, obtaining the following values: $\beta_{PV} = 2.1 \times 10^{-4}$ and $\gamma_{PV} = 1 \times 10^{-5}$ and a $w_{PV} = 0.0641$.

The PV agent is activated in response to solar radiation (ϕ) (see Equation (4)) and consumer demand (P_L) as there are no ESSs to store the produced energy. Figure 3 exhibits the key variables involved in the PV agent, with stable sunny days (as shown at the top of Figure 3).

Figure 3A depicts the consistent solar power available throughout the study, providing a foundational context for the PV agent's operation. This consistent availability of solar radiation is crucial for the agent's activation. In addition, it is observed how the interaction between the stimulus variables (context-dependent, such as solar radiation) and threshold intertwine to exhibit a dynamic (see Figure 3B,E,F) which results in the probability of activation or deactivation of the PV agent (see Figure 3C,D) that comes into operation when conditions are favorable, which will be explained below.

Figure 3E shows a day (24 h) during which the PV agent must be used to meet the demand from sunrise to sunset. Figure 3F illustrates how the stimulus (pink line) exhibits a decreasing behavior at night and an increasing behavior with sunrise, which is expected. In addition, the threshold (blue line) decreases when the solar panels contribute to satisfying the demand, translating into an increase in probability when there is solar radiation (see Figure 3F). At approximately 7:30 am, the PV agent switch is activated to supply power to the load, and remains activated for almost 8 h, until its deactivation at 3:20 p.m. when the power generated by the panel is no longer sufficient. This reveals that the activation mechanism is based on the probability signal (stimulus due to the sun), which allows the switches to be automated.

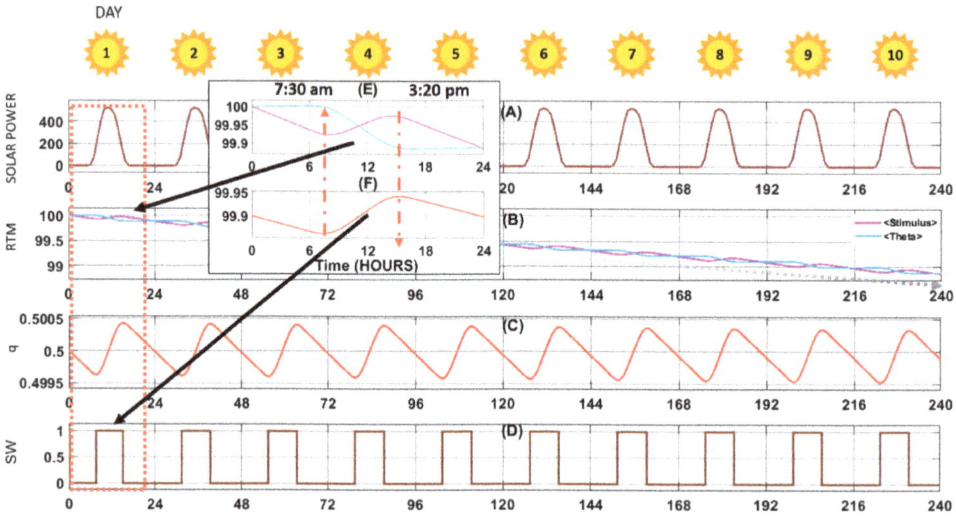

Figure 3. Solar energy in remote rural locations. From top to bottom: (**A**) Solar power. (**B**) Relationship between stimulus and threshold. (**C**) Probability of activation of the solar panels. (**D**) Smart switch activation. Figures (**E**,**F**) are the detailed and bounded views in a period of 24 h.

The simulation was conducted 20 times, and the results, including the averages and worst results for the ISE and IAE metrics, are summarized in Table 3. The standard deviation (S), trending towards zero, indicates the control system's stable performance. This pattern of stability, as reflected in the ISE and IAE results, is consistent with the results achieved in the other case studies (see Sections 5.2.2–5.2.8). This pattern of stability, as reflected in the ISE and IAE results, is consistent with the results achieved in the other case studies (see Sections 5.2.2–5.2.8).

Table 3. Performance metrics for case study 1.

Metrics	$\overline{\Delta P}$	Worst Cases ΔP	%	S
ISE	1.01×10^7	1.5×10^7	67.47%	0.0
IAE	4.03×10^4	6×10^4	67.17%	0.0

Table 3 provides a comprehensive overview of the system's behavior by presenting a set of metrics. In the worst-case scenario, where the power supply–demand discrepancy (ΔP) is at its maximum, ISE and IAE values reach their highest. These values are compared to the average ISE (1.01×10^7) and IAE (4.03×10^4) recorded during the experiment. The difference in magnitude between the worst-case and average scenarios is approximately 67.47% for ISE and 67.17% for IAE. This implies that over the 10-day (or 240-h) study period, the agent remained deactivated 67% of the time, translating to the PV agent being active for about 33% of the time. In terms of power supply, this equates to approximately 79 h or 3.3 days of activation. Table 4 details the parameters derived from the hyperparameter optimization process specific to this case study.

Table 4. Optimization of hyperparameters for functional control mode.

Parameter	Symbol	PV Agent	Battery Agent Producer	Battery Agent Consumer	Exchange Agent	Load Agent
Initial stimulus s_i	s	100	100	100	100	100
Initial response threshold θ_{ij}	θ	100	100	100	100	100
Attenuation factor	w	10.1×10^{-6}	0.005	5×10^{-8}	5×10^{-8}	5×10^{-8}
Learning factor	β	1.06×10^{-6}	1×10^{-6}	1×10^{-6}	1×10^{-6}	1×10^{-6}
Forgetting factor	γ	3×10^{-5}	3.9×10^{-5}	3.9×10^{-5}	3.9×10^{-5}	3.9×10^{-5}

The most relevant parameters of our model are the initial values of the stimuli and thresholds, as well as the factors used by each model, which significantly impact the speed and effectiveness of the reinforcement process. Notably, for each parameter, the forgetting factor outweighs the learning factor. This characteristic enables the agents to effectively utilize and build upon the knowledge they have acquired. Additionally, the attenuation factor, which plays a crucial role in modulating the stimulus, is kept relatively small. This careful calibration ensures that the stimulus is adjusted in a controlled manner, contributing to the overall precision and responsiveness of the model.

5.2.2. Case Study 2: Variable Solar Power in Remote Rural Areas

In a typical scenario encountered in rural or remote areas, such as farms without electricity service, inhabitants often turn to alternatives like solar systems to harness daylight radiation. This case study, inspired by the research of Yousif et al. [39], examines renewable energy sources without ESS, focusing on the quality and stability of the power delivered. Over a 10-day period, marked by variability in weather conditions and demand, the study explores these dynamics. The findings are presented in Table 5 and Figure 4.

Table 5. Performance metrics for case study 2.

Metrics	$\overline{\Delta P}$	Worst Cases ΔP	%	S
ISE	1.40×10^7	1.5×10^7	93.33%	0.0
IAE	4.56×10^4	5.52×10^4	82.6%	0.0

In Figure 4A, there are two significant shifts in energy demand. For the initial five days, the demand remains low at 100 W, allowing for optimal use of the available solar power. However, from day 6 onwards, the demand surges to 400 W. Consequently, solar energy is only sufficient to meet this increased demand on days 6 and 7. Starting from day 8, the solar radiation drops to levels too low to activate the Photovoltaic (PV) agent, as illustrated in Figure 4D.

Figure 4E highlights the energy deficit resulting from the lack of alternative energy sources to fulfill the demand. This deficit is corroborated by the data in Table 5, where $\overline{\Delta P}$ is close to the worst of the scenarios, with 93.33% for ISE and 82.6% for IAE, which means the percentage that fails to satisfy the demand. As a result, the PV agent's contribution is limited to 17.4%, with the shortfall leading to a blackout.

These outcomes, derived from 20 simulations as shown in Table 5, also reveal a standard deviation (S) tending towards zero. This indicates that despite the challenging conditions and the resultant shortfall in energy supply, the control system's performance remains stable.

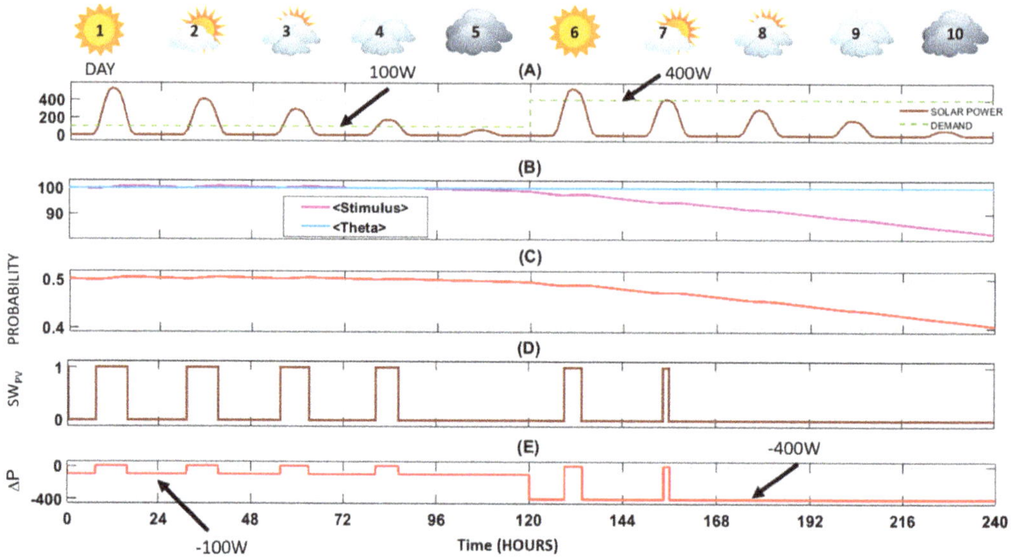

Figure 4. Variable solar power in remote rural areas with load demand. From top to bottom: (**A**) Continuous line represents solar power as a function of irradiance and the dotted line represents the real demand. (**B**) Relationship between stimulus and threshold. (**C**) Probability of activation of the solar panels. (**D**) Digital control signal for activation/deactivation of the switch by the PV agent. (**E**) Difference between the power delivered and consumed.

5.2.3. Case Study 3: Coordination between Energy Supply Agents

The objective of this case is to evaluate the coordination of the energy supply agents (solar energy and main grid) over a 10-day period characterized by variable sunny conditions. This scenario, typical of a residential setting, assumes a constant demand of 250 W with unvarying costs. The study, inspired by the findings in the work of Hu et al. [11], aims to assess how photovoltaic systems and the main electrical network collaborate to meet the energy needs of residential consumers.

In this case study, cloudiness significantly impacts the solar energy output, as observed in Figure 5A. The reduction in solar radiation due to cloud cover, particularly on days 2–5 and 7–10, leads to a decreased contribution from solar energy, which is clearly demonstrated in Figure 5B. The impact is most pronounced on days 4–5 and 9–10, where the cloudiness is so substantial that the Photovoltaic (PV) agent remains inactive, unable to generate adequate power due to insufficient radiation.

During these days, with the PV agent being inactive, the residential demand for power cannot be met by solar energy alone. Consequently, as depicted in Figure 5C, the electricity demand is fulfilled by an external power source. This reliance on an external grid during periods of reduced solar output underscores the importance of having a diversified energy supply, particularly in residential settings where consistent power availability is crucial.

Figure 5D illustrates the disparity between the energy supplied and the energy consumed. This difference is quantitatively depicted in Table 6, which shows exceptional demand satisfaction ($\Delta P = 0$), marked by a singular peak in both the IAE and ISE, each at 0.01%. The distribution of contributions from various components of the energy ecosystem is as follows: solar panels account for 13.13%, the battery contributes 0.0%, and the external power supply provides the majority at 86.88%. This distribution underscores the predominant role of the external power supply in this energy mix.

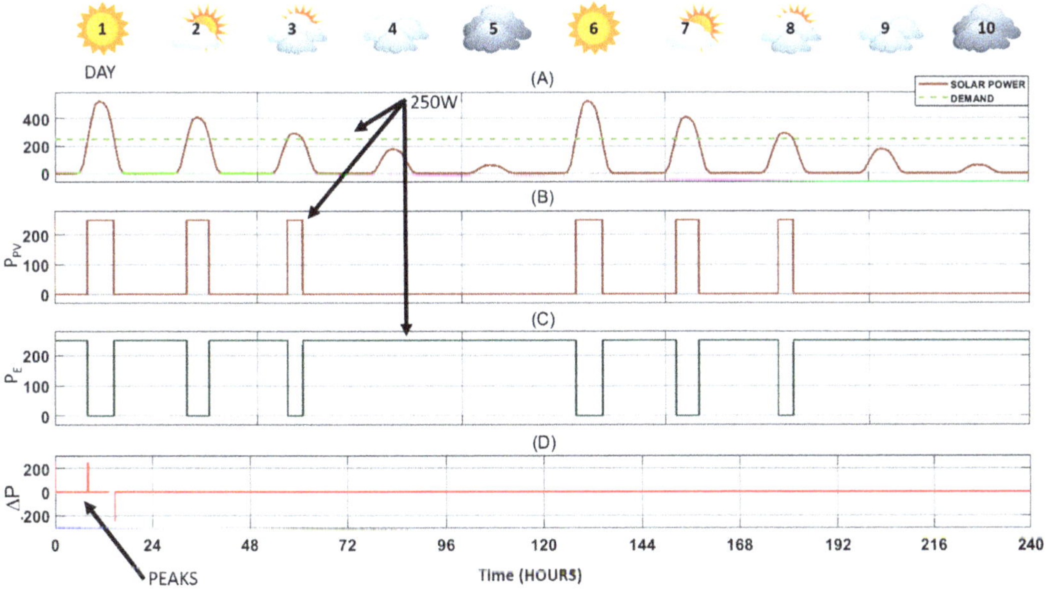

Figure 5. Failure in the energy storage system. From top to bottom: (**A**) Continuous line represents solar power as a function of irradiance and the dotted line represents the real demand. (**B**) Real power delivered by the solar panels. (**C**) Power from the external electrical grid. (**D**) Difference between the power delivered and consumed.

Table 6. Performance metrics for case study 3.

Metrics	$\overline{\Delta P}$	Worst Cases ΔP	%	S
ISE	1513	1.5×10^7	0.01%	0.0
IAE	6.05	6×10^4	0.01%	0.0

5.2.4. Case Study 4: Storage System with Failures

This case study centers on examining the coordination among PV agents, the battery, and the main electrical network in a situation where the battery, initially 100% charged, experiences a deep discharge due to a malfunction in its recharging mechanism, rendering its recovery impossible. Uniformity in energy costs and demand is maintained in this typical residential scenario. However, the battery charge and discharge controller encounter an issue in its electrical system, preventing proper charging of the storage system. This study is grounded in the research presented by Hu et al. [11].

Figure 6A displays solar power alongside the constant demand, highlighting days (such as 4, 5, 9, and 10) where solar power is insufficient due to weather conditions. In addition, Figure 6B shows how the photovoltaic agent is activated when there is solar radiation, which influences the time of the power delivered by the solar panels due to the level of radiation. Also, it is observed between the period of absence of solar radiation or cloudiness (see Figure 6A) that the battery comes into operation to supply the demand of consumers (Figure 6C). On day 4, with inadequate irradiation, solar panels fail to activate, and the storage device must supply energy all day. However, a failure in the battery agent leads to its complete discharge. (Figure 6D). The exchange agent intermittently compensates until it fully assumes supply due to the battery's complete discharge (SOC = 0) and insufficient solar radiation from day 9 onwards (Figure 6E).

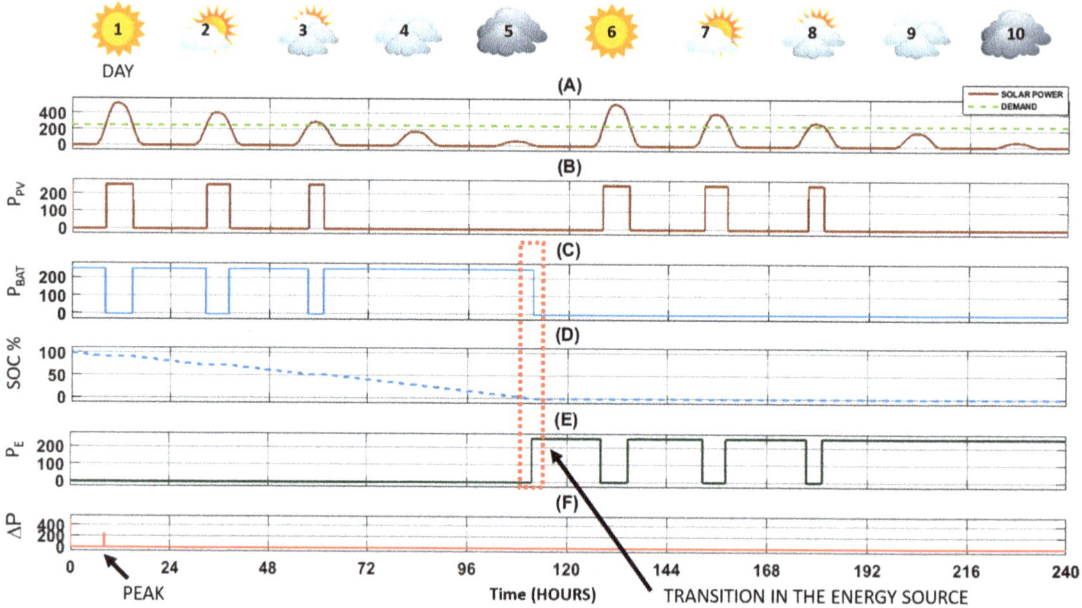

Figure 6. Storage system charging process failure. From top to bottom: (**A**) Continuous line represents solar power as a function of irradiance and the dotted line represents the real demand. (**B**) Real power delivered by the solar panels. (**C**) Battery power. (**D**) Dotted line represents the SOC battery charge percentage. (**E**) Power of the external electrical grid. (**F**) Difference between the power delivered and consumed.

Figure 6F depicts the balance achieved in the electrical network, fulfilling demand through the joint coordination of the agents and compensating for the deficiencies due to failures. Notably, the utilization of the storage device improves the metrics compared with case study 2 by reducing peaks caused by the entry of the external electrical network. The figure shows the difference between the energy delivered and consumed, as quantitatively represented in Table 7, with excellent demand satisfaction ($\Delta P = 0$). A single peak is noted in the IAE (0.02%) and ISE (0.04%) metrics, with the latter being higher. The contributions from each component of the energy ecosystem are as follows: solar panels at 13.13%, the battery at 39.95%, and the external power supply at 46.96%. This distribution highlights the significant influence of climatic conditions on renewable energy production (solar energy) and the increased reliance on external energy sources due to the battery's failure.

Table 7. Performance metrics for case study 4.

Metrics	$\overline{\Delta P}$	Worst Cases ΔP	%	S
ISE	6013	1.5×10^7	0.04%	0.0
IAE	12.05	6×10^4	0.02%	0.0

5.2.5. Case Study 5: Storage System Failure and Restoration

Case Study 5 examines the ecosystem's response to a battery failure in a typical residential setting, where the battery, initially fully charged, encounters a communication failure. This failure prevents the SOC variable from being detected by any system agent during the first five days. After 120 h, the communication issue is resolved. The scenario assumes variable solar radiation due to weather conditions and a constant demand over

10 days. This study, based on the insights from Hu et al. [11], analyzes the impact of a battery agent problem and its subsequent restoration on the overall energy supply system.

Figure 7A displays solar power in relation to irradiance alongside the actual demand. It highlights days, such as 4, 5, 9, and 10, where solar power is insufficient due to weather conditions. In addition, Figure 7B illustrates the activation of the PV agent and its influence on the power delivered by the solar panels, dictated by the level of solar radiation. Figure 7C shows that for the first five days, the battery does not operate due to the failure in communicating the SOC variables. Once this issue is resolved, the battery resumes operation and supplies the required power for the subsequent days. In Figure 7D, the dotted line represents the SOC battery charge percentage throughout the period, and Figure 7E indicates the periods when the power from the external electrical grid is utilized, particularly after the battery is fully discharged on the 10th day.

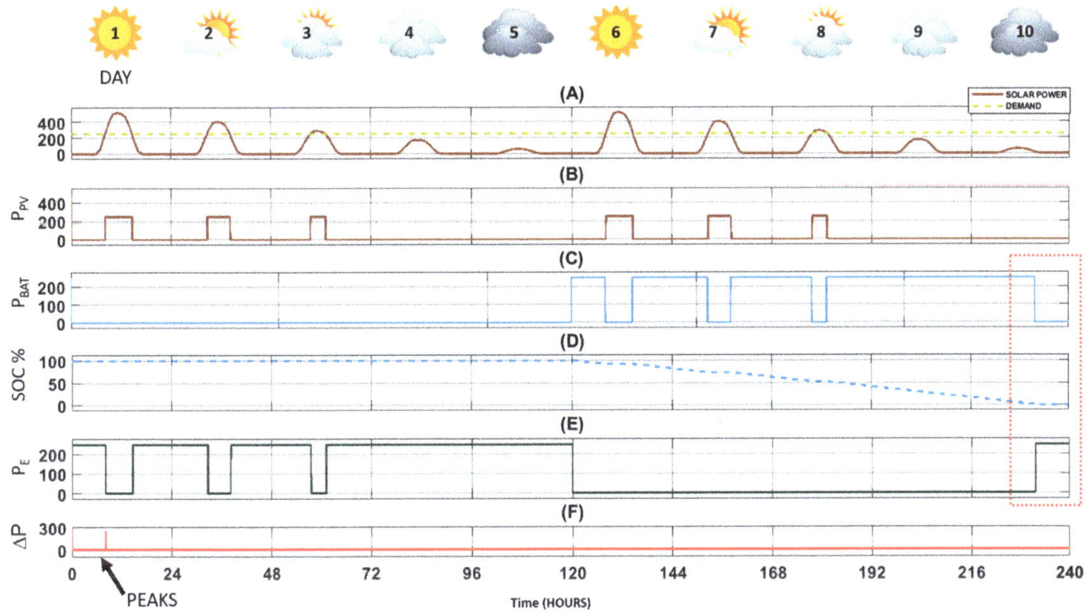

Figure 7. Storage system failure and restoration. From top to bottom: (**A**) Continuous line represents solar power as a function of irradiance and the dotted line represents the real demand. (**B**) Real power delivered by the solar panels. (**C**) Period where the battery power is supplied. (**D**) Dotted line represents the SOC battery charge percentage. (**E**) Period where the power of the external electrical grid is used. (**F**) Difference between the power delivered and consumed.

Figure 7F reveals the difference between the energy delivered and that consumed, whose behavior is represented quantitatively through Table 8, with excellent demand satisfaction ($\Delta P = 0$), and with a single peak expressed in the metric IAE (0.03%) and ISE (0.09%), the latter being higher due to the presence of the peak. The contribution of each of the components of the energy ecosystem is distributed as follows: 13.13% the solar panels, 39.94% the battery, and 46.96% the external power supply, which demonstrates the high influence of climatic conditions on production from renewable sources (solar energy), and the greater participation of the external energy source.

Table 8. Performance metrics for case study 5.

Metrics	$\overline{\Delta P}$	Worst Cases ΔP	%	S
ISE	1.35×10^4	1.5×10^7	0.09%	0.0
IAE	18.05	6×10^4	0.03%	0.0

5.2.6. Case Study 6: Normal Operation of the MG with Constant Demand

The aim of this case study is to evaluate the performance of the energy storage agent in managing energy supply and consumption processes under constant demand. It operates under the assumption of stable energy costs over time ($cost_{comfort} > cost_{real}$), and varying solar radiation, as depicted in see Figure 8A. This scenario is typical of a residential environment powered by various sources, with the study focusing on the coordination of its components. The foundational research for this case is drawn by Zhou et al. [10].

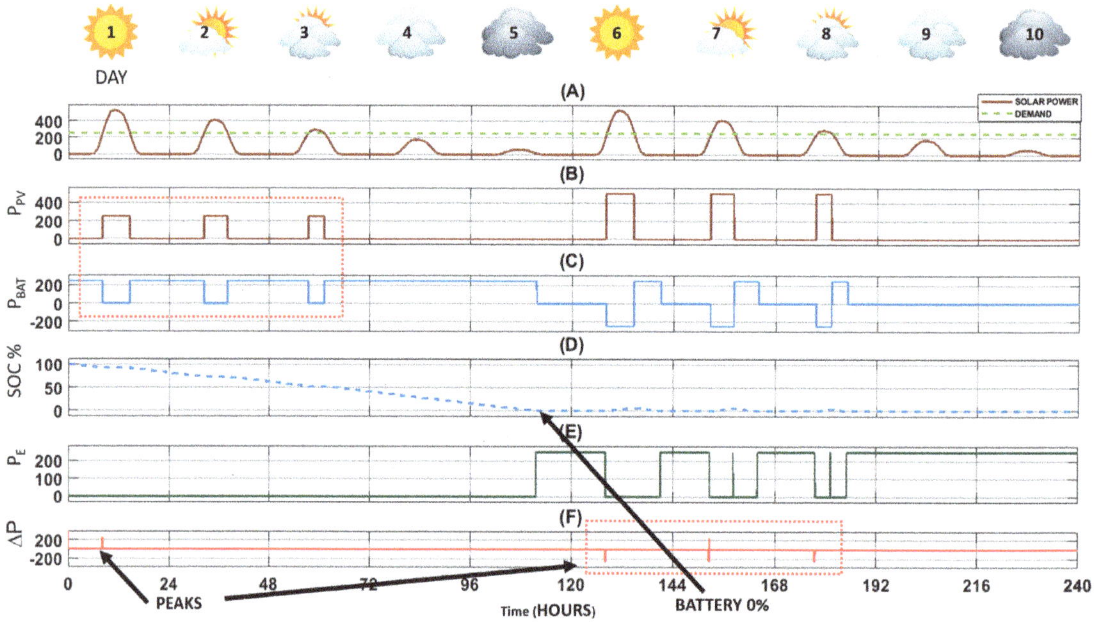

Figure 8. Normal operation of the MG with constant demand. From top to bottom: (**A**) Continuous line represents solar power as a function of irradiance and the dotted line represents the real demand. (**B**) Real power delivered by the solar panels. (**C**) Period where the battery power is supplied. (**D**) Dotted line represents the SOC battery charge percentage. (**E**) Period where the power of the external electrical grid is used. (**F**) Difference between the power delivered and consumed.

Figure 8A shows the solar power with constant demand, evidencing that there are days where the solar power is not enough to satisfy the demand, such as days 4, 5, 9, and 10, due to weather conditions. In addition, Figure 8B shows how the PV agent is activated according to the solar radiation and load demand. On the one hand, it is evident that in the absence of solar energy, the storage device starts to work and is activated again when the power of the panel is not enough, which is highlighted between days 1 and 3 of Figure 8B,C. On the other hand, the battery is completely discharged on day 5, and tries to recover slowly, but, motivated by the demand and variability of the climatic conditions, the SOC tends to zero (see Figure 8D). Thus, the external power system comes into operation intermittently from day 5, and becomes permanent from day 9, due to insufficient solar radiation and battery discharge.

Figure 8F presents the difference between the energy delivered and consumed. This behavior, quantitatively depicted in Table 9, shows excellent demand satisfaction ($\Delta P = 0$), with occasional peaks in the IAE (0.09%) and ISE (0.16%) metrics, the latter being higher due to the quantity and magnitude of the peaks. Meanwhile, the IAE remains lower than the ISE due to the infrequent occurrence of these peaks. The energy contributions are distributed as follows: solar panels at 13.11%, the battery at 46.58%, and the external power supply at 40.31%. This distribution highlights the significant impact of climatic conditions on renewable energy production (solar energy) and the subsequent effects on the battery charging processes.

Table 9. Performance metrics for case study 6.

Metrics	$\overline{\Delta P}$	Worst Cases ΔP	%	S
ISE	2.328×10^4	1.5×10^7	0.16%	0.0
IAE	57.11	6×10^4	0.09%	0.0

5.2.7. Case Study 7: Normal Operation of the MG with Variable Consumption

The focus of this case study is to analyze the coordination between the components of an MG—specifically the solar panels and external power source—in a scenario where there is variable consumption against a backdrop of constant solar radiation over a 10-day period. The battery in this scenario is fully functional and charged (charge and discharge). This residential scenario, drawing on various energy sources, examines the coordination between the MG components in response to demand variations. The study is based on the research conducted by Zhou et al. [10].

Figure 9A shows the solar power and the variation in the demand of the electrical network, highlighting the excess energy that will allow the battery to be charged. The power generated by P_{PV} (solar panels) is used depending on the demand needs and availability of storage in the battery (SOC << 1). For example, on days 1 and 2, only what is required by the load is supplied, while for the rest of the days, both the demand and the battery charge are covered (see Figure 9B). It can also be seen in Figure 9C,D, the power delivered by the battery and the percentage (SOC), respectively, where it is fully discharged on day 6 and begins a gradual recovery process, due to the demand decreases at the end of day 7. Between days 6 and 8, the battery is completely discharged, so the external power source comes into operation (see Figure 9E).

Figure 9F presents the difference between the energy delivered and consumed. This behavior, as quantified in Table 10, indicates excellent demand satisfaction ($\Delta P = 0$), with occasional peaks expressed in the metrics IAE (0.73%) and ISE (0.19%). The ISE is lower due to the frequency and magnitude of the peaks, whereas the IAE is higher, reflecting an increase in magnitude due to demand variations on days 6 and 7. The energy contributions are distributed as follows: solar panels at 27.92%, the battery at 58.38%, and the external power source at 13.72%. This distribution signifies that 86.26% of the energy comes from renewable sources, underscoring the MG's effective management and utilization of renewable energy.

Table 10. Performance metrics for case study 7.

Metrics	ΔP	Worst Cases ΔP	%	S
ISE	2.84×10^4	1.5×10^7	0.19%	0.0
IAE	73.26	6×10^4	0.73%	0.0

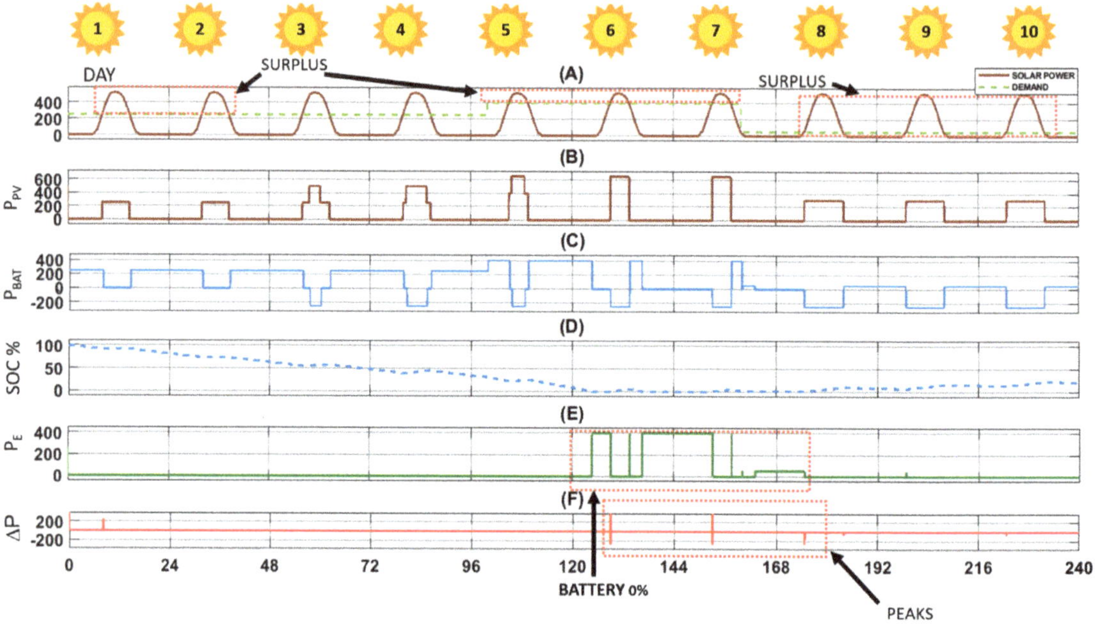

Figure 9. Normal operation of the MG with variable consumption. From top to bottom: (**A**) Continuous line represents solar power as a function of irradiance and the dotted line represents the real demand. (**B**) Real power delivered by the solar panels. (**C**) Period where the battery power is supplied. (**D**) Dotted line represents the SOC battery charge percentage. (**E**) Period where the power of the external electrical grid is used. (**F**) Difference between the power delivered and consumed.

5.2.8. Case Study 8: Energy Price Variability

This case study investigates the functionality of the entire energy ecosystem amidst energy price fluctuations over a 10-day period, focusing on how these changes affect behavior around the cost of comfort. The scenario assumes constant solar radiation, a fully charged (100%) battery at the start, and normal operation for charging and discharging processes. The non-controllable loads consume 200 W, and the controllable loads consume 50 W, representing a scenario with diverse energy sources and fluctuating energy costs.

The dotted line in Figure 10A indicates solar power related to irradiance, while the solid line shows the actual potential delivered by the solar panels. For the initial two days, the energy supply is solely from the battery and solar agents, as shown in Figure 10B. Despite high battery charge levels and solar panel energy surpluses, these surpluses are not utilized until the third day.

The battery has a high percentage of load, and the energy surpluses of the solar panels are not used, but from the third day on, the excess energy from the solar panels is gradually used to charge the battery, which is evidenced by the overlapping of both signals (see Figure 10A) This charge is very tenuous (See Figure 10C). Figure 10B shows the three states of the battery: when $P_{BAT} > 0$ then it supplies power; when $P_{BAT} = 0$ then it does nothing; and when $P_{BAT} < 0$ then it charges. On day 7th, the batteries are completely discharged for the first time, and that is when the external electrical energy comes into operation to supply the energy deficiencies (see Figure 10D). It should be noted that the batteries manage to partially charge themselves in periods of excess solar energy, but are not enough to achieve 100% charge. So, coordinated control is obtained between the solar panels, the batteries and the external electrical network.

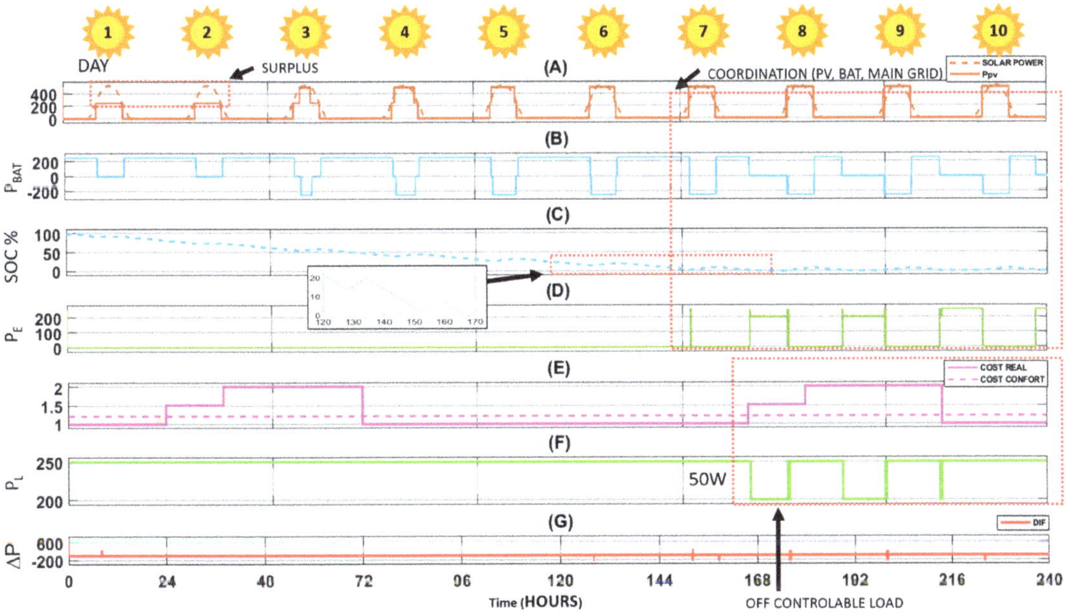

Figure 10. Energy price variability. From top to bottom: (**A**) The dotted line represents solar power as a function of irradiance and the solid line represents the real potential delivered by the solar panels. (**B**) Continuous line is the battery power supplied. (**C**) Dotted line represents the SOC battery charge percentage. (**D**) Period where the power of the external electrical grid is used. (**E**) The dotted line represents the cost of comfort and the continuous line the variations in the price of energy. (**F**) Demand or power consumed by the apartment. (**G**) Difference between the power delivered and consumed.

Figure 10E shows the variations in the price of energy. When the batteries are charged in the first 6 days, despite the cost increase, it is not relevant because the external source does not act. But then, in the second cycle of variations in the price of energy that occurs between days 8 and 9, it affects the operation of the electrical network, which causes the deactivation of controllable loads to reduce consumption. Most of the time (91.46%), full energy supply is guaranteed to both controllable and non-controllable loads, but only 8.54% of the time, controllable loads are turned off due to high energy prices.

Figure 10G quantitatively represents the difference between the energy delivered and consumed. In other words, a measure of satisfaction of consumers who seek to have the trend as close to zero as possible, which is shown in Table 11, when the peaks and transitions of the energy sources are minimal with values of 0.10% in the case of the IAE and 0.17% in the ISE. The latter is higher because it penalizes the peaks that appear. To calculate the above measurements, the worst case is considered when the error is maximum (250 W) during the 10 days.

Table 11. Performance metrics for case study 8.

Metrics	$\overline{\Delta P}$	Worst Cases ΔP	%	S
ISE	2.54×10^4	1.5×10^7	0.17%	0.0
IAE	66.53	6×10^4	0.10%	0.0

5.3. General Discussion

The results of the case studies, reflecting the contributions of the MG generation sources, are summarized in Table 12. The data are expressed in hours, days, and percentages relative to a period of 240 h/10 days/100 percent. Each case study explores a variety of scenarios that can occur in electrical networks. These scenarios include weather variations, component failures, market price fluctuations, and changes in consumer demand, among others.

Table 12. Summary of the case studies.

		Case 1	Case 2	Case 3	Case 4	Case 5	Case 6	Case 7	Case 8
P_{PV}	Hours	78.02	41.75	31.52	31.5	31.52	31.47	67	65.17
	Day	3.25	1.74	1.31	1.31	1.31	1.31	2.79	2.72
	%	32.51%	17.40%	13.13%	13.13%	13.13%	13.11%	27.92%	27.15%
$P_{BAT,S}$	Hours	0	0	0	95.87	95.86	111.8	140.1	141.4
	Day	0.00	0.00	0.00	3.99	3.99	4.66	5.84	5.89
	%	0.00%	0.00%	0.00%	39.95%	39.94%	46.58%	58.38%	58.92%
P_E	Hours	0	0	208.5	112.7	112.7	96.75	32.93	33.5
	Day	0.00	0.00	8.69	4.70	4.70	4.03	1.37	1.40
	%	0.00%	0.00%	86.88%	46.96%	46.96%	40.31%	13.72%	13.96%

Figure 11 graphically represents the percentage contribution of each component. In cases 1 and 2, only the solar panels are involved, with differences between the two cases due to climatic variability.

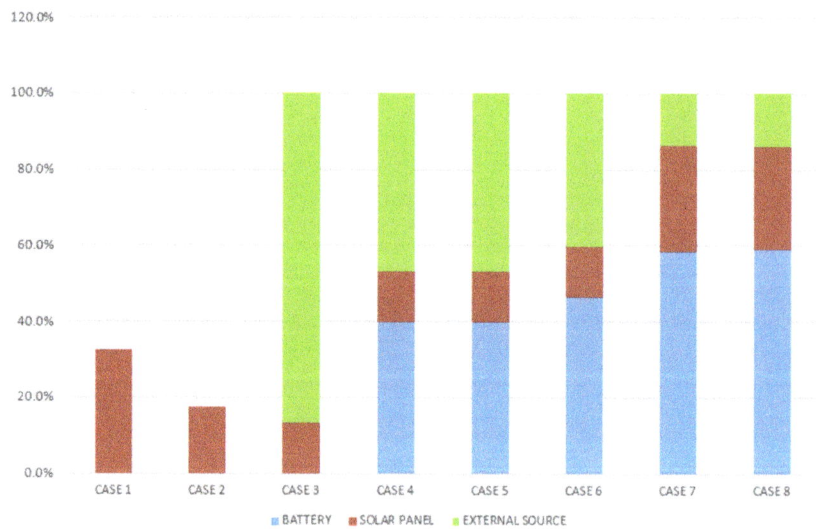

Figure 11. Contribution of supply sources in each case study.

According to Figure 11, blackouts are caused by an energy deficit when other supply sources are absent, leading to a vulnerable and deficient electricity supply. Conversely, from cases 3 to 8, there is evident perfect coordination among the various agents to ensure energy supply under any condition. Cases 4–6 exhibit similar behaviors, as do cases 7 and 8, where most of the energy supply is from renewable sources. This situation is advantageous for consumers, as renewable energy is generally cheaper and less polluting than fossil fuel-based sources such as gas and oil.

This paper has demonstrated the technical feasibility of our proposal through simulations. However, for real-world implementation, other technological aspects must be considered. Specifically, the emerging agent-oriented RTM-based controller would operate at the tertiary level of an MG control architecture, necessitating coordination, control, and communication processes with higher-level agents. This would require communication capabilities depending on the distributed nature of the MG components. For wireless communication needs, technologies like GSM, GPRS, 3G, 4G, and Wi-Max could be utilized; ZigBee, WiFi, or Bluetooth are suitable for shorter distances. Wired solutions might include Power Line Communication (PLC), fiber optics, and DSL for longer distances, or RS485 and Ethernet for shorter distances [9].

5.4. Qualitative Comparison with Previous Works

In comparing our work with previous studies (refer to Table 13), several criteria have been considered:

1. Context of Distributed Control: one criterion is whether the study is within the context of distributed control.
2. Energy Problem Involving Both Supply and Demand: another criterion is whether the study addresses the energy problem involving both supply and demand simultaneously.
3. Consideration of New Agents: lastly, we consider if the study incorporates dynamically new agents, such as new renewable energy sources.

The x symbol on Table 13 indicates that this work meets that criterion.

Table 13. Comparison with previous works.

Criteria vs. Works	1	2	3
[14]			x
[15]	x		x
[16]	x		
[20]	x		x
[40]			x
[10]	x	x	
[12]		x	
[13]	x	x	
Our approach	x	x	x

Following is a qualitative comparison with some of the previous works: Harmouch et al. [14] suggest a decentralized MA-EMS with the goal of minimizing power exchange with the main grid. While it shares a similar architecture with our paper, it does not search into the coordination between different elements or variations in market prices, and the study time periods are shorter. Additionally, its objective is to minimize grid reliance rather than satisfy consumer needs.

Kofinas et al. [15] recommend a cooperative MAS for managing the energy of a stand-alone MG. The MAS learns to control the MG components, but it requires previous experience to define the control rules. Moreover, it does not consider the participation of controllable loads on the demand side. Zhou et al. [9] propose an autonomous coordination control strategy for MGs, and Areekkara et al. [13] study a MAS approach to model an MG EMS with diverse forecasting agents and a real-time correction agent.

Chouikhi, et al. [16] present electricity management mechanisms in smart buildings to reduce total energy cost and the peak to the average consumption ratio, and maximize the exploitation of renewable energy. Unlike our proposal, which focuses on controllable loads, this work analyzes various loads within building apartments but does not detail their distribution within the building.

Zheng et al. [40] examine self-organization algorithms that are appropriate for task distribution in industrial manufacturing processes with significant task dependencies. They propose two algorithms based on the RTM for task dependencies. They only rely on local interactions between the relevant agents and the environment for decision-making. It is a different application of the RTM compared to our approach as an emergent control strategy.

The rest of the works [9,11,17,18] fall along the same lines as the previous ones: they do not simultaneously consider both sides (producer and consumer), or they do not allow distributed control or the natural integration of new agents. These aspects are covered by our proposal. It is worth highlighting the works [11,12,19] that address trends in coordination strategies around energy management systems as an MG. In these works, it can be verified that the RTM has not been previously used, showing that our RTM-based approach is an innovative proposal. Additionally, this paper indicates the way to specify an MG using the MAS paradigm with an emergent coordination approach between its agents (see Section 4.2).

6. Conclusions

Based on the emergent control paradigm, this paper proposes a distributed control architecture for MGs. Specifically, it is based on the RTM, which was inspired by how ants react to internal or external stimuli of local variables in their environment. The paper suggests an emergent control approach for MGs composed of heterogeneous agents, including batteries, solar panels, controllable loads, and external energy sources. These agents must coordinate to satisfy consumer demand.

One of the main features of our emergent control method is its robustness. It adapts to component failures, changes in climatic conditions, or economic variables, ensuring the achievement of global goals. The activation mechanisms derived from this approach significantly improve performance, with IAE metrics as low as 0.01% (in case study 3), indicating minimal difference between supplied and demanded power.

The research results demonstrate the potential of this approach to address several challenges in the energy ecosystem: (i) MG component decentralized coordination based on local environmental knowledge; (ii) Robustness in distributed control actions in the face of different failures (see cases 3, 4 and 5); (iii) RES prioritization. From cases 3 to 8, there is evidence of perfect coordination between the various agents to guarantee the energy supply in any condition $\Delta P = 0$. Cases 4–6 show similar behaviors, as well as cases 7 and 8, where most of the energy comes from renewable energy sources, 59.7% and 86.1%, respectively. The main findings of the research work focus on the definition of an autonomous distributed control based on the RTM, and the definition of a methodology to apply our emergent approach to control an MG. Our approach allows maximizing the use of renewable energy, uses controllable charging when it is least expensive, and optimizes the use of batteries, among other things. On the other hand, one of the main limitations of this work is the implementation in a reduced number of energy agents.

Future work aims to integrate additional RES such as wind and geothermal and consider their joint performance. The study will also dig into controllable loads and other agents like electric vehicles, exploring various user comfort levels for load adjustment. The emergent control proposal will be integrated into a MAPE-K (Monitor-Analyze-Plan-Execute plus Knowledge) model [41] for smart energy management in smart grids. Furthermore, the approach will be implemented in an energy simulator like ESDL-based Energy System Simulator (ESSIM) [42] and investigated for compatibility with advanced RL models, including Deep RL.

Additionally, other conditions, such as penalties stipulated in electricity supply contracts for interruptions to consumer supply, can be incorporated into the design of our emerging controller. Under these conditions, maintaining the stability of the MG requires action on the electricity supply networks rather than on the consumers. Another aspect for future studies is the monitoring of frequency values to control the balance between the electrical energy produced and consumed. Finally, while future work will concentrate on distributed control issues, there is also a potential to explore more complex component models and apply MG design methods to size components appropriately according to various needs and operational requirements.

Author Contributions: Conceptualization, M.G. and J.A.; Methodology, J.A.; Validation, M.D.R.-M.; Formal analysis, M.G. and J.A.; Investigation, M.G., J.A. and M.D.R.-M.; Writing—original draft, M.G.; Writing—review & editing, J.A. and M.D.R.-M. All authors have read and agreed to the published version of the manuscript.

Funding: This project received funding from the European Union's Horizon 2020 research and innovation programme under the Marie Skłodowska-Curie grant agreement No 754382 GOT ENERGY TALENT. M.D.R.-M. is supported by the JCLM project SBPLY/19/180501/000024 and the Spanish Ministry of Science and Innovation project PID2019-109891RB-I00, both under the European Regional Development Fund (FEDER).

Data Availability Statement: The data presented in this study are available on request from the corresponding author.

Conflicts of Interest: The authors declare no conflict of interest.

Disclaimer: The content of this publication does not reflect the official opinion of the European Union. Responsibility for the information and views expressed herein lies entirely with the authors.

Nomenclature and Abbreviations

Subscripts	
i	For agents/individuals
j	For tasks, j = {PV, ESS, UG}
Variables	
θ	Response threshold
s	Stimulus
q	Activation probability
N_{act}	Quantity of active entities
N	Quantity of entities that may be active in the colony
f_s	Instant stimulus
y_{ij}	Portion of entities of type i doing task j
P_{PV}	Maximum active power output PV [W]
ϕ	Solar radiation [kw/m^2]
T_a	Ambient temperature [°C]
V_t	Terminal voltage of the battery [V]
SOC	Battery state of charge at t [%].
I	The effective value of the discharge current [A]
P_T	Total active power generated [W]
ΔP	Error in the supply-demand [W]
P_H	Generated active power by the renewable systems [W]
P_E	Main grid active power [W]
P_{BAT}	Active power in the energy storage system [W]
P_L	Total energy consumption or demand [W]
P_{LC}	Power consumed by controllable loads [W]
P_{LNC}	Power consumed by uncontrollable loads [W]
$cost_{comfort}$	Comfort cost
$cost_{real}$	Market cost
Abbreviations	
DER	Distributed energy resources
DRL	Deep Reinforcement Learning
EP	Energy pool
PEU	Power exchange unit
EMS	Energy Management System
ES	Energy system

ESS	Energy storage system
HES	Hybrid energy systems
LC	Controllable loads
LNC	Non-controllable loads
MAS	Multiagent Systems
MASINA	Multiagent Systems for Integrated Automation
MG	Microgrid
PV	Photovoltaic
RES	Renewable Energy Systems
RTM	Response Threshold Model
RL	Reinforcement Learning
UG	Utility grid
WECS	Wind energy conversion systems
Parameters	
α	Scale factor that measures the efficiency in performing the task
δ	Increase in the stimulus's intensity per unit of time
β	Learning rate
γ	Forgetting rate
η	Conversion efficiency of the PV array
S	Area of the PV array [m^2]
V_{OC}	Circuit voltage [V]
R_{int}	Internal resistance of the battery [Ω]
K	Polarization constant
w_j	Attenuation factor of the instantaneous stimulus function (fs)

References

1. Ray, P.K.; Mohanty, S.R.; Kishor, N. Small-signal analysis of autonomous hybrid distributed generation systems in presence of ultracapacitor and tie-line operation. *J. Electr. Eng.* **2010**, *61*, 205–214. [CrossRef]
2. Moghaddas-Tafreshi, S.; Jafari, M.; Mohseni, S.; Kelly, S. Optimal operation of an energy hub considering the uncertainty associated with the power consumption of plug-in hybrid electric vehicles using information gap decision theory. *Int. J. Electr. Power Energy Syst.* **2019**, *112*, 92–108. [CrossRef]
3. Anvari-Moghaddam, A.; Rahimi-Kian, A.; Mirian, M.S.; Guerrero, J.M. A multi-agent based energy management solution for integrated buildings and microgrid system. *Appl. Energy* **2017**, *203*, 41–56. [CrossRef]
4. Ahmethodzic, L.; Music, M. Comprehensive review of trends in microgrid control. *Renew. Energy Focus* **2021**, *38*, 84–96. [CrossRef]
5. Aguilar, J.; Garcès-Jimènez, A.; Gallego-Salvador, N.; de Mesa, J.G.; Gomez-Pulido, J.; Garcìa-Tejedor, À. Autonomic Management Architecture for Multi-HVAC Systems in Smart Buildings. *IEEE Access* **2019**, *7*, 123402–123415. [CrossRef]
6. Rahman, M.S.; Orchi, T.; Saha, S.; Haque, M.E. Cooperative Multiagent Based Distributed Power Sharing Strategy in Low-Voltage Microgrids. *IEEE Trans. Ind. Appl.* **2020**, *56*, 3285–3296.
7. Bonabeau, E. Fixed Response Thresholds and the Regulation of Division of Labor in Insect Societies. *Bull. Math. Biol.* **1998**, *60*, 753–807. [CrossRef]
8. Shahgholian, G. A brief review on microgrids: Operation, applications, modelling, and control. *Int. Trans. Electr. Energy Syst.* **2021**, *31*, e12885. [CrossRef]
9. Liu, Z.; Xiang, T.; Wang, T.; Mu, C. Cooperative Optimization Strategy for Distributed Energy Resource System using Multi-Agent Reinforcement Learning. In Proceedings of the IEEE Symposium Series on Computational Intelligence, Orlando, FL, USA, 5–7 December 2021; pp. 1–6.
10. Zhou, Q.; Shahidehpour, M.; Paaso, A.; Bahramirad, S.; Alabdulwahab, A.; Abusorrah, A. Distributed Control and Communication Strategies in Networked Microgrids. *IEEE Commun. Surv. Tutor.* **2020**, *22*, 2586–2633. [CrossRef]
11. Hu, M.; Xiao, F.; Wang, S. Neighborhood-level coordination and negotiation techniques for managing demand-side flexibility in residential microgrids. *Renew. Sustain. Energy Rev.* **2021**, *135*, 110248. [CrossRef]
12. Zhou, X.; Zhou, L.; Chen, Y.; Guerrero, J.; Luo, A.; Wu, W.; Yang, L. A microgrid cluster structure and its autonomous coordination control strategy. *Int. J. Electr. Power Energy Syst.* **2018**, *100*, 69–80. [CrossRef]
13. Areekkara, S.; Kumar, R.; Bansal, R. An Intelligent Multi Agent based Approach for Autonomous Energy Management in a Microgrid. *Electr. Power Compon. Syst.* **2020**, *49*, 18–31. [CrossRef]
14. Harmouch, F.; Krami, N.; Hmina, N. A multiagent based decentralized energy management system for power exchange minimization in microgrid cluster. *Sustain. Cities Soc.* **2018**, *40*, 416–427. [CrossRef]
15. Kofinas, P.; Dounis, A.I.; Vouros, G.A. Fuzzy Q-Learning for multi-agent decentralized energy management in microgrids. *Appl. Energy* **2018**, *219*, 53–67. [CrossRef]
16. Chouikhi, S.; Merghem-Boulahia, L.; Esseghir, M.; Snoussi, H. A Game-Theoretic Multi-Level Energy Demand Management for Smart Buildings. *IEEE Trans. Smart Grid* **2019**, *10*, 6768–6781. [CrossRef]

17. Xu, X.; Jia, Y.; Xu, Y.; Xu, Z.; Chai, S.; Lai, C.S. A Multi-Agent Reinforcement Learning-Based Data-Driven Method for Home Energy Management. *IEEE Trans. Smart Grid* **2020**, *11*, 3201–3211. [CrossRef]
18. Arnone, D.; Croce, V.; Paternó, G.; Rossi, A.; Emma, S.; Miceli, R.; Di Tommaso, A.O. Energy Management of Multi-carrier Smart Buildings for Integrating Local Renewable Energy Systems. In Proceedings of the 5th International Conference on Renewable Energy Research and Applications, Birmingham, UK, 20–23 November 2016; Volume 5, pp. 20–23.
19. Aguilar, J.; Giraldo, J.; Zapata, M.; Jaramillo, A.; Zuluaga, L.; Moreno, M.R. Autonomous Cycle of Data Analysis Tasks for Scheduling the Use of Controllable Load Appliances using Renewable Energy. In Proceedings of the 2021 International Conference on Computational Science and Computational Intelligence (CSCI), Las Vegas, NV, USA, 15–17 December 2021.
20. García, M.; Aguilar, J. Emergent control in the context of industry 4.0. *Int. J. Comput. Integr. Manuf.* **2021**, *35*, 247–262. [CrossRef]
21. Du, Y.; Lu, X.; Wang, J.; Lukic, S. Distributed secondary control strategy for microgrid operation with dynamic boundaries. *IEEE Trans. Smart Grid* **2019**, *10*, 5269–5282. [CrossRef]
22. Chai, J.; Peng, Y. Coordinated Power Control for Islanded DC Microgrids Based on Bus-Signaling and Fuzzy Logic Control. In Proceedings of the 2nd IEEE Conference on Energy Internet and Energy System Integration, Beijing, China, 20–22 October 2018.
23. Arwa, E.O.; Folly, K.A. Reinforcement Learning Techniques for Optimal Power Control in Grid-Connected Microgrids: A Comprehensive Review. *IEEE Access* **2020**, *8*, 208992–209007. [CrossRef]
24. Zheng, Y.; Li, S.; Tan, R. Distributed Model Predictive Control for On-Connected Microgrid Power Management. *IEEE Trans. Control Syst. Technol.* **2018**, *26*, 1028–1039. [CrossRef]
25. Kyriakou, D.G.; Kanellos, F.D. Optimal Operation of Microgrids Comprising Large Building Prosumers and Plug-in Electric Vehicles Integrated into Active Distribution Networks. *Energies* **2022**, *15*, 6182. [CrossRef]
26. Escobar, L.M.; Aguilar, J.; Garcés-Jiménez, A.; De Mesa, J.G.; Gomez-Pulido, J. Advanced Fuzzy-Logic-Based Context-Driven Control for HVAC Management Systems in Buildings. *IEEE Access* **2020**, *8*, 16111–16126. [CrossRef]
27. Bonabeau, E.; Sobkowski, A.; Theraulaz, G.; Deneubourg, J. Adaptive Task Allocation Inspired by a Model of Division of Labor in Social Insects. *Biocomput. Emergent Comput.* **1997**, 36–45.
28. Yamada, K. Specialization in Swarm Robotics using Local Interactions. *Int. J. Eng. Res. Technol.* **2016**, *5*, 103–108. [CrossRef]
29. Teruya, D.; Indurkhya, B.; Maksaki, T.; Nakajo, H. Autonomous Distributed System Based on Behavioral Model of Social Insects. In Proceedings of the International Conference on Parallel and Distributed Processing Techniques and Applications, Las Vegas, NV, USA, 30 July–2 August 2018; pp. 289–295.
30. Ding, Z.; Huang, Y.; Yuan, H.; Dong, H. Introduction to Reinforcement Learning. In *Deep Reinforcement Learning*; Springer: Berlin/Heidelberg, Germany, 2020; pp. 47–123.
31. Antonanzas-Torres, F.; Urraca, R.; Polo, J.; Perpiñán-Lamigueiro, O.; Escobar, R. Clear sky solar irradiance models: A review of seventy models. *Renew. Sustain. Energy Rev.* **2019**, *107*, 374–387. [CrossRef]
32. Mousavi, S.M.; Nikdel, M. Various battery models for various simulation studies and applications. *Renew. Sustain. Energy Rev.* **2014**, *32*, 477–485. [CrossRef]
33. Shen, J.; Jiang, C.; Li, B. Controllable Load Management Approaches in Smart Grid. *Energies* **2015**, *8*, 11187–11202. [CrossRef]
34. Aguilar, J.; Cerrada, M.; Hidrobo, F. A Methodology to Specify Multiagent Systems. In Proceedings of the Agent and Multi-Agent Systems: Technologies and Applications, Wroclaw, Poland, 31 May–1 June 2007; Volume 4496, pp. 92–101.
35. Aguilar, J.; Besembel, I.; Cerrada, M.; Hidrobo, F.; Narciso, F. Una Metodología para el Modelado de Sistemas de Ingeniería Orientado a Agentes. *Intel. Artif. Rev. Iberoam. Intel. Artif.* **2008**, *12*, 39–60.
36. Domański, P.D. *Control Performance Assesment: Theoretical Analyses and Industrial Practice*; Springer: Berlin/Heidelberg, Germany, 2020.
37. Bergstra, J.; Bengio, Y. Random Search for Hyper-Parameter Optimization. *J. Mach. Learn. Res.* **2012**, *13*, 281–305.
38. Morales, L.; Ouedraogo, C.; Aguilar, J.; Chassot, C.; Medjiah, S.; Drira, K. Experimental comparison of the diagnostic capabilities of classification and clustering algorithms for the QoS management in an autonomic IoT platform. *Serv. Oriented Comput. Appl.* **2019**, *13*, 199–219. [CrossRef]
39. Yousif, M.; Ai, Q.; Wattoo, W.; Jiang, Z.; Hao, R.; Gao, Y. Least cost combinations of solar power, wind power, and energy storage system for powering large-scale grid. *J. Power Sources* **2019**, *412*, 710–716. [CrossRef]
40. Zheng, K.; Schmitt, J.; González, A.; Fettweis, G. Self-organized Allocation of Dependent Tasks in Industrial Applications. In Proceedings of the IEEE International Conference on Autonomic Computing and Self-Organizing Systems (ACSOS), Washington, DC, USA, 27 September–1 October 2021; pp. 170–176.
41. Sánchez, M.; Aguilar, J.; Cordero, J.; Valdiviezo-Díaz, P.; Barba-Guamán, L.; Chamba-Eras, L. Cloud Computing in Smart Educational Environments: Application in Learning Analytics as Service. In *New Advances in Information Systems and Technologies*; Rocha, Á., Correia, A., Adeli, H., Reis, L., Mendonça Teixeira, M., Eds.; Springer: Berlin/Heidelberg, Germany, 2016; Volume 444, pp. 993–1002.
42. Wetzels, W.; van Dam, K.; Geerdink, T.; Meerman, H. *Decision Making on Regional Energy Transition. Models, Tools and Approaches*; Technical Report: TNO2020P10053; ESTRAC: Guiche, France, 2019. Available online: https://research.hanze.nl/ws/files/35101462/TNO2020_P10053_Decision_making_on_regional_energy_transition.pdf (accessed on 1 September 2023).

energies

Article

Adaptive Machine-Learning-Based Transmission Line Fault Detection and Classification Connected to Inverter-Based Generators

Khalfan Al Kharusi *, **Abdelsalam El Haffar** and **Mostefa Mesbah**

Electrical and Computer Engineering, Sultan Qaboos University, P.O. Box 33, Muscat 123, Oman; a.elhaffar@squ.edu.om (A.E.H.); m.mesbah@squ.edu.om (M.M.)
* Correspondence: s49626@student.squ.edu.om

Abstract: Adaptive protection schemes have been developed to address the problem of behavior-changing power systems integrated with inverter-based generation (IBG). This paper proposes a machine-learning-based fault detection and classification technique using a setting-group-based adaptation approach. Multigroup settings were designed depending on the types of power generation (synchronous generator, PV plant, and type-3 wind farm) connected to a transmission line in the 39-Bus New England System. For each system topology, an optimized pretrained ensemble tree classifier was used. The adaptation process has two phases: an offline learning phase to tune the classifiers and select the optimum subset of features, and an online phase where the circuit breaker (CB) status and the active output power of the generators are continuously monitored to identify the current system topology and to select the appropriate setting group. The proposed system achieved an average accuracy of 99.4%, a 99.5% average precision, a 99.9% average specificity, and a 99.4% average sensitivity of classification. The robustness analysis was conducted by applying several fault scenarios not considered during training, which include different transmission network configurations and different penetration levels of IBGs. The case of incorrect selection of the appropriate setting group resulting from selecting the wrong topology is also considered. It was noticed that the performance of developed classifiers deteriorates when the transmission network is reconfigured and the incorrect setting group is selected.

Keywords: adaptive protection; machine learning; fault detection and classification; inverter-based generators; groups setting

Citation: Kharusi, K.A.; Haffar, A.E.; Mesbah, M. Adaptive Machine-Learning-Based Transmission Line Fault Detection and Classification Connected to Inverter-Based Generators. *Energies* **2023**, *16*, 5775. https://doi.org/10.3390/en16155775

Academic Editor: Chunhua Liu

Received: 10 May 2023
Revised: 13 July 2023
Accepted: 21 July 2023
Published: 2 August 2023

1. Introduction

Adaptive protection schemes have emerged in the last few decades to address the problem of behavior-changing power systems with IBGs. Examples of system-changing problems include the fault level contribution, the fault characteristics of the IBGs, and system reconfigurations [1]. Analyzing the fault characteristics becomes difficult because the IBGs have unique electricity-generating principles due to the integration of power electronic converters [2]. Adaptive protection was suggested to automatically adjust protection functions to make them more attuned to prevailing power system conditions [3]. These adaptive protection schemes grant the power system protection the ability to identify, categorize, and localize faults in power systems that have been penetrated by inverter-based generators (IBGs), which cause ongoing changes in the direction of the power flow, the source impedance, and the system inertia.

Several adaptive protection methods were proposed in the literature to mitigate the issues of integrating renewable resources and changing system topologies. One of the more recent methods for creating adaptable protection systems is adaptation utilizing machine learning (ML) algorithms. As a substitute method for fault detection, classification, and localization, ML techniques for power system protection have emerged. The main advantages of ML techniques are accuracy, self-adaptiveness, and robustness to parameter variations [4]. Classical ML techniques face several challenges, such as updating the training dataset for newly discovered faults and continuously tuning the classifier's parameters [5]. Adaptive ML algorithms have been introduced to address the power systems' changing behavior. The adaptation may occur at feature extraction, feature selection, or classification levels. Continuous data streaming for training should be supported by the adaptive classifier, which should be able to adjust to the changing behavior of the power system thanks to integrated renewables. This process is known as incremental or online learning [6].

An adaptive microgrid protection strategy employing traditional protection relay settings and a machine learning classifier was proposed by Hengwei et al. [7]. The rule-based method was able to choose the appropriate setting group of overcurrent and distance relays relevant to the current system topology, while the ML classifier was utilized to detect the faults. A support vector machine (SVM) classifier was used in [8] to predict the remote CBs' status, identify the circuit topology, and select the correct relay setting. Marín-Quintero et al. [9] developed an adaptive protection scheme for an active distribution network with distribution generators (DGs) using an ML algorithm. Different system topologies were considered, and each topology used a different ML classifier for fault detection. For each system topology, the relays in the network were equipped with the classifier that achieved the best classification accuracy during offline training. In this application, the different classifiers used a common set of features, which may not be an optimal subset. Yavuz et al. [10] proposed an adaptative algorithm that computes the optimum weights using particle swarm optimization for different ML classifiers to achieve a high performance independent of the system topology and the type of connected generators. The adaptation was made online using only the voltage, frequency, and phase angle signals obtained from the PMUs. Tang and Yang [11] developed an adaptive protection scheme by extracting features from the continuous wavelet transformation of the positive, negative, and zero sequences of voltage and current and the three-phase currents and voltages. Two system topologies were considered, namely fixed topology and changing topology. Features were selected using Pearson's correlation coefficient. For the fixed topology case, the relay operation thresholds, obtained using the DT classifier, were embedded in the relays to enable fault detection. For the changing topology, a neural network was adopted.

This paper proposes an adaptive transmission line protection connected to three generations: synchronous generators, PV plants, and wind farm. The adaptation design depends on online system topology identification using a combination of the CBs' status and real-time measurement of the active output power of the connected generating units behind the protected transmission line and the ML model for fault detection classification.

The combination of CB status and power measurement ensures more reliable topology identification. It is worth noting that the definition of the system topology in this paper differs from the definition in the literature. Our definition of the system topology is the generation mixture available in the busbar behind the protected line, whereas the topology defined by others was the lines or buses connected/disconnected in the power system network.

The fault detection and fault type classification for each system topology were designed using an ML model in the offline process. After appropriate topology identification, the designed ML model parameters were saved in the setting groups. The correct setting group is selected based on the relative topology in the online process. More specifically, the proposed method:

- defines the system topology as the generation mixture available at the busbar behind the protected line;
- considers two types of IBGs, namely large-scale PV and DFIG wind farms (WF);
- combines ML-based classifiers with setting group selection based on circuit breaker statuses and real-time active power measurements;

- tunes the hyperparameters of the classifier associated with each system topology using the Bayesian optimization algorithm to achieve the best classification performance;
- uses the ensemble feature method, an embedded-type feature selection method, to select the optimal feature subset associated with each system topology;
- selects the appropriate setting group online according to the system topology obtained from the lookup table.

The adaptive scheme in this study refers to the online topology identification from CB states and active power measurements of the connected generating units, followed by the appropriate selection of the setting groups equipped with pretrained ML classification models to detect and classify the faults.

This paper is organized as follows. Section 2 identifies different system topology identification methods in the literature. Section 3 describes the methodology used to identify the generation topology and to build the datasets for the adaptive fault detection and classification technique. It also describes the ML protection setting method. Section 4 presents and discusses the obtained results. It evaluates the performance of the classification models using different classification metrics: accuracy, specificity, precision, and sensitivity. The performance of the proposed adaptive scheme system was evaluated using new fault events, different IBG penetration levels, and transmission system configurations. Section 5 provides the conclusion.

2. System Topology Identification Methods

Identifying or detecting system topology changes is crucial in selecting the appropriate adaptive protection scheme. Several topology identification approaches have been proposed in the literature.

Identification with CB status: The CB status approach has been widely used for adaptive protection scheme design. Poudel et al. [8] developed an adaptive protection scheme for medium-voltage feeders by collecting the CB status of that feeder in the substation computer, along with the statuses of the loads, DGs, and autoreclosers. The computer substation saves predefined settings corresponding to each system topology. An adaptive overcurrent was established and proposed in [12] to detect the faults in two different setting groups, the islanded mode and the grid-connected mode, by defining the CBs of the network circuits and the CB status of the distributed generators using the IEC61850 communication protocol. The setting groups were changed after loss of mains, loss of DG, or islanding. Lin et al. [13] proposed lookup tables for the circuit breaker (CB) and relay events. The protection settings for different states were calculated offline and stored in the settings table (action table).

Identification using dynamic state estimation (DSE): When the operating system point changes more frequently and quickly, it becomes crucial to keep track of the system's dynamic state variables, including voltage and current magnitude and angle, current magnitude and angle, and real and reactive power [14]. The data were collected from PMUs, merging units (MUs), and digital fault recorders (FRs). The dynamic variables of the system equipment could be estimated using differential–algebraic equations [15]. Adaptive protection can be developed by estimating the dynamic states to identify the system topology. Korres et al. [16] utilized the state estimation algorithm to define the IEEE RTS-96 substation configuration via the circuit breaker status identification using the active and reactive power flows as continuous state variables. The authors in [16] tested two algorithms for topology identification: the recursive Bayesian estimation (RBE) and generalized state estimation (GSE). The GSE provided good model identification accuracy, even when the number of possible network configurations was increased. The state estimation approach, however, faces several challenges [17] such as dependency on the communication system to transfer the data from measuring devices, considering the network bandwidth and capacity, and limiting the accuracy and rate of data exchange. In addition, the higher penetration of renewable power resources introduces a higher level of uncertainty.

Identification using machine learning: ML can be used to detect and identify the system configuration changes by collecting measurements at different locations of the power system, such as lines, generators, and supplementary devices (fault current limiters, reactive power compensators, and others). The SVM classifier was used in [18], where the three-phase voltage and current, RMS values, and the zero-sequence current were the input features measured at different locations in the IEEE 123-node distribution test system. The authors in [19] proposed using several ML classifiers (SVM, k-NN, and ensemble algorithms) to identify the system configurations of a simulated standard power distribution system. The SVM outperformed the other classifiers by achieving 100% detection accuracy. Rajendra et al. [20] estimated the system configuration of the tie lines in the modified IEEE 123-bus distribution system using deep learning (CNN) and compared it with the SVM. The CNN outperformed the SVM.

Identification using data-driven approaches: The system topology can be identified by collecting data from different locations of the power system network and recognizing the available system components (lines, generators, and loads). The required data are voltage, current, real power, reactive power, or frequency signals and can also be extracted features from these signals. Razmi et al. [19] used transient voltage signals at different system-switching devices to identify the system topology and circuit status. The transient voltage was obtained from instantaneous voltage signals measured at each end of the distribution lines in an ANSI standard distribution system and extracted the maximum, minimum, and rate of change as features. The dataset was classified using SVM, k-NN, DT, and ensemble tree classifiers. The SVM classifier outperformed the others with its 100% achieved accuracy. In [8], The circuit status was identified according to the measurements of RMS voltage, current, real and reactive power collected from the lines of the modified IEEE 34-bus system and two-bus power system with the PV system. Then, the overcurrent protection settings were adopted for each circuit topology. The topology identification was achieved using the SVM classifier, and the results showed that using all measured signals resulted in a classification accuracy of 100%, whereas using current and voltage measurements only resulted in lower performance (98%). The authors in [21] designed an ML data-driven approach to identify the system topology by constructing a connectivity matrix that showed the status of switches with voltage and current phasors recorded by PMUs. The proposed method approximates network parameters using an ensemble-based deep learning model for a modified IEEE 33-bus network and real feeder in Queensland, Australia. Their design performance was a detection error rate of only 1.2%. For transmission system topology identification (IEEE 39-bus and 118-bus systems), the authors in [22] identified the line outage using the phasor angles at buses. Logistic regression (LR), random forest (RF), and graph convolutional network (GCN) were the models used for identification. The proposed classifiers were evaluated with two performance metrics (precision and recall). The results showed that logistic regression outperformed the others with 99% precision and recall.

3. Methodology

In this section, the proposed methodology is described in detail. It consists of three parts: topology identification procedure, machine learning design requirement, and the overall proposed adaptive scheme using ML-based protection setting approach.

3.1. Topology Identification Procedure

The proposed topology identification in this paper depends on the circuit breaker status of the generating units connected at one end of the protected transmission line and the active output power of these units. Depending on the circuit breaker status alone is not always reliable. The circuit breaker may fail due to mechanical or electrical reasons and the intermittency behavior of renewable resources where the output power ranges from zero to maximum, depending on weather conditions. There are cases where the output power is zero while the circuit breaker is in a closed position. Figure 1 shows a real case of PV plant and wind farm output power that varies from zero to maximum during the day's hours.

As a result, the circuit breaker status and the active power measurements are used for the reliability of topology identification. If the power measurement exceeds zero, the output is one; otherwise, it is set to zero. Considering the three generators connected at one bus (synchronous generator, PV plant, and wind farm), the topology lookup table is shown in Figure 2. Individual plant availability status is obtained using the AND gate between the CB status and the output power, and then the topology can be identified with the lookup table using the three statuses of the generating plants. The resulting number of topologies is eight, noting that the 'No generation' topology means that none of the three generation plants are in service. However, other generators connected at different system parts feed the faults incepted in the protected lines.

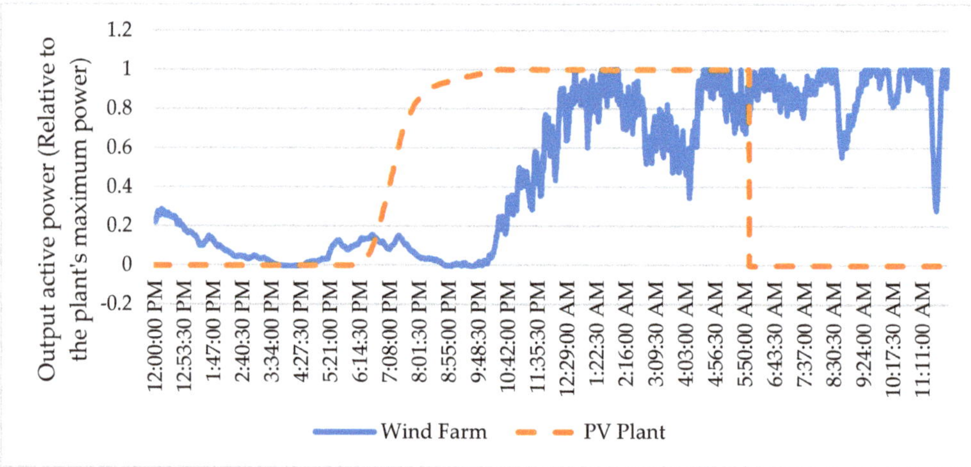

Figure 1. Measurements of the active power output of the PV plant and wind farm during the day hours.

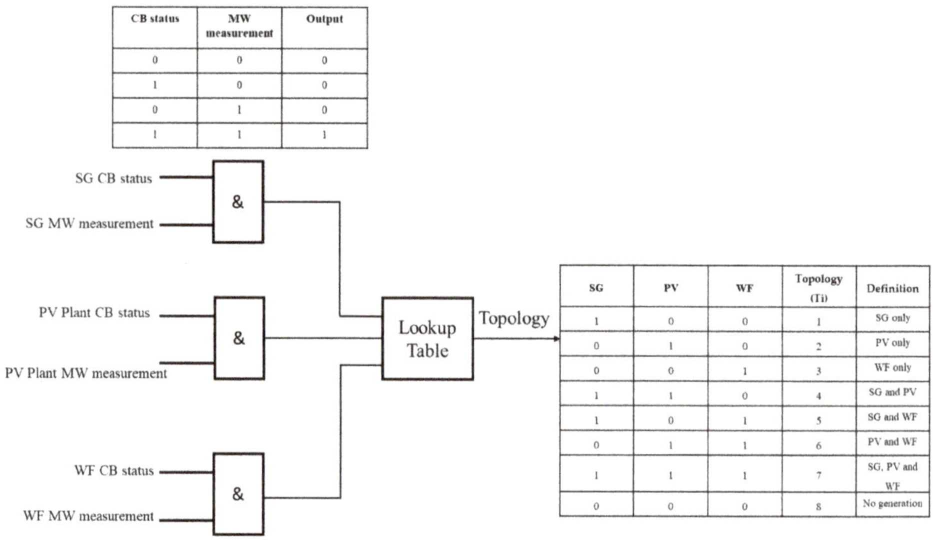

Figure 2. Lookup table for topology identification.

3.2. Machine Learning Design Requirement

This section explains the dataset construction phases, the feature selection method, the classification model, and the classification performance metrics.

3.2.1. Data Collection and Preparation

The datasets were simulated using the 39-Bus New England System model [23] shown in Figure 3. The data were simulated using the Power Factory DigSilent software package [24]. The details related to the parameters of the synchronous generators, transmission lines, power transformers, and loads can be found in [25]. The protected transmission line was line 1-2. The following signals were measured at bus -2: the instantaneous three-phase voltage, instantaneous three-phase current, and the angle between voltage and current. The signals included fault and nonfault events. The PV plant and wind farm, whose characteristics are shown in Table 1, were introduced at bus 2.

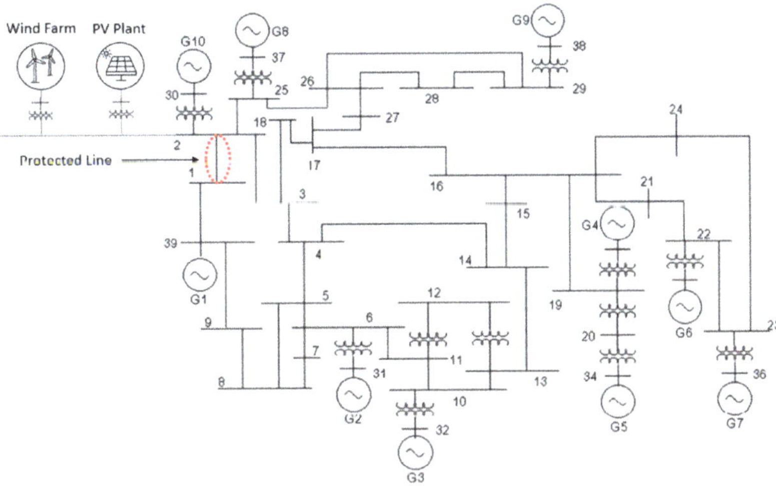

Figure 3. The 39-Bus New England System.

Table 1. PV plant and wind farm characteristics.

IBG	Characteristics	Dynamic Model Type
PV Plant	10 kVA per inverter, local controller: constant Q, Short circuit model: Dynamic voltage support, Sub-transient short circuit: 1.21 kVA, R to X" ratio: 0.1, K Factor: 2, Max. current: 1.1 pu, Td"= 0.03 s, Td' = 1.2 s	WECC Large-scale Photovoltaic Plant model
Wind Farm (Type 3: Doubly fed induction generator)	2MVA, 1.0 power factor, local controller: constant Q, Short circuit model: Dynamic voltage support, Sub-transient short circuit: 2.39 MVA, R to X" ratio: 0.1, K Factor: 2, Max. current: 1.1 pu, Td"= 0.03 s, Td' = 1.2 s	WECC Wind Turbine Model Type 3

Swing conditions and normal system behavior make up the nonfault events. The swing conditions were incorporated into the nonfault class because they should prevent the protection device from operating during power swing situations. Different fault types, locations, and resistances were simulated as part of the fault events. Power swing was detected using the swing center voltage (SCV) signal, as suggested in [26]. The magnitude of the SCV changes between 0 and 1 per unit of system nominal voltage. The SCV's magnitude remains constant under typical load situations. The voltage magnitude at the relay point is multiplied by the local voltage and current angle difference to determine it. The fault scenarios contain combinations of the following fault types: A-G, B-G, C-G, A-B, A-C, B-C, A-B-G, A-C-G, B-C-G, and A-B-C. They also feature fault resistances of 0 and 100 ohms, fault locations of 10%, 50%, and 90%, and three-phase faults occurring during power swing situations.

Eight datasets were created for this investigation that represented the generators attached to bus 2 behind the protected transmission line. Three different generator types might be coupled, as shown in Figure 3. (G10: synchronous generator, PV plant: connected to the system with inverters, and wind farm: doubly fed induction machines).

There are eight different system topologies: T1 (SG only), T2 (PV only), T3 (WF only), T4 (SG and PV), T5, T6, and T7 (SG, PV, and WF) (No generation). Seventy percent of the total observations were utilized in the training dataset to fine-tune the hyperparameters, and the remaining observations were used to test the ML classification model.

Eight balanced classes make up each dataset: "0" for normal and swing conditions, "1" for A-G fault events, "2" for B-G fault events, "3" for C-G fault events, "4" for A-B and A-B-G fault events, "5" for A-C and A-C-G fault events, "6" for B-C and B-C-G fault events, and "7" for three-phase fault events. The dataset's balanced classes have an equal amount of observations for each class.

3.2.2. Feature Extraction and Selection

According to Table 2, the features were derived from three domains: time, frequency, and time–frequency. Each dataset contained 343 characteristics in total (7 Signals × 49 features). These elements were taken into account in our previously published research [27].

Table 2. Extracted features.

Domain	Features	Number of Features for Each Signal
Time	Statistical features of the squared signals	9
	Statistical features of first-order difference of the squared signals	9
Time–frequency	Statistical features of spectrogram	9
	Statistical features of wavelet decomposition of first and second detail coefficients [28]	18
	Estimated instantaneous frequency [29]	1
Frequency	Spectral entropy [30]	1
	Mean and median frequency [31]	2

This research suggests using the ensemble-based feature selection strategy because it maximizes classification accuracy. For the same dataset, the ensemble tree classifier outperformed the k-nearest neighbor, support vector machine, and decision tree classification models, as described in [27]. The ensemble approach is a strategy for embedded feature selection that makes use of weak learners to choose the ideal subset of features that optimize classification accuracy and reduce error. The construction of a linear prediction model using embedded techniques aims to decrease the number of input features while simultaneously maximizing the goodness of fit of the model [32]. The rationale behind utilizing decision trees to assess the significance of a feature is that they perform splits that optimize impurity reduction. Calculating the mean reduction in impurity for each feature across all trees yields the feature significance [33]. Impurity-based feature importance is another name for this technique. The importance calculation follows the following procedure:

For each given feature
For each tree

1. Compute the impurity decrease (Gini or entropy)
2. Weight by the number of examples at that node
3. Average overall trees (i.e., average impurity decrease)
4. Normalize importance values so that the sum of feature importance values equals one

The feature that experiences a greater impurity reduction at each split is given more weight. This strategy's mathematical model can be found in [34].

3.2.3. Classification Model

There are several ML-based classification models, each with advantages and limitations. In our study [27], the results demonstrated that the ensemble classifier outperformed decision trees, k-nearest neighbors, and support vector machines. The ensemble tree classification model is, therefore, adopted in this paper. Additionally, the ensemble tree was also used in feature selection. The classifier's hyperparameters were tuned using the Bayesian optimization (BO) algorithm.

3.2.4. Classification Performance Metrics

The accuracy, sensitivity, specificity, and precision are used to describe classification performance indicators in this paper. Their mathematical definitions, derived from the confusion matrix, can be found in [27].

3.2.5. Machine Learning Protection Setting Method

As explained in Section 3.1, the suggested adaptive technique is based on gathering the CB statuses of these units (1: closed position, 0: open position) and the active power measurements of these units in order to track the availability of various generating types connected in bus 2.

Following the procedures shown in Figure 4, the ML classifier setting associated with each system topology was created offline. This graphic explains how to identify the ideal hyperparameters for each system topology using ensemble-based feature selection and an optimized ensemble classifier customized by Bayesian optimization. Each system topology's setup group is made up of the classifier hyperparameters and the chosen subset of features. This procedure could be regarded as offline.

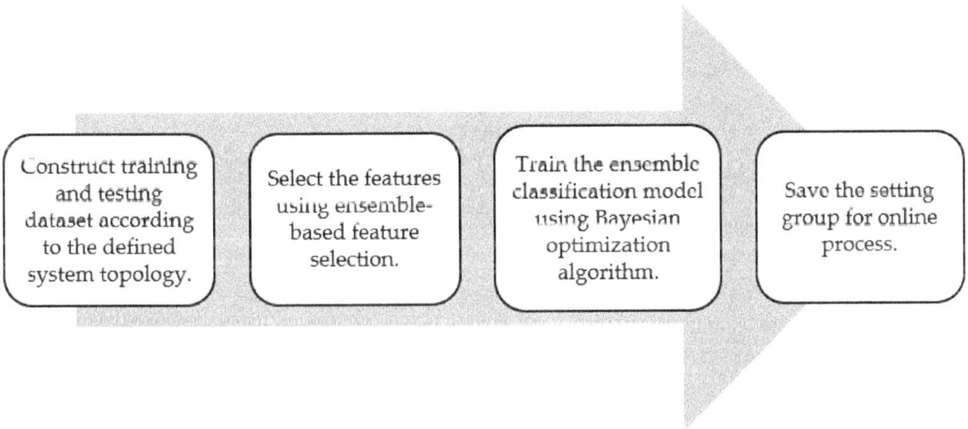

Figure 4. Steps to find out the setting groups.

On the other hand, the online procedure was accomplished by performing the following steps: (1) checking the information pertaining to the current topology; (2) choosing the classifier associated with the current topology; and (3) identifying and classifying new fault events. Figure 5 depicts the offline and online processes.

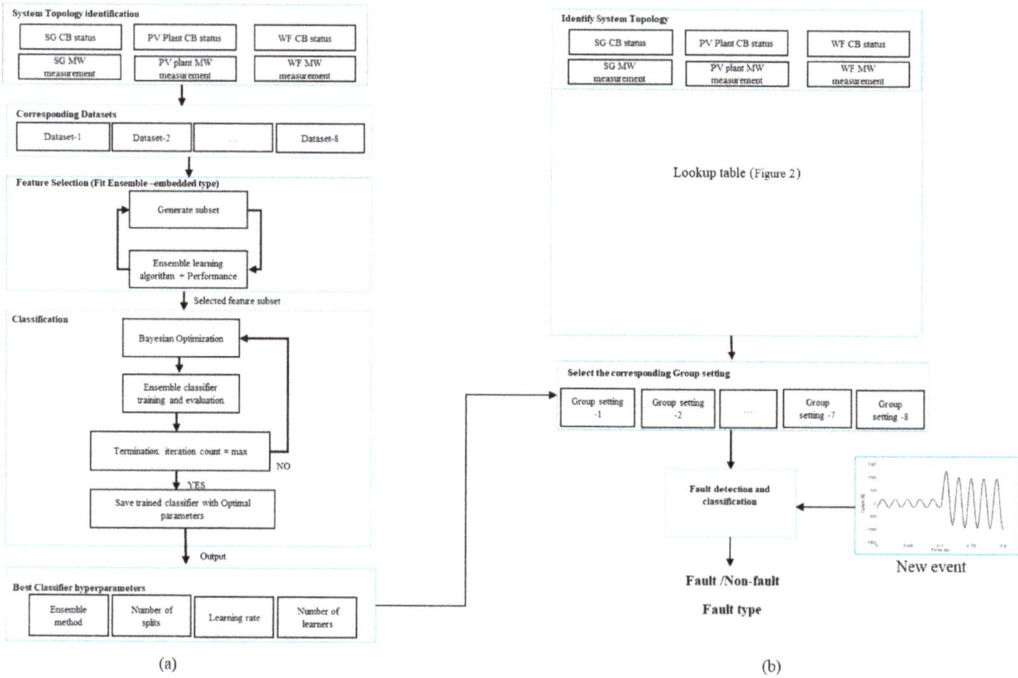

Figure 5. The proposed adaptive ML-based fault detection and classification workflow. (**a**) Offline process, (**b**) online process.

4. Results and Discussion of Results

The offline settings in this part begin with creating the best classifiers and choosing the best collection of features for each preselected topology using the training data. The system is then tested using test data, and the results are presented in terms of the performance metrics indicated earlier. The system is then tested for robustness utilizing faults on the protected line, faults at various degrees of IBG penetration (10, 50, and 100 percent of their maximum output power), and faults at various transmission system configurations (line outages). Finally, the classifier's performance under incorrect topology identification is assessed.

4.1. *Offline Settings*

In the offline settings, the system topologies are identified beforehand, the best feature subsets are chosen, and the various ensemble-based classifiers are trained and optimized. The number of features that were chosen, the ensemble classifier model hyperparameters for each system topology, and the performance metrics for validation and testing data are all displayed in Table 3. The training data represent 70% of the data (five cross-validation folds), and the testing data represent 30%. Figure 6 shows the importance estimation of the features for each system topology using the impurity-based feature importance.

Table 3. Results of the offline settings.

Topology (T)	Classification Model Hyperparameters	Number of Selected Features	Performance Using Training Data (%)	Performance Using Testing Data (%)
T1 SG only	Ensemble method: Adaboost Maximum number of splits: 5 Number of learners: 13 Learner rate: 10	39	Precision: 98.93 Recall: 98.84 Accuracy: 98.62 Specificity: 99.79	Precision: 97.56 Recall: 97.76 Accuracy: 96.77 Specificity: 99.51
T2 PV only	Ensemble method: Bag Maximum number of splits: 128 Number of learners: 131 Number of predictors to sample: 79	190	Precision: 99.17 Recall: 98.88 Accuracy: 98.8 Specificity: 99.82	Precision: 99.66 Recall: 99.46 Accuracy: 99.60 Specificity: 99.94
T3 WF only	Ensemble method: Bag Maximum number of splits: 13 Number of learners: 18 Number of predictors to sample: 82	155	Precision: 99.31 Recall: 98.80 Accuracy: 99.14 Specificity: 99.87	Precision: 99.34 Recall: 99.46 Accuracy: 99.60 Specificity: 99.95
T4 SG + PV	Ensemble method: RUSboost Maximum number of splits: 159 Number of learners: 25 Learner rate: 0.873	20	Precision: 98.35 Recall: 98.23 Accuracy: 97.58 Specificity: 99.63	Precision: 97.28 Recall: 97.12 Accuracy: 96.47 Specificity: 99.48
T5 SG + WF	Ensemble method: Bag Maximum number of splits: 168 Number of learners: 45 Number of predictors to sample: 18	157	Precision: 99.77 Recall: 99.57 Accuracy: 99.66 Specificity: 99.95	Precision: 99.31 Recall: 98.81 Accuracy: 99.20 Specificity: 99.88
T6 PV + WF	Ensemble method: Bag Maximum number of splits: 46 Number of learners: 10 Number of predictors to sample: 10	204	Precision: 99.52 Recall: 99.53 Accuracy: 99.65 Specificity: 99.95	Precision: 98.98 Recall: 99.17 Accuracy: 99.19 Specificity: 99.89
T7 SG + PV + WF	Ensemble method: Bag Maximum number of splits: 28 Number of learners: 16 Number of predictors to sample: 10	163	Precision: 99.89 Recall: 99.76 Accuracy: 99.90 Specificity: 99.99	Precision: 99.21 Recall: 99.02 Accuracy: 99.52 Specificity: 99.93
T8 No generation	Ensemble method: Bag Maximum number of splits: 40 Number of learners: 10 Number of predictors to sample: 3	168	Precision: 99.45 Recall: 98.74 Accuracy: 99.12 Specificity: 99.86	Precision: 100 Recall: 100 Accuracy: 100 Specificity: 100

4.2. Performance Evaluation with New Fault Events

Applying new fault events could further assess the resulting classifiers' performance. At 70 percent of the protected line (line 1-2) from the measurement point, three cascading within-the-line faults were simulated. A-G fault (class 1) was the first fault, followed by A-B fault (class 4) and A-B-C fault (class 7) at 1.0, 2.0, and 3.0 s, respectively. The fault durations were 100 msec. The faults were created for all previously defined generation topologies, each with hyperparameters and a subset of features.

As shown in Figure 7, the proposed classifiers for each generation topology were successful in precisely identifying and classifying the incepted faults. The proposed classifiers can detect and classify the faults accurately for each generation topology determined by the CB statuses and active power measurements, with the exception of a two-phase fault in topology T6 (PV and WF), where the classifier's output was classified as class 7 (three-phase fault). In addition, after clearing the three-phase fault in T6, there was an output of fault detection as A-B fault. The misclassification occurred due to the percentage error of the classification model with this topology, which was reported as 0.81%.

4.3. Performance Evaluation with Different IBG Penetration Levels

The training and testing datasets used to train the classifiers in the offline mode and the testing dataset were thus far simulated assuming either 0% penetration (not connected or zero output power) and 100% penetration of IBGs (wind farm and PV plant). The present section aims to evaluate the proposed classifiers' performance at different penetration levels other than zero or 100% (i.e., 10% and 50%) for T2 and T3 generation topologies. Only these two topologies were taken into account because the synchronous generator's (G10) fault contribution was frequently dominating and the IBGs' (PV and wind turbines') fault contributions were constrained by the controller parameters of the inverters.

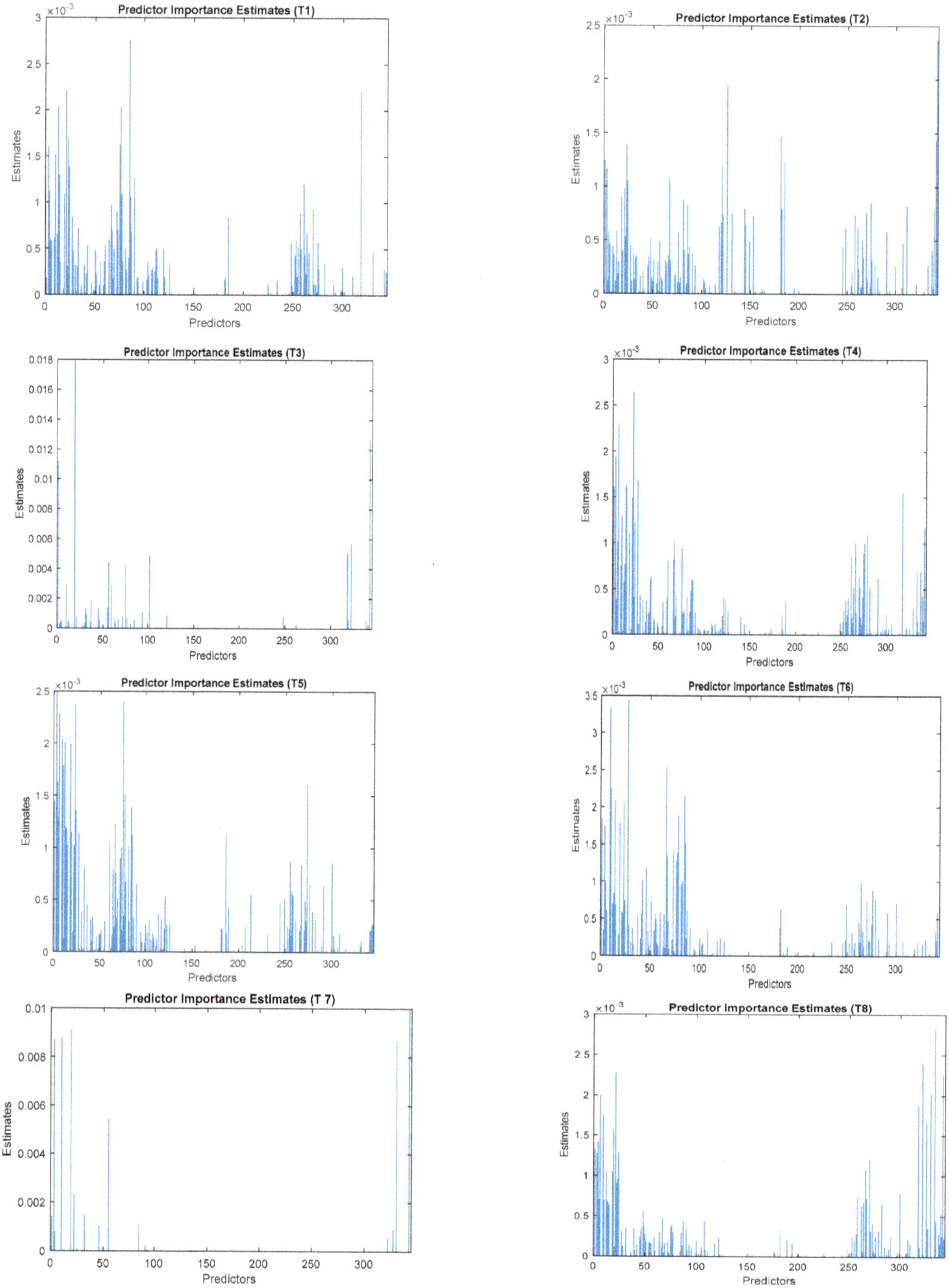

Figure 6. Feature importance estimation using ensemble tree feature selection algorithm for each system topology.

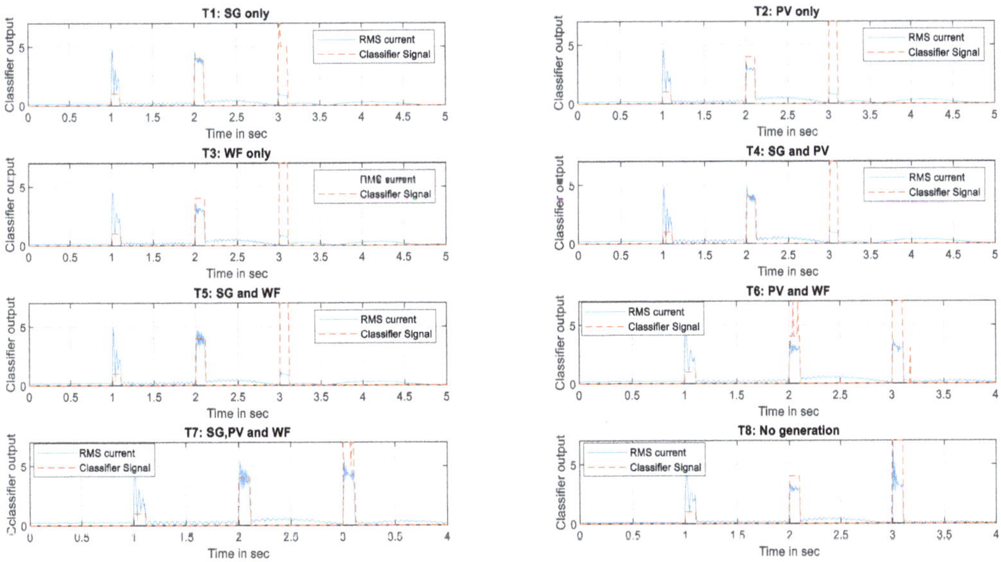

Figure 7. Classifiers' outputs for different line faults at each generation topology.

The contributions of the PV plant and the wind farm for various faults and locations are given in Figure 8 and Figure 10, respectively, at each penetration level (10%, 50%, and 100%). The minimum fault current contribution provided by the PV plant was the three-phase fault at the end of the protected line (near bus 1), and the maximum was for a single-phase fault at the beginning of the protected line near the PV plant (Figure 8). This observation can also be made for the wind farm connected to bus 2 (Figure 10).

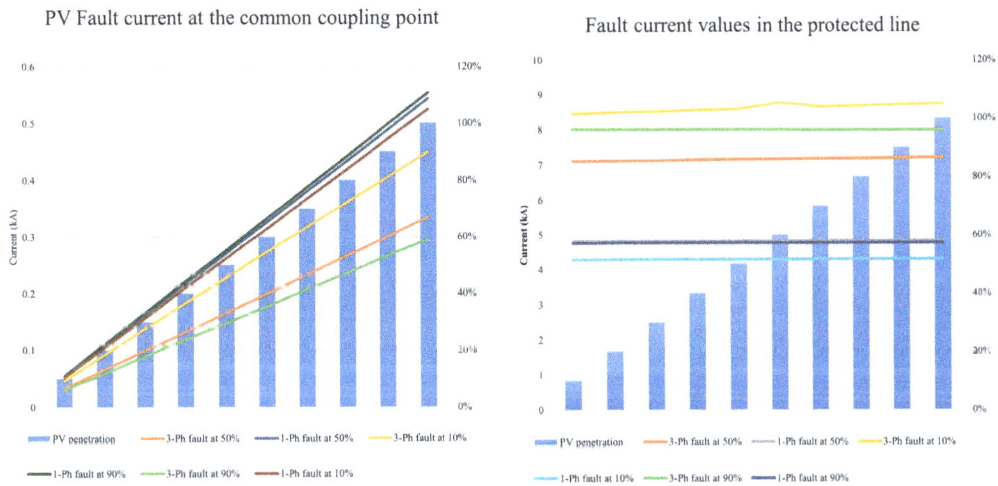

Figure 8. PV plant fault contribution at different output power.

Figures 9 and 10 display the topological performance of the T2 classification model for two distinct PV plant penetration levels and two different fault types (single- and three-phase faults) at the 10% and 50% fault locations. As can be seen, the classifier detected the faults accurately. For A–G and three-phase faults, the classifiers' output was classes

4 and 7, respectively. The results can be generalized for other PV penetration levels and fault types.

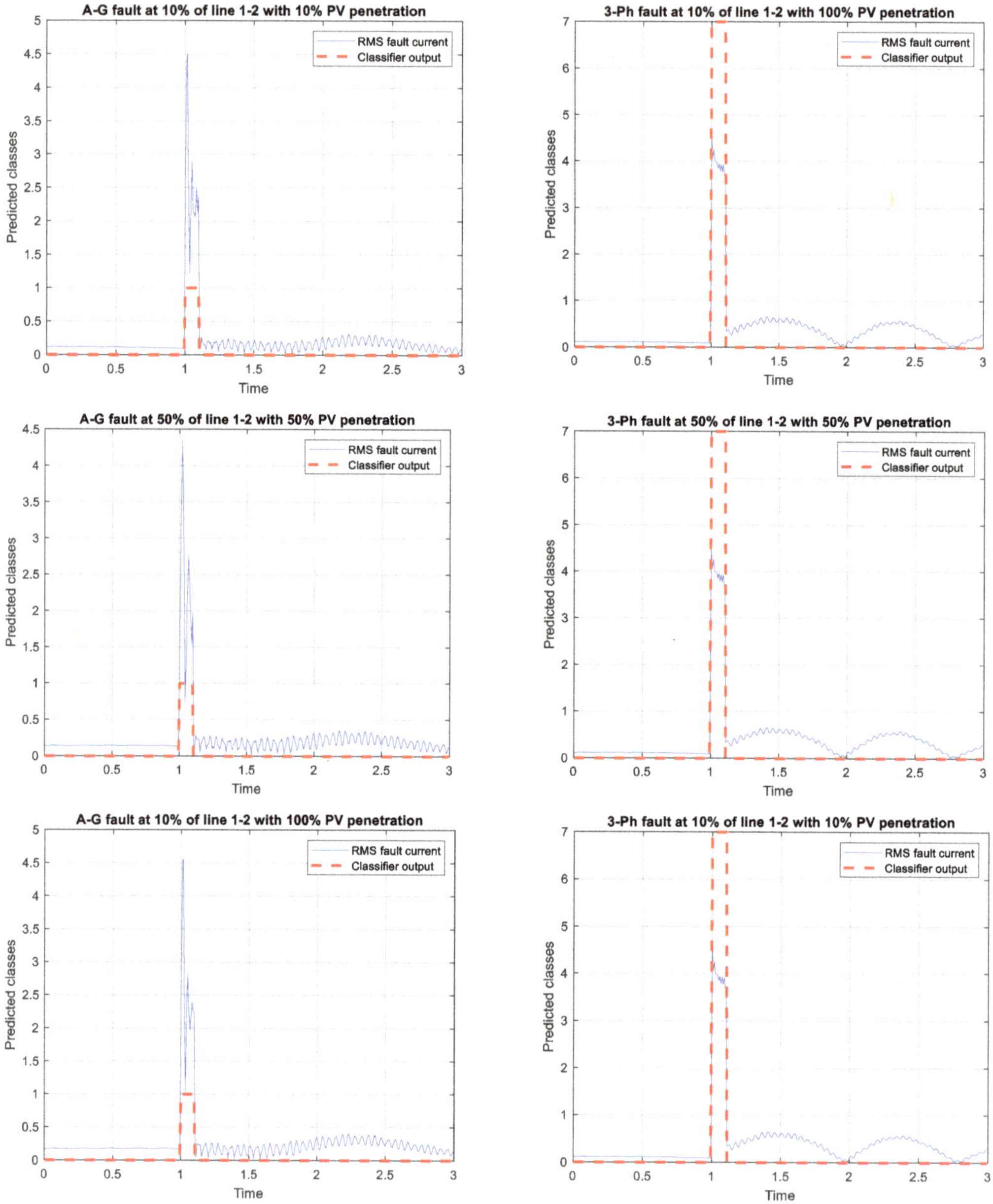

Figure 9. T2 classification model performance at different PV plant penetration levels.

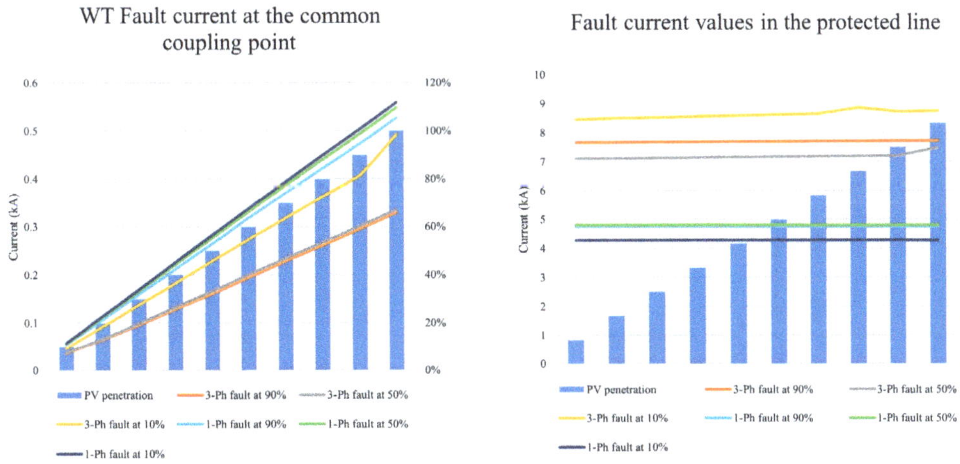

Figure 10. Wind farm fault contribution at different output power.

Similarly, the classifier's performance for the T3 setting topology was investigated for different wind farm penetration levels. Figure 11 shows that the classifier proposed for T3 could also detect all types of faults at different locations and for two levels of the wind farm output power (10% and 50%) of its maximum power. For A-G and three-phase faults, the classifiers' output was classes 4 and 7, respectively. The results can be generalized for other wind farm penetration levels and fault types.

4.4. Performance Evaluation with New Transmission System Configurations

In the previous results, the investigation was achieved considering the same transmission system configuration with different generation topologies at bus 2. This section examines the performance of the developed classifiers at different transmission system configurations. Three scenarios are considered, which are shown in Figure 12. The first case is to cut off the supply from G1 by disconnecting the line 1-39. The second is to limit the contribution of G8 by switching the line 2-25 to the OFF position. The third is the disconnection of both lines. This line selection will impact different sources' contribution to faults that occurred in the protected line (the line 1-2).

Case 1: line 1-39 outage

The outage of line 1-39 prevents the contribution of G1 to faults that occurred in line 1-2, but still, the faults are fed through bus 2 generators, line 2-25, and line 3-2. Figure 13 shows the outputs of the classifiers for each generation topology for the following faults Phase A fault (Class 1), A-B fault (Class 4), and three-phase fault (Class 7). Topologies 1, 2, 3, 4, 5, and 8 classifiers performed well in detecting and classifying these faults, which predicted them as classes 1, 4, and 7. However, the effect of line 1-39 outage was clear on topologies 6 and 7. For topology T6 (wind farm and PV plant were connected), the classifier still had outputs after a three-phase fault detection, although the fault was cleared. On the other hand, the T7 classifier detected faults after each clearing time, which should be reset to zero as the faults were cleared. By retraining them using these new batches of data, the classifier's hyperparameters can be updated, which will enhance detection and classification performance.

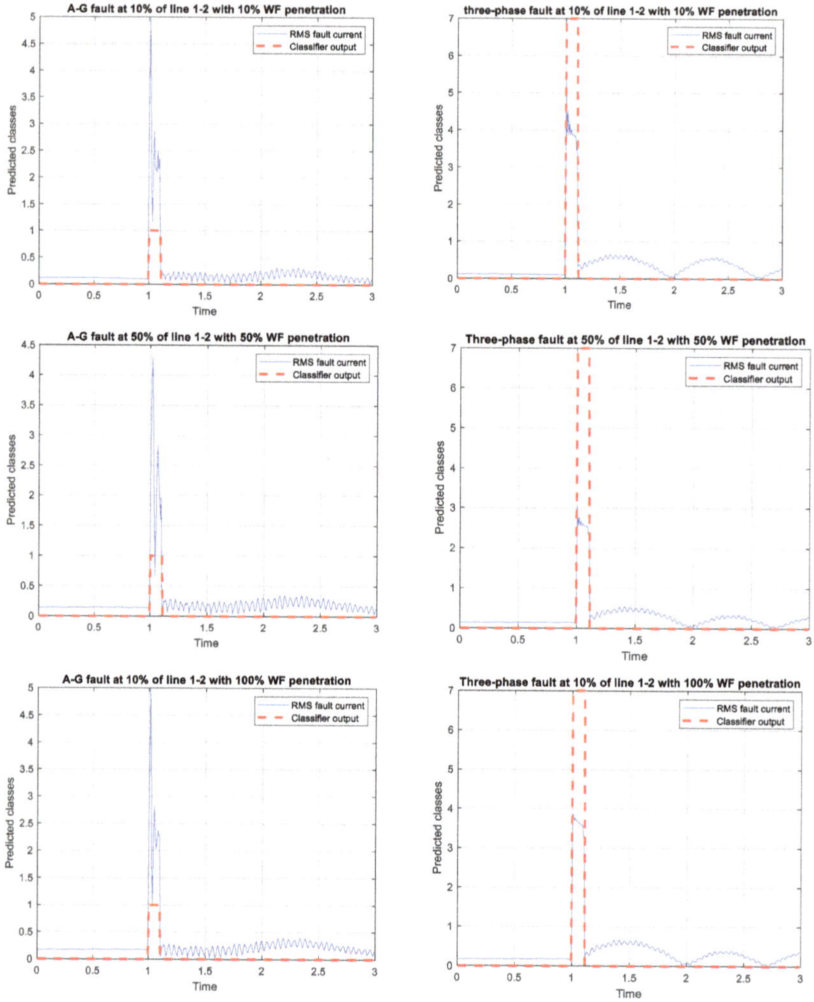

Figure 11. T3 classification model performance at different wind farm penetration levels.

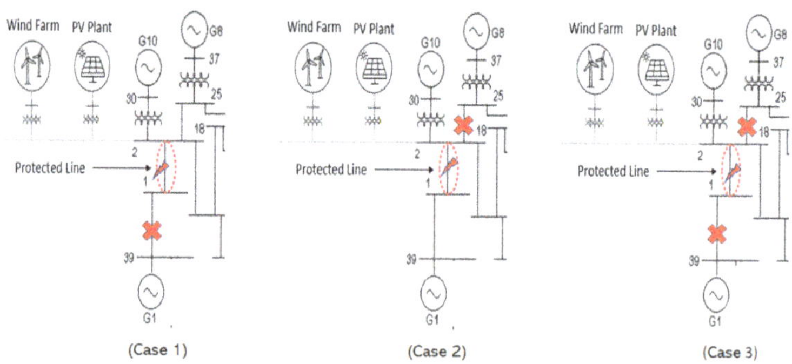

Figure 12. Transmission system configurations.

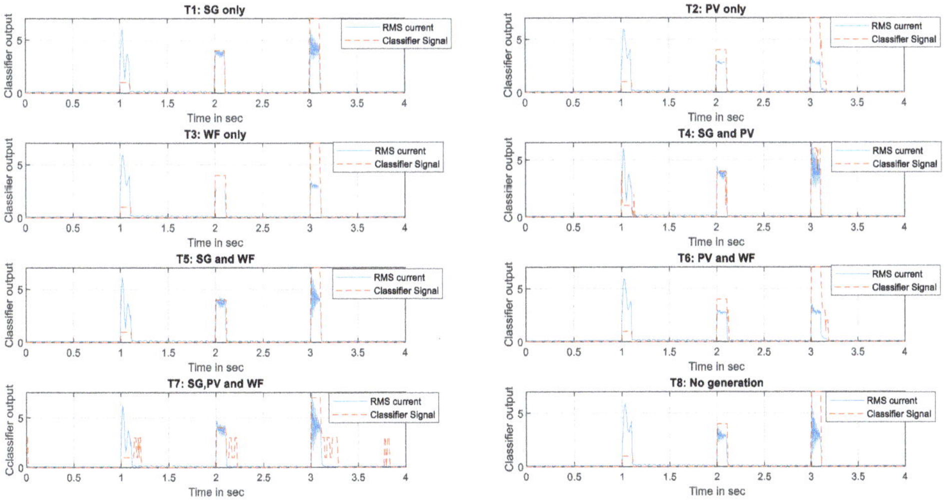

Figure 13. Classifiers' outputs for different line 1-2 faults with the outage of line 01-39 (Case 1).

Case 2: line 2-25 outage

The outage of line 2-25 prevents the contribution of G8 to faults in line 1-2, so that the faults were fed through bus 2 generators, line 1-39, and line 3-2. Figure 14 shows the outputs of the classifiers for each generation topology for the following faults: Phase B fault (Class 2), Phase B to C fault (Class 5), and three-phase fault (Class 7). It could be noted that although the magnitude of the RMS current for single- and two-phase faults was minimum, the classifiers were able to detect these faults in topologies 1, 2, 3, 4, 5, and 8. The misclassification rate was high in the case of topologies 6 and 7, where the PV plant and WF were connected to bus 2. Retraining the classifiers with this new dataset or taking into account the new setting group for each transmission network configuration could increase the detection and classification performance. The classifiers performed worse in this case than they did in the first.

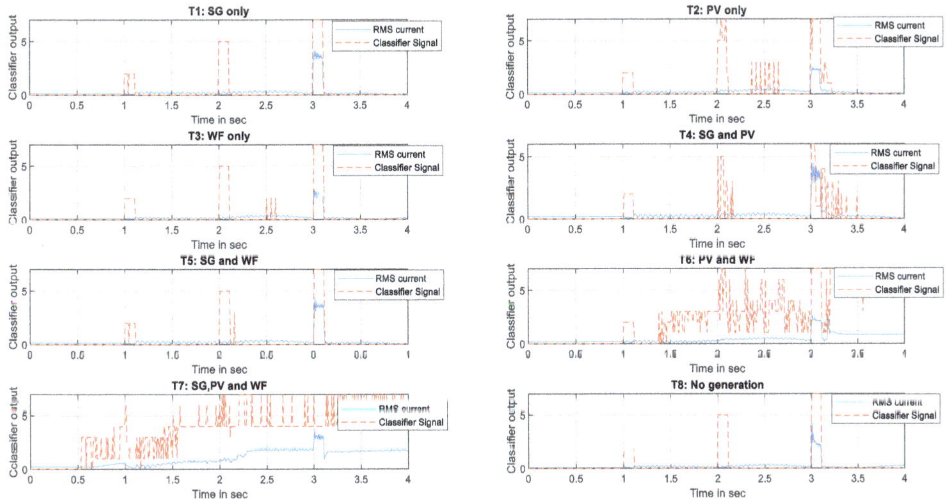

Figure 14. Classifiers' outputs for different line 1-2 faults with the outage of line 2-25 (Case 2).

Case 3: line 2-25 and line 1-39 outages

The combination outage of lines 2-25 and 1-39 allows the fault to be fed through bus 2 generators and lines 3-2. Figure 15 shows the outputs of the classifiers for each generation topology for the following faults: Phase B fault (Class 2), B-C fault (Class 5), and three-phase fault (Class 7). As in case 2, the magnitude of the RMS current of single-phase and phase-phase faults is low, but the classifiers were able to detect these two faults in most cases using other proposed features in different domains. The classifiers at each generation topology can efficiently detect the three types of faults, except for topologies 6 and 7. The misclassification events in topology 6 were more, and an update of the classifier's hyperparameters was required to include this transmission topology with the PV system connection. Generation topology 7 had an issue with fault detection after each fault-clearing event, with misclassification between classes 5 and 7.

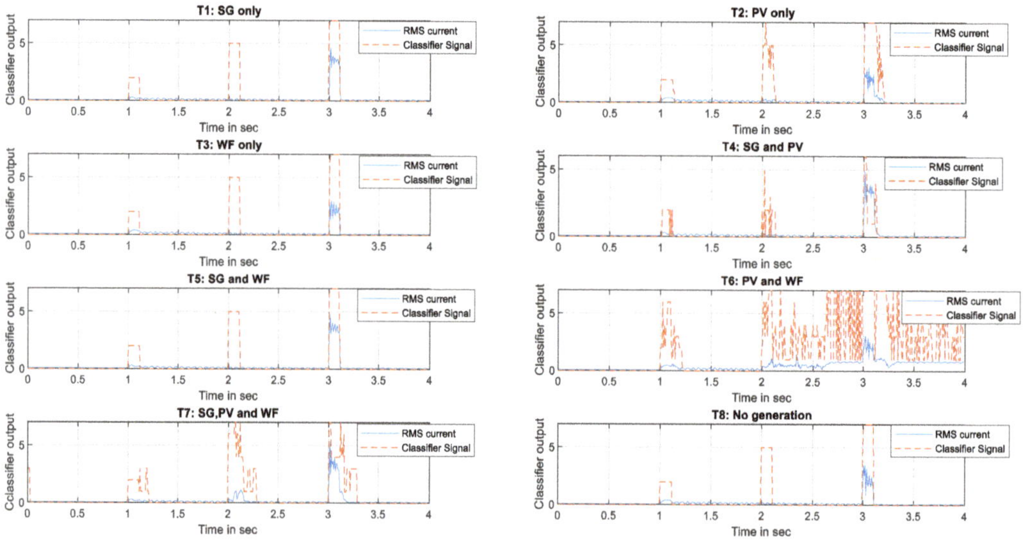

Figure 15. Classifiers' outputs for different line 1-2 faults with the outage of lines 1-39 and 2-25 (Case 3).

By implementing a new system topology and following the instructions in Figure 4 to identify the setting groups, as well as by converting the existing classifiers' models into incremental learning to update hyperparameters by retraining them with any new data stream, the misclassification events in cases 2 and 3 can be reduced.

4.5. Performance Evaluation for Incorrect Topology Identification

The previous analysis assumes the correct identification of generation topology. However, there is a possibility of incorrect identification of the topology, and, hence, the setting group is inappropriately selected. To investigate this case, the performance of the classifiers is assessed by creating faults in the protected line at a specific generation topology with the different selected setting groups.

Fault scenario: Three cascaded in-zone faults at 70% of the protected line (Line 1-2) from the measurement point were simulated. The first fault was an A-G fault (class 1) at 1.0 s, the second was followed by an A-B fault (class 4) at 2.0 s, and the third was a three-phase fault (class 7) at 3.0 s. The fault durations were 100 ms. The faults were created when the PV plant was only connected to bus 2. This means that T2 should be selected as the setting group. The selected setting group (wrong selection): topology 1 (SG only).

Results: Figure 16 depicts the RMS current signal for each of the three faults: A-G fault (Class 1), A-B fault (Class 4), and three-phase fault (Class 7) and the prediction of the classifiers in setting group 1 (wrong selection) and setting group 2 (correct selection). It is evident that the prediction of a single-phase fault was correct in the case of correct and wrong topology identification, but the classifier of setting group 1 misclassified the two-phase fault (A-G fault), which was predicted as classes 4 and 7, and the three-phase fault, that was also predicted as classes 5 and 7. Moreover, there were incorrect predictions of normal events after clearing the second fault using the classification model of setting group 1. This result can be generalized for other setting groups. Correct identification of the generation topology resulted in the appropriate selection of the setting group and, hence, correct detection and classification of the faults. One way to mitigate this issue is to convert the classification models into incremental learning models where they are retrained and the hyperparameters are updated to fit the new system events.

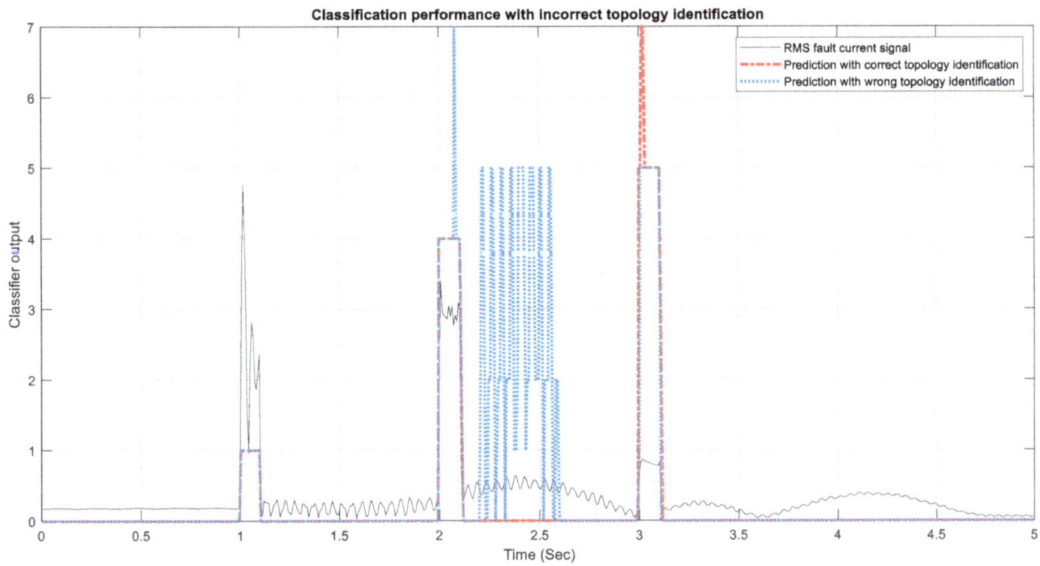

Figure 16. Classification performance with incorrect topology identification.

4.6. Comparative Analysis of Different Methods in the Literature

The proposed adaptive scheme in this paper is compared with the previous methods of adaptive approaches using system topology identification and machine learning. Table 4 presents four research studies along with our proposed approach. It could be noticed that each method follows different approaches to define the topology, identify the system configuration, and utilize ML procedures. Our approach's classification accuracy outperformed the approach in [7] for the 39-bus IEEE power system. This could be due to the dataset having more valuable features, including features from different domains, as well as the performance of the feature selection method, which selects features that maximize classification accuracy. The references [10,11] depend on deep learning techniques to identify the system topologies, which require lots of data gathered at different locations in the power system.

Moreover, it is evident that the definition of system topology in this research was specific to the types of generation units connected to one bus, while others were related to the connection and disconnection of lines, buses, and other system elements.

In comparison to [7], which used the PSO algorithm to optimize the hyperparameters of the classifier, the proposed approach used a Bayesian optimization algorithm, resulting in better performance.

Table 4. Comparative analysis of selected methodologies from the literature.

Reference	Power System Network	Topology Definition	Topology Change Identification Method	Number of Topologies	ML Method			Performance
					Feature Extraction Method	Feature Selection	Classifier	
[11]	Distribution network: modified IEEE 30-bus system	DG availability, FCL, and load varying	ANN	Two	CWT	Nil	DT	Failure rate = 0%
[7]	Medium-voltage network: Aalborg microgrid and transmission network: IEEE 9-bus model	Meshed or radial network configuration, grid-connected or islanded modes, and load variations.	ANN-SVM algorithm	Not defined	Real-time measurements (no feature extraction)	Nil	SVM for fault location	The error of ANN = 0%, Average error of SVM = 0.215%
[10]	Transmission networks: Standard IEEE 14-bus and standard IEEE 39-bus	Add/drop new bus or transmission lines	PSO detects structural changes	Unlimited	Measurements of frequency and phase values of all buses in the time domain (no feature extraction)	Nil	PSO-based weighted ensemble method of k-NN, LDA, LR, NB, DT, boosting algorithm	Accuracy = 97.93% for the IEEE classical model. Accuracy = 96.68% for the modified system (PV added) Accuracy = 96.61% for the IEEE-39 bus model.
Proposed	Transmission (39-Bus New England System)	Type of generators connected behind the relay point (synchronous machine, PV plant, and DFIG wind farm)	Circuit breaker statuses and the active power of these generating plants	Eight	Table 2	Ensemble trees (embedded-type)	Optimized ensemble trees using Bayesian optimization	Average accuracy = 98.79% Average precision = 98.92% Average specificity = 98.76% Average sensitivity = 99.82%

5. Conclusions

This paper proposes an adaptive ML-based fault detection and classification approach for transmission lines connected to inverter-based renewables like PV plants and wind farms with type-3 wind turbines. The adaptative scheme tracked the availability of a synchronous generator, PV plant, and wind farm behind the protected transmission line. The generation topology was identified using two field data: circuit breaker status and active output power measurement.

The setting groups were selected for the eight system topologies, including optimized ensemble tree classification models' hyperparameters for fault detection and classification. The reported averaged classifiers' performance was 98.79% accuracy, 98.92% precision, 98.76% recall, and 99.82% specificity.

Several system events were evaluated for the robustness of the classifiers: fault events at the protected line, different IBG penetration levels, and new transmission system configurations. The proposed classifiers can efficiently detect and classify faults incepted in the protected lines and for different IBG penetration levels (10%, 50%, and 100%). Changing the transmission system configuration and incorrect selection of the setting group degraded the performance of the developed classifiers in several cases. One of two methods can be used to overcome this. The first is creating new setting groups, and the ML models are used for each one. This approach has limitations to the number of available setting groups equipped with protective devices and the difficulty of assuming all expected scenarios of the system topologies for large-scale power systems. The second is retraining the same classifiers with new system events or converting the existing classifiers into incremental models. Incremental learning updates the models without ignoring the previously accumulated knowledge, and adapts to any new system event at each topology. Incremental learning algorithms will be considered in future studies.

Furthermore, a practical implementation of the scheme is suggested for proof of concept (POC) using a real-time digital simulator. The scheme can also be improved by facilitating fault localization and fault direction. The improvement requires more data samples for fault localization and adding features to indicate the fault direction. In addition, the design framework of this research was limited to allocating protection at only one end of the transmission line in the power system. The scheme may be developed similarly to other transmission lines with appropriate coordination procedures. Moreover, advanced methods of incipient fault diagnosis analysis, such as [35,36], can be studied further to improve detectability and speed.

Author Contributions: Conceptualization, K.A.K.; methodology, K.A.K.; software, K.A.K.; validation, K.A.K., A.E.H. and M.M.; formal analysis, K.A.K., A.E.H. and M.M.; investigation, K.A.K.; resources, K.A.K.; data curation, K.A.K.; writing—original draft preparation, K.A.K.; writing—review and editing, A.E.H. and M.M.; supervision, A.E.H. and M.M. All authors have read and agreed to the published version of the manuscript.

Funding: This research received no external funding.

Acknowledgments: The authors are thankful to the Department of Electrical Engineering, Sultan Qaboos University, for providing facilities to conduct this research.

Conflicts of Interest: The authors declare no conflict of interest.

References

1. Mukherjee, S.; Marshall, M.; Smith, T.; Piesciorvosky, E.; Snyder, I.; Sticht, C. Adaptive Protective Relay Settings—A Vision to the Future. In Proceedings of the 2022 IEEE Rural Electric Power Conference (REPC), Tulsa, OK, USA, 23–25 April 2022; pp. 25–30. [CrossRef]
2. Liu, S.; Bi, T.; Liu, Y. Theoretical analysis on the short-circuit current of inverter-interfaced renewable energy generators with fault-ride-through capability. *Sustainability* **2017**, *10*, 15. [CrossRef]
3. Voima, S.; Kauhaniemi, K. Adaptivity of Protection in Smart Grids. In Proceedings of the PAC World Conference, Budapest, Hungary, 25–28 June 2012. Available online: http://sgemfinalreport.fi/files/P024.pdf (accessed on 9 August 2022).

4. Ray, P.; Mishra, D.P. Support vector machine based fault classification and location of a long transmission line. *Eng. Sci. Technol. Int. J.* **2016**, *19*, 1368–1380. [CrossRef]
5. Ali, S.F.A.A.M.; Kumar, M.; Muthukaruppan, V. Utility Perspective Towards Machine Learning Techniques in Power System Protection. *IRJET* **2021**, *8*, 4140–4146.
6. Lawal, I.A.; Abdulkarim, S.A. Adaptive SVM for data stream classification. *S. Afr. Comput. J.* **2017**, *29*, 27–42. [CrossRef]
7. Lin, H.; Sun, K.; Tan, Z.H.; Liu, C.; Guerrero, J.M.; Vasquez, J.C. Adaptive protection combined with machine learning for microgrids. *IET Gener. Transm. Distrib.* **2019**, *13*, 770–779. [CrossRef]
8. Poudel, B.; Garcia, D.R.; Bidram, A.; Reno, M.J.; Summers, A. Circuit Topology Estimation in an Adaptive Protection System. In Proceedings of the 2020 52nd North American Power Symposium, NAPS 2020, IEEE, Tempe, AZ, USA, 11–13 April 2021; pp. 1–6. [CrossRef]
9. Marín-Quintero, J.; Orozco-Henao, C.; Percybrooks, W.S.; Vélez, J.C.; Montoya, O.D.; Gil-González, W. Toward an adaptive protection scheme in active distribution networks: Intelligent approach fault detector. *Appl. Soft Comput.* **2021**, *98*, 106839. [CrossRef]
10. Yavuz, L.; Soran, A.; Önen, A.; Muyeen, S.M. An adaptive fault detection scheme using optimized self-healing ensemble machine learning algorithm. *CSEE J. Power Energy Syst.* **2021**, *8*, 1145–1156. [CrossRef]
11. Tang, W.J.; Yang, H.T. Data Mining and Neural Networks Based Self-Adaptive Protection Strategies for Distribution Systems with DGs and FCLs. *Energies* **2018**, *11*, 426. [CrossRef]
12. Memon, A.A.; Kauhaniemi, K. An adaptive protection for radial AC microgrid using IEC 61850 communication standard: Algorithm proposal using offline simulations. *Energies* **2020**, *13*, 5316. [CrossRef]
13. Lin, H.; Guerrero, J.M.; Vasquez, J.C.; Liu, C. Adaptive distance protection for microgrids. In Proceedings of the IECON 2015—41st Annual Conference of the IEEE Industrial Electronics Society, Yokohama, Japan, 9–12 November 2015; pp. 725–730. [CrossRef]
14. Zhao, J.; Netto, M.; Huang, Z.; Yu, S.S.; Gómez-Expósito, A.; Wang, S.; Kamwa, I.; Akhlaghi, S.; Mili, L.; Terzija, V.; et al. Roles of dynamic state estimation in power system modeling, monitoring and operation. *IEEE Trans. Power Syst.* **2021**, *36*, 2462–2472. [CrossRef]
15. Eisa, S.A.; Stone, W.; Wedeward, K. Mathematical analysis of wind turbines dynamics under control limits: Boundedness, existence, uniqueness, and multi time scale simulations. *Int. J. Dyn. Control* **2018**, *6*, 929–949. [CrossRef]
16. Korres, G.N.; Katsikas, P.J.; Chatzarakis, G.E. Substation topology identification in generalized state estimation. *Int. J. Electr. Power Energy Syst.* **2006**, *28*, 195–206. [CrossRef]
17. Dehghanpour, K.; Wang, Z.; Wang, J.; Yuan, Y.; Bu, F. A survey on state estimation techniques and challenges in smart distribution systems. *IEEE Trans. Smart Grid* **2019**, *10*, 2312–2322. [CrossRef]
18. Poudel, B.P.; Bidram, A.; Reno, M.J.; Summers, A. Zonal Machine Learning-based Protection for Distribution Systems. *IEEE Access* **2022**, *10*, 1–12. [CrossRef]
19. Razmi, P.; Asl, M.G.; Canarella, G.; Emami, A.S. Topology identification in distribution system via machine learning algorithms. *PLoS ONE* **2021**, *16*, e0252436. [CrossRef] [PubMed]
20. Kurup, A.R.; Martinez-Ramon, M.; Summers, A.; Bidram, A.; Reno, M.J. Deep learning based circuit topology estimation and fault classification in distribution systems. In Proceedings of the 2021 IEEE PES Innovative Smart Grid Technologies Europe: Smart Grids: Toward a Carbon-Free Future, ISGT Europe, Espoo, Finland, 18–21 October 2021. [CrossRef]
21. Amoateng, D.O.; Yan, R.; Mosadeghy, M.; Saha, T.K. Topology Detection in Power Distribution Networks: A PMU Based Deep Learning Approach. *IEEE Trans. Power Syst.* **2022**, *37*, 2771–2782. [CrossRef]
22. He, J.; Cheng, M.X. Machine learning methods for power line outage identification. *Electr. J.* **2021**, *34*, 106885. [CrossRef]
23. DIgSILENT. *GmbH 39 Bus New England System*; DIgSILENT: Gomaringen, Germany, 2015; pp. 1–16.
24. DIgSILENT GmbH. *PowerFactory 2019, Manual, User*; DIgSILENT: Gomaringen, Germany, 2019.
25. González-Longatt, F.M. The P.M. Anderson Test System. Available online: https://www.fglongatt.org/Test_Systems/PM_Anderson_PF.html (accessed on 5 February 2019).
26. ENTSO-E. *System Protection Behavior and Settings during System Disturbances*; ENTSO-E: Brussels, Belgium, 2018.
27. Al Kharusi, K.; El Haffar, A.; Mesbah, M. Fault Detection and Classification in Transmission Lines Connected to Inverter—Based Generators Using Machine Learning. *Energies* **2022**, *15*, 5475. [CrossRef]
28. Aliyu, I.; Lim, C.G. Selection of optimal wavelet features for epileptic EEG signal classification with LSTM. *Neural Comput. Appl.* **2021**, *35*, 1077–1097. [CrossRef]
29. Taheri, B.; Salehimehr, S.; Razavi, F.; Parpaei, M. Detection of power swing and fault occurring simultaneously with power swing using instantaneous frequency. *Energy Syst.* **2020**, *11*, 491–514. [CrossRef]
30. Kłosowski, G.; Rymarczyk, T.; Wójcik, D.; Skowron, S.; Cieplak, T.; Adamkiewicz, P. The use of time-frequency moments as inputs of lstm network for ecg signal classification. *Electronics* **2020**, *9*, 1452. [CrossRef]
31. Phinyomark, A.; Thongpanja, S.; Hu, H.; Phukpattaranont, P.; Limsakul, C. The Usefulness of Mean and Median Frequencies in Electromyography Analysis. In *Computational Intelligence in Electromyography Analysis—A Perspective on Current Applications and Future Challenges*; IntechOpen: London, UK, 2012; pp. 195–220. [CrossRef]
32. Liu, H.M.H. *Computational Methods of Feature Selection*; Taylor & Francis: Abingdon, UK, 2007.
33. Kazemitabar, S.J.; Amini, A.A.; Bloniarz, A.; Talwalkar, A. Variable importance using decision trees. In Proceedings of the 31st Conference on Neural Information Processing Systems (NIPS 2017), Long Beach, CA, USA, 4–9 December 2017; pp. 426–435.

34. Nembrini, S.; König, I.R.; Wright, M.N. The revival of the Gini importance? *Bioinformatics* **2018**, *34*, 3711–3718. [CrossRef] [PubMed]
35. Wu, Y.; Liu, X.; Zhou, Y. Deep PCA-Based Incipient Fault Diagnosis and Diagnosability Analysis of High-Speed Railway Traction System via FNR Enhancement. *Machines* **2023**, *11*, 475. [CrossRef]
36. Han, J.; Miao, S.; Li, Y.; Yang, W.; Yin, H. Fault Diagnosis of Power Systems Using Visualized Similarity Images and Improved Convolution Neural Networks. *IEEE Syst. J.* **2022**, *16*, 185–196. [CrossRef]

energies MDPI

Article

Categorizing 15 kV High-Voltage HDPE Insulator's Leakage Current Surges Based on Convolution Neural Network Gated Recurrent Unit

Wen-Bin Liu [1], Phuong Nguyen Thanh [1,2], Ming-Yuan Cho [1] and Thao Nguyen Da [3,4,*]

1 Department of Electronic Engineering, National Kaohsiung University of Science and Technology, Kaohsiung 800, Taiwan
2 Department of Electronic—Electrical Engineering, Nha Trang University, Nha Trang 650000, Vietnam
3 Department of Business Intelligence, National Kaohsiung University of Science and Technology, Kaohsiung 800, Taiwan
4 Faculty of Economics and Management, Thai Binh Duong University, Nha Trang 650000, Vietnam
* Correspondence: thao.nd@tbd.edu.vn

Abstract: The leakage currents are appropriate for determining the contamination level of insulators in the power distribution system, which are efficiently cleaned or replaced during the maintenance schedule. In this research, the hybrid convolution neural network and gated recurrent unit model (CNN-GRU) are developed to categorize the leakage current pulse of the 15 kV HDPE insulator in the transmission towers in Taiwan. Many weather parameters are accumulated in the online monitoring system, which is installed in different transmission towers in coastal areas that suffer from heavy pollution. The Pearson correlation matrix is computed for selecting the high correlative features with the leakage current. Hyperparameter optimization is employed to decide the enhancing framework of the CNN-GRU methodology. The performance of the CNN-GRU is completely analyzed with other deep learning algorithms, which comprise the GRU, bidirectional GRU, LSTM, and bidirectional LSTM. The developed CNN-GRU acquired the most remarkable improvements of 79.48% CRE, 83.54% validating CRE, 14.14% CP, 20.89% validating CP, 66.24% MAE, 63.59% validating MAE, 73.24% MSE, and 71.59% validating MSE benchmarks compared with other methodologies. Therefore, the hybrid CNN-GRU methodology provides comprehensive information about the contamination degrees of insulator surfaces derived from the property of leakage currents.

Keywords: classify 15 kV HDPE insulator's leakage current; convolutional neural network; gated recurrent unit; deep learning machine; hyperparameter optimization

Citation: Liu, W.-B.; Nguyen Thanh, P.; Cho, M.-Y.; Nguyen Da, T. Categorizing 15 kV High-Voltage HDPE Insulator's Leakage Current Surges Based on Convolution Neural Network Gated Recurrent Unit. *Energies* **2023**, *16*, 2500. https://doi.org/10.3390/en16052500

Academic Editors: Veerapandiyan Veerasamy, Shailendra Singh and Sunil Kumar Singh

Received: 12 February 2023
Revised: 26 February 2023
Accepted: 1 March 2023
Published: 6 March 2023

1. Introduction

The insulators installed in severe environments near the sea or industrial areas commonly cause flashovers and disruptions in the transmission power line. These insulators are contaminated by salt fog or dust, which increase the electrical conductivity on the surface because of equivalent salt deposit density (ESDD) and non-soluble deposit density (NSDD). Many electrical power companies have developed insulator monitoring systems to evaluate the pollution severity of insulators, which provide the maintenance procedures for periodical washes [1–4]. The insulator monitoring technologies combined with leakage current prediction are the most appropriate methods to continuously measure without interruption of the transmission lines [5–8].

Many laboratory experiments and research proved that the prediction of the insulator leakage current could be utilized to determine the process of contaminative levels on the insulator surface [9–11]. The insulator leakage currents are affected by the environmental parameters and the contaminative levels on the insulator surface [12–14]. Research on leakage current prediction has been accomplished using many neural networks (NN) and mathematical models with environmental features. Li et al. enhanced the neural network model to predict the ESDD in the security stage [15]. The 10-set training and validating data were recorded from the laboratory experiments, and the accurate predictions were reliable for practical applications. Zhao et al. employed linear and nonlinear regressive methods to predict the leakage current of the insulator from the environmental variables [16]. The eight-month data were utilized to evaluate the regressive model, which needed to adjust the coefficients in the actual applications. Ali et al. implemented the leakage current prediction on silicone rubber insulators by a neural network [17]. The predicted leakage current with the feed-forward backpropagation neural network obtained less than 12% in the tested cases. Ayman et al. employed the Gaussian radial basis function NN to forecast the leakage current of non-ceramic insulators in the salt fog condition [18]. The proposed algorithm acquired a 3.5–5.3% accuracy with a random optimizing algorithm in the identical insulators. Felix et al. proposed the linear stochastic and statistical flashover prediction with a signature analysis tool [2]. The digitized samples of leakage current were collected offline for statistical investigation of the leakage current envelope with a new twist method [19]. In the previous research, Phuong et al. employed the particle-swarm-optimization-based backpropagation NN to forecast the leakage current of 69 kV and 161 kV high-voltage insulators [20]. The enhanced algorithm combined with the surface spark discharge data improved the effectiveness and accuracy of the predicting application, which were 8.35×10^{-4} MSE, 8.92% MAPE, and 0.0857% R square. Santos and Holtzhausen proposed a novel method to classify the leakage current with different weather variables and the cumulative pollution index [21]. The authors utilized the random forest methodology to predict the leakage current of ceramic and RTV silicone-coated insulators in twenty-two consecutive months, which obtained a higher accuracy. Jahromi et al. proposed an NN to forecast the leakage current in the early aging interval [22]. The two-layer feed-forward backpropagation achieved a higher performance with a 15% maximum error with the data from 15 kV commercial silicon rubber insulators. Pinotti and Meyer proposed the mathematical models with the validation of the least squares method and statistical hypothesis testing to predict the leakage current with reasonable accuracy [23]. The prediction method utilized the weather data, which included temperature, pressure, humidity, wind speed, and rain for 25 kV distribution insulators. Hadi et al. proposed a statistical method to predict flashover occurrence by utilizing the harmonic components of leakage current [24]. The ratio between the fifth and third harmonic components combined with geometric distribution was proposed to approximately predict the contaminative level of insulators. Sheik et al. proposed the salt contaminant accumulation prediction by utilizing the feed-forward backpropagation neural network and statistical data of the leakage current [25]. The research utilized the statistical character of leakage current to predict the contaminant level and provide the pre-warning of flashover. Ali et al. extracted the ESDD, NSDD, wetting rate and pollution level to effectively evaluate the insulator condition [26]. These proposed indicators could provide more critical information about the insulator condition and the leakage current characteristic, which can be implemented in the monitoring system or prototype application. Volat et al. deployed the ANN to evaluate the leakage current development through the melting session [27]. The ANN was integrated into the monitoring system to forecast the appearance of the discharge phenomenon for protection with an average delay of 9 min. Zahra et al. developed the real asymmetric contamination to evaluate the leakage current of the 20 kV polymeric insulator [28]. The experiment illustrated that the various parameters of contamination layers were crucial for the electrical conductivity of insulators. The data from the non-uniform fan-shaped contamination were utilized in ANN to predict the leakage current, which achieved less than 5% relative error. Hui proposed the prediction theory with an extreme value risk function to forecast the leakage current by calculating the probability density function [29]. The developed method could predict the evolutional trend of the leakage current, which provided more information on flashover prevention procedures. Li et al. extracted three main statistical characteristics of leakage current, which were utilized in the ANN for predicting the ESDD level on the insulator surface [30]. The predicted performance was less than 0.035 mg/cm^2, which was appropriate for the pre-warning system before the flashover. In another study, the leakage current was predicted by utilizing the multi-regression model with the weather parameters in 10-min intervals [31]. Nameer and Ayman proposed the Bayesian neural network to predict the first and third harmonic components in the salt fog environment [32]. The research results proved that the nonlinear autoregressive neural network achieved the highest performance in terms of prediction error. The developed method was integrated into the condition monitoring system to prevent flashover accumulation. Wang et al. proposed a deeper mathematical model to predict the leakage current of 110 kV insulators based on the relative humidity [33]. The experiment results proved the strong correlation between the leakage current and the environmental features, especially the temperature and relative humidity. Suhas et al. classified the contamination levels based on the short-time modified Hilbert transform, which combined with local characteristic current waveform [34]. The comparable performance was better than the literature review, which included ANN, neuro Fuzzy classifier, and Rough set classifier, and could be implemented in many applications with 95% overall classification accuracy. Abouzeid et al. developed a non-intrusive algorithm to predict the ESDD on silicone rubber insulators [35]. The proposed method achieved an overall predicting improvement from 68% to 95% compared with KNN, polynomial, and fuzzy neuro methodologies. Sun et al. developed the least square SVM (LSSVM) and exploratory factor analysis (EFA) for the contamination level of the distribution insulator [36]. The research utilized the air pollution parameters and meteorological parameters as input factors to predict the ESDD and NSDD values. The effectiveness of EFA and LSSVM outperformed other traditional models in different evaluation indexes. Therefore, many researchers have proved the accuracy and effectiveness of utilizing the weather parameter to predict the insulator leakage current for evaluating the contaminative level.

Recently, with the development of computational resources and accurate sensor devices, deep learning methods which require more accumulated data are widely proposed for many applications, especially computer vision, language processing, and sequential data. The deep learning algorithms outperform the traditional methodologies, which obtain more accuracy and higher stability. Medeiros et al. utilized different neural network models to predict the increase in leakage current and compared the performance between them [19]. The experiment results proved that the RNN outperformed other methodologies and achieved an accuracy of up to 97.25%. Yeh et al. proposed the bidirectional long short-term memory (Bi-LSTM) to categorize the leakage current of insulators [37]. The accumulated data from the outdoor insulator were utilized in the predicting model, which obtained an improvement of 49.529%, 72.736% error, and 12.761%, 36.641% accuracy in the training and validating operations. Nguyen et al. developed the hybrid model, which combined the convolution neural network (CNN) and Bi-LSTM to categorize the insulator leakage current [38]. The experiment proved that the CNN-Bi-LSTM outperformed other deep learning methods, which achieved improvements of 71.331% and 91.250% errors in the training and validating data with a 24-sequential window size. Therefore, the deep learning algorithms proved better performance in predicting the insulator leakage current compared with the traditional methodologies.

Although many researchers conducted many deep learning algorithms, the novel hybrid model, which combines the CNN and gated recurrent unit (GRU) algorithm, has not been evaluated and compared in predicting the leakage current of the insulator. In this research, the authors propose the novel CNN-GRU to more accurately and effectively categorize the leakage current of 15 kV HDPE distribution insulators. The innovative contributions in this study are as follows: (1) calculating the Pearson correlation between different climate features and the 15 kV high-voltage insulator leakage current, (2) developing the novel hybrid CNN and GRU to distinguish the leakage current levels of 15 kV HDPE insulators, (3) employing the hyperparameter optimization to establish the framework of the developed model, (4) analyzing and comparing the effectiveness and accuracy with other deep learning models, which include the recurrent neural network (RNN), the LSTM, the Bi-LSTM, and GRU. The methodologies are presented in the next section. The third section illustrates the hyperparameter results, and the fourth section presents the comparative tables between different deep learning methodologies.

2. Algorithm Flow and Methodology

The high frequent leakage currents come from the pollution particles in the environment, which accumulate on the insulator surface. The accumulated layer is influenced by many features, including air pollutants, wind direction, temperature, humidity, etc. In this research, many weather parameters were accumulated every hour for more than a two-year period, which is illustrated in Figure 1. The general timelines of this study are illustrated in the following aspects:

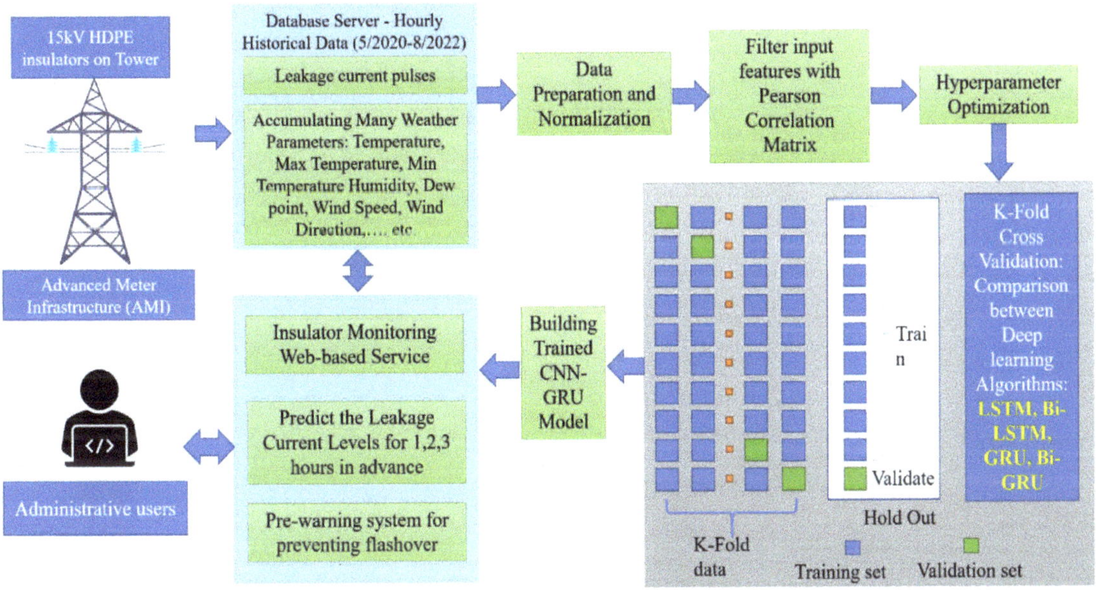

Figure 1. The algorithm flow for implementing the predicting methodology.

1. Collecting data: The weather parameters were gathered for more than two years. The advanced meter infrastructure (AMI) with accurate sensors was installed on the different transmission towers in Taiwan. The AMI systems transferred all the collected information to the database server. Moreover, the leakage current pulses were also accumulated and categorized based on their values in four locations. This research utilizes the weather parameters as input features and leakage current level as the target vector. All the collected information needs to be normalized by the min-max methodology, as in Equation (1), where in_i, in_{min}, and in_{max} are the variable, the minimum value, and maximum of the input data, respectively.

$$In\_nor_i = \frac{in_i - in_{min}}{in_{max} - in_{min}} \quad (1)$$

2. Selecting input features: To completely evaluate the influence of accumulated features on the insulator leakage current, this research calculates the Pearson correlation matrix between variables. The Pearson values present the correlation between two factors which vary from −1 to 1, as calculated in Equation (2), where In_i, $\overline{In}$, Out_i, and $\overline{Out}$ are the input, average value of input, output, and average value of output,

respectively [39,40]. The correlation coefficients are calculated for each collected environmental factor with the leakage current pulses. The influences of weather parameters are evaluated based on the interaction with the leakage current levels. The environmental indexes with high correlation are selected as dominant input features for utilization in the predicting model.

$$\mathrm{Corr}_{\mathrm{In-Out}} = \frac{\sum(\mathrm{In_i} - \overline{\mathrm{In}})\sum(\mathrm{Out_i} - \overline{\mathrm{Out}})}{\sqrt{\sum(\mathrm{In_i} - \overline{\mathrm{In}})^2 \sum(\mathrm{Out_i} - \overline{\mathrm{Out}})^2}} \quad (2)$$

3. Performing the hyperparameter optimization to analyze the effectiveness of setting parameters on the deep learning models: This research identifies the most appropriate parameter which significantly improves the performance of classifying models. The simple grid search method is deployed to define the optimized setting parameter of the proposed models [41]. Moreover, the simple grid search analyzes the effects of setting variables on the predicting accuracy.
4. Comparing the performance and accuracy between the proposed algorithm and other deep learning methodologies with the optimized architecture: The CNN-GRU is completely analyzed with the RNN, LSTM, Bi-LSTM, and GRU models by utilizing the K-fold cross-validation [42,43]. This method could effectively evaluate the performance of deep learning algorithms on the collected data.
5. The collected data come from the natural insulators which were installed in Kaohsiung City, a coastal area in Taiwan. The test site is the 15 kV transmission line which is normally operated, as shown in Figure 2. The performance between predicting models is analyzed through different evaluation indexes, which include the mean square error (MSE), category cross entropy (CRE), cosine proximity (CP), and mean absolute error (MAE), as in the following Equations (3)–(6), where $\mathrm{out_i}$, $\hat{\mathrm{out}}_\mathrm{i}$, $\overline{\mathrm{out}}$, and K are the actual, predictive, average vectors, and the number of testing set, respectively. The collected leakage current surges are transformed into one-hot vectors for the category purpose, which are utilized as target vectors in the forecasting algorithms, as in Figure 3. A total of 5 leakage current levels are categorized with different leakage current surges.

$$\mathrm{CRE} = -\log\left(\frac{e^{s_p}}{\sum_j^C e^{s_j}}\right) \quad (3)$$

$$\mathrm{MSE} = \frac{1}{K}\sum_{i=1}^{K}(\mathrm{out_i} - \hat{\mathrm{out}}_\mathrm{i})^2 \quad (4)$$

$$\mathrm{CP} = \frac{\sum_{i=1}^{n} \mathrm{out_i}\, \hat{\mathrm{out}}_\mathrm{i}}{\sqrt{\sum_{i=1}^{n} \mathrm{out_i^2}}\sqrt{\sum_{i=1}^{n} \hat{\mathrm{out}}_\mathrm{i}^2}} \quad (5)$$

$$\mathrm{MAE} = \frac{1}{K}\sum_{i=1}^{K}|\mathrm{out_i} - \hat{\mathrm{out}}_\mathrm{i}| \quad (6)$$

Figure 2. Collecting the leakage current surges of 15 kV HDPE insulator in the natural environment.

Time	Level 1 100uA~500uA	Level 2 500uA~1 mA	Level 3 1mA~5mA	Level 4 5mA~10mA	Level 5 >10mA
11/10/2020 07:00	75139	2432	24	0	0
11/10/2020 08:00	78710	1821	14	0	0
11/10/2020 09:00	88039	48	0	0	0
11/10/2020 10:00	84668	693	0	0	0
11/10/2020 11:00	85111	976	0	0	0

Transfer to One Hot Vector

Time	LV0	LV 1	LV 2	LV 3	LV4	LV5
11/10/2020 07:00	0	0	0	1	0	0
11/10/2020 08:00	0	0	0	1	0	0
11/10/2020 09:00	0	0	1	0	0	0
11/10/2020 10:00	0	0	1	0	0	0
11/10/2020 11:00	0	0	1	0	0	0

Figure 3. Transforming the leakage current surges to one-hot vector.

2.1. Primary Influence of Environmental Factors on Insulator Leakage Current

Previous research proved that the weather indexes correlate critically with the leakage current. These weather factors accumulated the pollution particles in the air, which gradually deposited on the surface of insulators. Moreover, high humidity and rainfall conditions intensify the wet deposition, which makes contamination accumulate obviously. However, the effect of other weather factors on the insulator leakage current is comparatively complex. In this research, the authors calculate the Pearson correlation matrix to effectively analyze the significant correlation with leakage current, as presented in Figure 4. Each collected insulator leakage current surge is computed in the correlation index with the weather parameters. The analysis of the influence parameters of the insulator leakage current is detailed as follows:

1. Level 3, level 4, and level 5 have a weak correlation with all of the collected weather parameters, which indicates a weak relationship between the high level of leakage current and other environmental parameters. However, level 1 and 2 prove substantial similarity with the weather factors. Therefore, the leakage current levels from 0 to 2 are utilized as target variables in the categorizing models.

2. The average, the maximum, and the minimum temperature during a 1-h interval have the most negative correlation with leakage current, which are −0.60223, −0.60086, and −0.60276, respectively. In addition, the dew point, wind direction, and maximum wind direction have a contradictory connection with the leakage current, which could be applied in the predicting model. The wind chill index, heat index, insulator temperature, and temperature humidity sunshine wind index are also implemented as input variables for effectively distinguishing the leakage current levels.
3. The air pressure has the most positive connection with the leakage current due to the installed high mountain area, which is the 0.440471 Pearson index. Moreover, the wind speed, wind path, maximum wind speed, and equilibrium moisture content have an ordinary relationship with the leakage current levels 1 and 2, and are employed as high correlation components in the predicting methodology. The increase in these environmental factors will enhance the number of surges of 15 kV HDPE insulators because of their positive correlation magnitudes.
4. Moreover, some weather factors have a weak correlation with the leakage current, but they have formidable relationships with other weather parameters and are also included in the predicting methodology. Solar illuminance, maximum solar illuminance, and solar energy are also engaged in the categorizing algorithm to enhance accuracy and stability effectively.

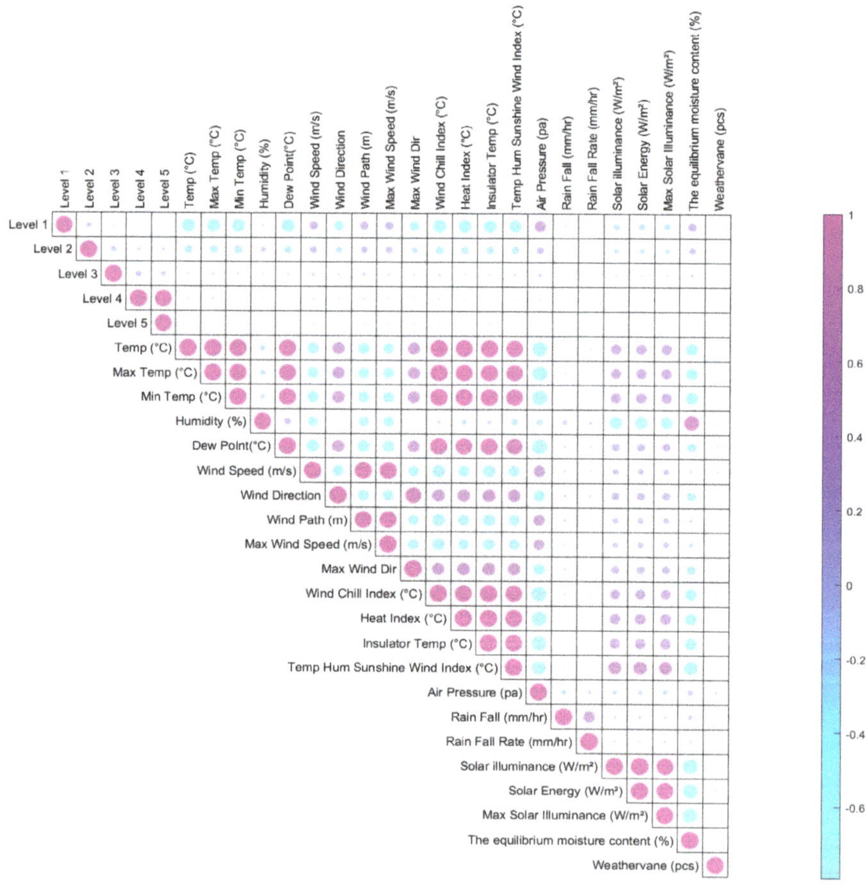

Figure 4. The correlation between different leakage current and various natural environmental features.

From the Pearson correlation matrixes, some natural environmental features which have formidable correlations with leakage current are selected as input parameters for categorizing the 15 kV HDPE insulators from level 0 to level 2.

2.2. The Series Sequential Predicting Methodologies

The leakage current requires an accurate categorization which is a non-linear function with other natural environmental factors. This study provides a better solution for this challenge, which utilizes many sequential predicting algorithms. The LSTM is an appropriate solution that could solve the gradient vanishing exploding issues in the recurrent neural network (RNN) [44,45]. The LSTM is the sequential dependency methodology that effectively reaches a better performance. The LSTM is generally composed of input, output, and forget gates which effectively memorize the sequential historical information for a long time, as in Figure 5. The forget gate, F_n, calculates the sequential input with a sigmoid activation function, which controls the output value between 0 and 1, as in Equation (7), where W_{FH}, W_{FX}, and b_F are the weight matrixes and bias vectors. If the output value is nearly zero, the achieving data will be forgotten, and on the contrary, more data could be retained. This gate will forget unnecessary previous information and check the reserved information in the memory cell. The sequential data navigate to the input gate I_n, which is computed by a sigmoid activation function, as in Equation (8), where W_{IH}, W_{IX}, b_I, H_{n-1}, and X_n are the weight matrixes, bias vector, the previous cell hidden state, and the input gate, respectively. The input gate takes responsibility for checking whether the information is worth reserving in the long memory. The candidate state, $\widetilde{C}_n$, is updated by a TANH operation which controls the output data in range −1 and 1, and the input vector, X_n, as in Equation (9). The cell state, C_n, is computed with the previous state, C_{n-1}, and the candidate state, as in Equation (10). Conclusively, the output gate, O_n, and the cell hidden state, H_n, are computed by the sigmoid and TANH activation function, as in Equations (11) and (12). The hidden layer cell state is computed based on the cell state and the output gate, which controls the long- and short-term dependency in the LSTM algorithm.

$$F_n = \sigma(W_{FH}H_{n-1} + W_{FX}X_n) + b_F \quad (7)$$

$$I_n = \sigma(W_{IH}.H_{n-1} + W_{IX}X_n) + b_I \quad (8)$$

$$\widetilde{C}_n = \tan h(W_{CH}.H_{n-1} + W_{CX}X_n) + b_C \quad (9)$$

$$C_n = F_nC_{n-1} + I_n\widetilde{C}_n \quad (10)$$

$$O_n = \sigma(W_{OH}.H_{n-1} + W_{OX}X_n) + b_O \quad (11)$$

$$H_n = O_n\tan h(C_n) \quad (12)$$

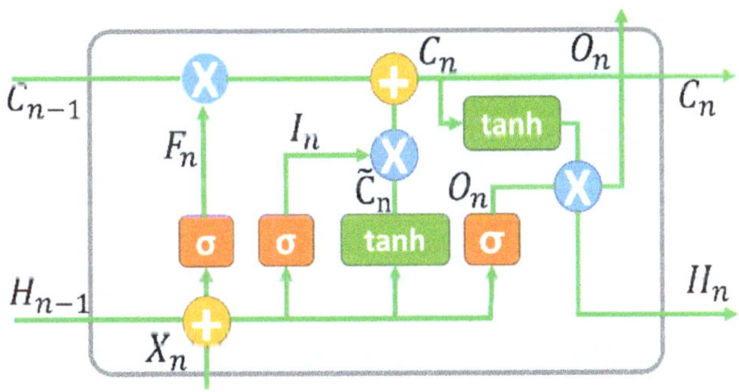

Figure 5. The diagrammatic representation of the LSTM structure.

Another LSTM variant is the gated recurrent unit (GRU), which consists of a simplified configuration and fewer training weights. In many applications, the GRU demonstrated its ability to obtain similar prediction accuracy to LSTM while utilizing a few training epochs. The GRU and LSTM algorithms could hold more periods of historical serial information to enhance the feature augmentation and improve the multistep ahead categorization ability [46,47]. The GRU cell includes the reset gate, R_n, and update gate, Z_n, as illustrated in Figure 6. The update gate, Z_n, controls the new update state of the hidden state, whereas the reset gate, R_n, is employed to ignore the information in the previous hidden state, as calculated in Equations (13) and (14), where W and b are the weight and bias matrixes and H_{n-1} and X_n are the hidden state and input vector. The cell state, $\widetilde{H}_n$, and hidden state, H_n, are combined into one hidden unit, representing the GRU algorithm's stored memory, as computed in Equations (15) and (16).

$$Z_n = \sigma(W_{ZH}.H_{n-1} + W_{ZX}X_n) + b_Z \tag{13}$$

$$R_n = \sigma(W_{RH}.H_{n-1} + W_{RX}X_n) + b_R \tag{14}$$

$$\widetilde{H}_n = \tanh(W_H[R_n \otimes H_{n-1} + X_n]) + b_C \tag{15}$$

$$H_n = (1 - Z_n) \otimes H_{n-1} + Z_n \otimes \widetilde{H}_n \tag{16}$$

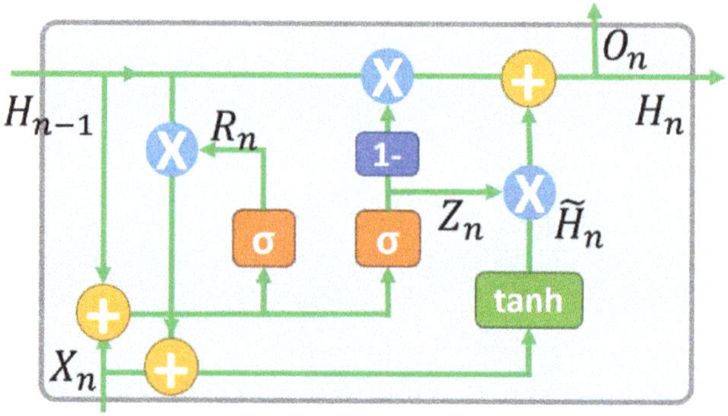

Figure 6. The basic configuration of GRU cell.

The LSTM and GRU are self-connected units that adapt to reduce error over time. These internal gates maintain the content in the memory cell, update necessary information, and forget unrelated contents. However, both LSTM and GRU only process the information in the forward direction, which could not reconstruct the memory in the backward direction. Therefore, the bidirectional models, which include the Bi-LSTM and Bi-GRU, are developed to procedure the sequential data in both directions, as in Figure 7. The bidirectional model consists of two dependent layers for inputting the forward and backward sequential information, which could better capture the long-term dependencies and facilitate more accurate predictions [48]. The forward and backward propagative processes are combined into one output by Equation (17), where $W_{\overrightarrow{H}}$, $\overrightarrow{H}_n$, $W_{\overleftarrow{H}}$, and $\overleftarrow{H}_n$ are the weight and hidden state of forward and backward directions, respectively.

$$Out_n = W_{\overrightarrow{H}}\overrightarrow{H}_n + W_{\overleftarrow{H}}\overleftarrow{H}_n + b_{Out} \tag{17}$$

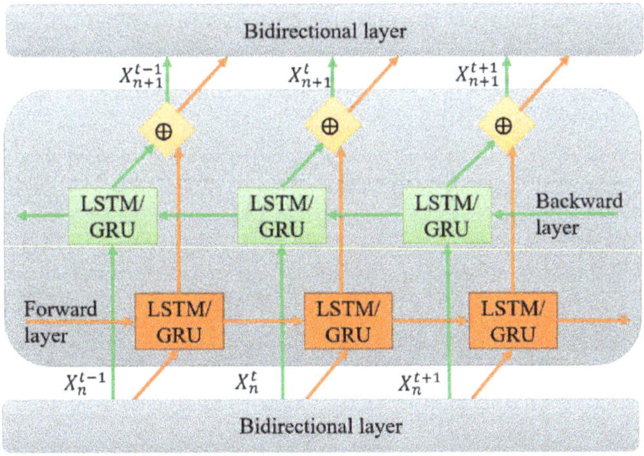

Figure 7. The structure of the Bi-GRU/Bi-LSTM model.

2.3. The Proposed Hybrid Convolution Neural Network and Gated Recurrent Unit

The convolution neural network (CNN) was inspired by the natural perception in the animal visual cortex, which was discovered by Fukushima in 1980 [49,50]. CNN is a well-known algorithm that has been successfully and effectively utilized in many applications. The CNN includes two main layers named convolutional and pooling layers which concentrate on learning feature representation of the information. The convolutional layers implement several kernel matrixes to calculate different feature maps and then deploy a nonlinear activation function to convolve the consequences. Equation (18) calculates the feature map value at the nth layer in the location (i, j) of the kth feature map, where f is the activation function and w and b are the weight term and bias vector. The activation function is the ReLU, which is generally desirable in stacked multilayer structures [51,52]. The pooling layer decreases the resolution of the feature map by employing shift invariance. The pooling layer is connected consequently to the convolution layers for extracting the dominant features by mathematical average and maximum operation [53]. By combining many CNN layers, the higher-level property representations are extracted for the subsequent layers. In this research, the max pooling layer is utilized in the CNN-GRU model for distinguishing the leakage current of 15 kV HDPE insulators.

$$\mathrm{out}^{n}_{i,j,k} = f\left(w^{n\ T}_{k} x^{n}_{i,j} + b^{l}_{k}\right) \quad (18)$$

The extracted dominant features from the CNN are transferred to the GRU layers, which perform high-level training operations. The last layers are fully connected layers with a SoftMax activation function for the classification task. The general architecture of CNN-GRU is illustrated in Figure 8. The proposed methodology is employed in distinguishing the leakage current level of the 15 kV HDPE insulator in the natural environmental experiment. The developed method first utilizes the CNN layers to decrease the dimension of the original matrix, which reduces the complexity while retaining most of the initial data. Then, the GRU is established to improve the predicting ability and accuracy during the optimizing process. The proposed CNN-GRU algorithm is analyzed and assessed with other deep learning methodologies, including LSTM, GRU, Bi LSTM, and Bi-GRU models. The number of sequential historical data is defined as window size, and the level of leakage currents is the output of the proposed model. The hyperparameter optimization for the proposed methodology is presented in the next section.

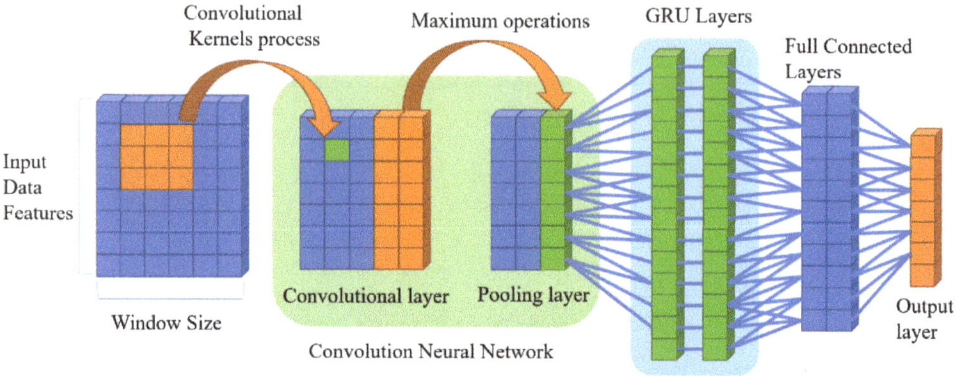

Figure 8. The hierarchically structured of CNN-GRU model.

3. Hyperparameter Optimization for CNN-GRU

Hyperparameter optimization (HPO) is implemented to enhance the performance of CNN-GRU with different configurations. This technique evaluates the CNN-GRU performance with different benchmark indexes and resource consumption. In this study, the grid search is developed to analyze the effect of variant setting configuration on the categorizing performance [41,54,55]. Table 1 describes the variant configuration of the proposed CNN-GRU. The number of window sizes alters from 24 to 36 to 48. In the CNN layers, the number of CNN layers adjusts from 1 to 2, and the number of kernel filters evolves from 16 to 32 to 64. In the GRU configuration, the number of GRU layers is modified from 2 to 3 to 4, and the number of GRU cells is 32, 64, and 128. Combined different configurations produce more than 162 CNN-GRU structures which are all simulated for analysis and evaluation. The summarized experiment results are illustrated in the radar maps in Figure 9 with different data collections from 1 to 4.

Table 1. Different configurations of HPO for CNN-GRU.

Hyperparameter Configurations	Value Ranges
Window size	24, 36, 48
CNN layers	1, 2
CNN filters	16, 32, 64
CNN kernel size	3
GRU layers	2, 3, 4
GRU neural nodes	32, 64, 128
Fully connected layers	2
Fully connected neural nodes	64
Learning rate	0.001
Train/validation ratio	90/10
Training epochs	50

For better evaluating the advantages and disadvantages of different configurations on the CNN-GRU model, different benchmark indexes of CRE, MSE, CP, and MAE are described in the radar maps with training and validating operations. The coordinate values become greater from the center to the outside, which illustrates a lower performance for the CRE, MSE, and MAE indexes. In contrast, the CP value gradually increases from inside to outside, which presents a better achievement.

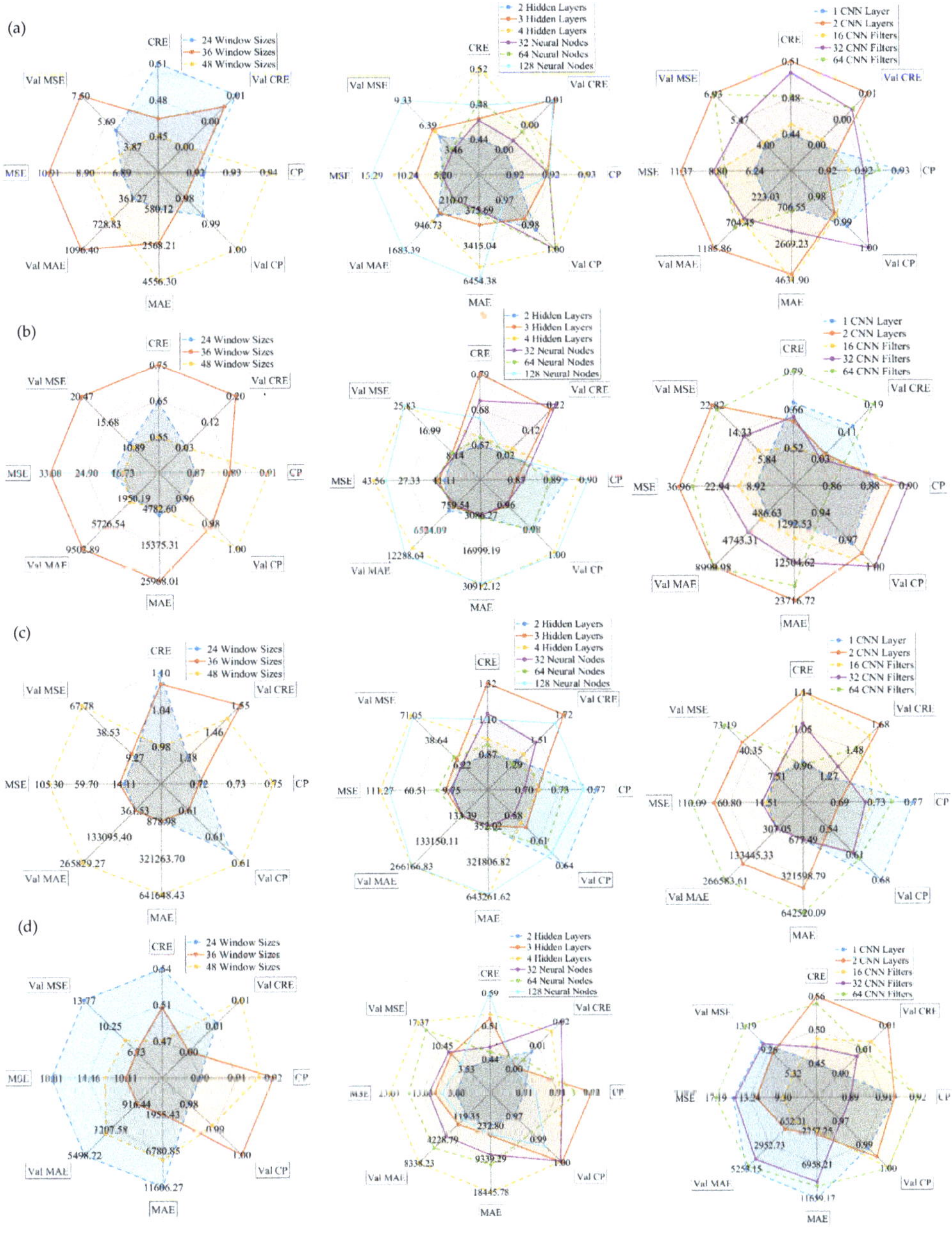

Figure 9. The effectiveness of varying window size, number of hidden layers and neural nodes, and number of CNN layers and filters on the CNN-GRU performance in different data collection; (**a**) the first row is for data collection 1; (**b**) the second row is for data collection 2; (**c**) the third row is for data collection 3; (**d**) the fourth row is for data collection 4.

For the first data collection, the 48-window size, 2 hidden layers, 32 neural nodes, 1 CNN layer, and 16 CNN filters achieve more robust and better performance compared with other configurations, as clearly illustrated in the first row of Figure 9. The effect of selecting these configurations is superior to other parameters in predicting the leakage current levels of data collection 1. The increasing window size tends to decrease CRE, validating CRE, MSE, and the validating MSE benchmarks, and increases the CP and validating CP indexes. In the GRU structures, the increase in hidden layers prefers raising the instability in predicting benchmarks, which extends the CRE, MSE, MAE, and validating MAE. Moreover, the addition of neural nodes gravitates the variability of predicting performance, which enlarges the MSE, validating MSE, MAE, validating MAE, and validating CRE benchmarks. For optimizing the CNN layers, the additional layers will extend all the error benchmarks and decrease the CP index, which illustrates the unreliability in categorizing the leakage currents. Furthermore, the growth of CNN filters gravitates toward the growth of validating MAE, validating SE, CRE, CP, and validating CRE benchmarks.

The varying performances of CNN-GRU in the second data collection are clearly described in the second row of Figure 9. The optimized window size, hidden layers, CNN layers, and CNN filters are identical to data collection 1. However, the 64 neural nodes achieved superior performance compared with the 32 and 128 configurations. The 128 neural nodes enlarge the MAE, validating MAE, MSE, and the validating MSE benchmarks, while the 32 neural nodes increase the CRE and validate CRE indexes. Therefore, the 48-window size, 2 hidden layers, 64 neural nodes, 1 CNN layer, and 16 CNN filter are the most optimized specification which obtains the highest achievement in distinguishing the leakage current of data collection 2.

For the third data collection, the decrease in window size prefers enhancing the predicting performance of the CNN-GRU model, which decreases the MAE, validating MAE, MSE, validating MSE, and validating CRE benchmarks. Therefore, the 24-window size obtains higher accuracy than other configurations in classifying the leakage current. In the GRU architecture, the 64 neural nodes obtain the lowest CRE, validating CRE, MAE, validating MAE, MSE, and validating MSE criteria compared with 32 and 128 neural nodes. For optimizing the CNN construction, the 32 filters achieve the highest accuracy with a better error benchmark and higher CP index compared with the 16 and 64 filters, as depicted in the third row of Figure 9. Therefore, the 24-window size, 2 hidden layers, 64 neural nodes, 1 CNN layer, and 32 CNN filters are considered optimized parameters for the CNN-GRU model in data collection 3.

In final data collection, 4, the 36-window size acquires the lower CRE, validating CRE, MAE, validating MAE, MSE, and validating MSE standards and secures higher CP and validating CP criteria, as represented in the fourth row of Figure 9. In the GRU structure, the 128 neural nodes obtain the most stable benchmarks compared with 32 and 64 neural nodes in validating CRE, MAE, validating MAE, MSE, and validating MSE. Consequently, the 36-window size, 2 hidden layers, 128 neural nodes, 1 CNN layer, and 16 CNN filters are selected as optimized structures in classifying the leakage current of data collection 4.

Hyperparameter optimization is utilized for evaluating the efficacity of varying setting parameters on the accuracy of CNN-GRU in classifying the leakage current with different data collection groups. The optimized specification of CNN-GRU for each data collection is obtained to achieve the highest performance. In the next section, the comparison between CNN-GRU and other deep learning methodologies is illustrated and analyzed.

4. Evaluating Performance between Algorithms

This section presents the comparative consequence between the CNN-GRU and another methodology, as illustrated in Figure 10. For the first data collection, the proposed CNN-GRU is demonstrated to obtain better precision with minor expansion in the MAE, validating MAE, MSE, and validating MSE. In the bar chart, the CNN-GRU acquires more maximum improvements compared with other methodologies, which attain 50.84% higher CRE, 8.58% higher CP, 5.88% higher validating CP, 30.42% higher MAE, and 71.59% higher validating MSE benchmarks than the GRU algorithm; 0.19% higher validating CRE and 63.59% higher validating MAE benchmarks than the LSTM methodology; and 65.96%

higher MSE criterion than the Bi-GRU technique. Therefore, in data collection 1, the proposed CNN-GRU achieves better prediction performance compared with other algorithms.

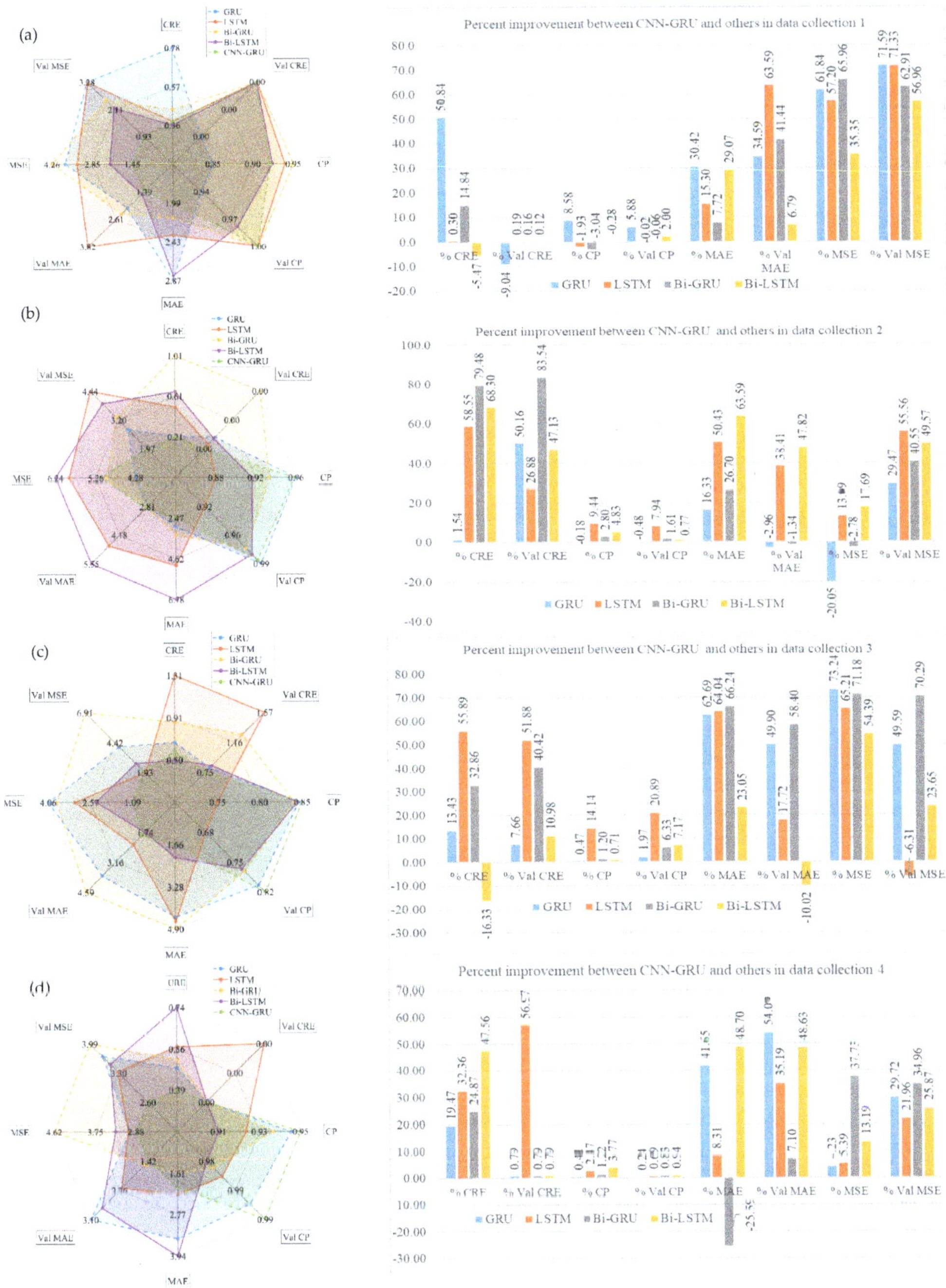

Figure 10. The comparative charts between the proposed algorithm and others: radar map and bar charts for (**a**) data collection 1; (**b**) data collection 2; (**c**) data collection 3; and (**d**) data collection 4.

In the second collection, the CNN-GRU proves the efficiency in categorizing the leakage current. In the radar figure, the CNN-GRU obtains a narrow expansion in the CRE, validating CRE, MAE, validating MAE, and validating MSE. The bar chart describes that the CNN-GRU accomplishes the most significant enhancements in comparison to other methodologies, which produces better improvement of 79.48% CRE and 83.54% validating CRE benchmarks than the Bi-GRU approach; 9.44% higher CP, 7.94% higher validating CP, and 55.56% higher validating MSE indexes than the LSTM algorithm; and 63.59% higher validating MAE, 47.82% higher validating MAE, and 17.69% higher MSE than the Bi-LSTM method. The combination between the CNN layers and the GRU method could retain the original information, which increases the accurate performance in categorizing the leakage current.

In the third data accumulation, the CNN-GRU produces a narrower expansion than other methodologies in the validating CRE, MAE, and MSE benchmarks and superior spread in the CP and validating CP indexes. Moreover, the developed algorithm obtains the most critical enhancements compared with other methods, which achieve 55.89% higher CRE, 51.88% higher validating CRE, 14.14% higher CP, and 20.89% higher validating CP benchmarks compared with the LSTM method; 66.24% higher MAE, 58.40% higher validating MAE, and 70.29% higher validating MSE compared with the Bi-GRU algorithm; and a more remarkable 73.24% MSE index compared with the GRU approach. Therefore, in the third accumulative data, the CNN-GRU outperforms the accurate indexes and performing benchmarks compared with GRU, LSTM, Bi-GRU, and Bi-LSTM algorithms.

In the last data accumulation, the developed CNN-GRU accomplishes the highest performance in the CRE, CP, validating CP, validating MAE, MSE, and validating MSE benchmarks with smaller spread areas. Furthermore, the CNN-GRU performs larger spreads in the CP and the validating CP indexes. In comparison to the Bi-LSTM method, the CNN-GRU accomplishes better improvements which are 47.56% CRE, 3.77% CP, 0.94 validating CP, and 48.70% MAE benchmarks. The developed algorithm also acquires a 56.97% improvement in validating CRE compared with the LSTM method and a 54.09% enhancement in validating MAE compared with the GRU algorithm. In the MSE and validating MSE metrics, the CNN-GRU outperforms other algorithms with significant improvements of 37.73% and 34.96%. Therefore, the effect of utilizing the hybrid CNN layers and GRU algorithms for the insulator's leakage current prediction is more accomplished than the other four methods.

Compared across different accumulated data, the proposed CNN-GRU algorithm defeats the GRU method in different error benchmarks, which acquires the most extreme enhancements of 50.84% CRE, 50.16% validating CRE, 8.58% CP, 5.88% validating CP, 62.69% MAE, 49.90% validating MAE, 73.24% MSE, and 71.59% validating MSE benchmarks. The additional CNN layers prove greater efficacity and higher precision in distinguishing the leakage current, which maintains the original information while decreasing the dimension of the input matrixes.

Compared through predicting technologies, the developed methodology surpasses the LSTM approach, which obtains significant improvements of 58.55% CRE, 56.97% validating CRE, 14.14% CP, 20.89% validating CP, 64.04% MAE, 63.59% validating MAE, 65.21% MSE, and 71.33% validating MSE metrics. Compared with the Bi-GRU algorithm, the CNN-GRU surpasses with supreme improvements of 79.48% CRE, 83.54% validating CRE, 2.8% CP, 6.33% validating CP, 66.24% MAE, 58.40% validating MAE, 71.18% MSE, and 70.29% validating MSE benchmarks. The additional CNN layers outperform the bidirectional features when combined with the GRU algorithm, which achieves higher accuracy and greater performance. In comparison with the Bi-LSTM approach, the CNN-GRU overwhelms all the evaluating metrics, which achieve maximum augmentations of 68.30% CRE, 47.13% validating CRE, 4.83% CP, 7.17% validating CP, 63.59% MAE, 48.63% validating MAE, 54.39% MSE, and 56.96% validating MSE metrics. Therefore, the proposed hybrid CNN and GRU outperform the original GRU and the Bi-GRU algorithms in distinguishing the insulator's leakage current.

In general, the developed CNN-GRU acquired the most remarkable improvements of 79.48% CRE, 83.54% validating CRE, 14.14% CP, 20.89% validating CP, 66.24% MAE, 63.59% validating MAE, 73.24% MSE, and 71.59% validating MSE benchmarks, which greatly enhances the predicting ability and modeling precision in distinguishing the insulator's leakage current. Therefore, the hybrid CNN-GRU provides the proper guideline for leakage current level in practical engineering applications. Figure 11 illustrates the CRE metric in the testing data with different predicting methods. For most data collections, the developed hybrid algorithm provides the most optimum CRE benchmark, which demonstrates the higher performance and greater performance in distinguishing the leakage current of 15 kV insulators. The limitation of this research is the application of the specific 15 kV insulators. The proposed algorithm needs more evaluation in different kinds of high-voltage insulators for more comprehensive reviews.

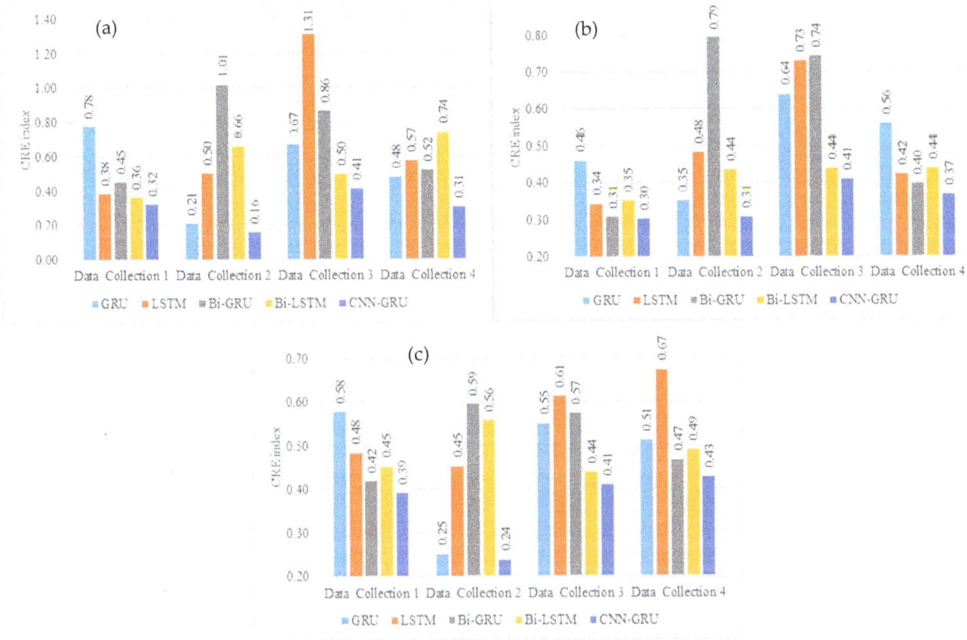

Figure 11. The comparative CRE index between predicting algorithms in testing data: (**a**) for 1 h in advance; (**b**) for 2 h in advance; (**c**) for 3 h in advance.

5. Conclusions

The categorizing leakage current for 15 kV insulators is critical for the safety and stability operation of the power system. An accurate prediction of leakage current is significant for contributing to appropriate maintenance schedules to prevent the flashover phenomenon in advance. This project develops the hybrid CNN and GRU methodology to categorize the leakage current levels through several group data collections. The simulation results validate the efficiency and precision of the proposed methods with other benchmark methodologies. This study evaluates the correlation between several weather parameters and the leakage current levels to maintain the most critical information as the input vector for the deep learning algorithms. The hyperparameters are employed with different evaluation indexes for constructing the higher forecasting accuracy and stability of the CNN-GRU algorithms. The optimized structures of the proposed CNN-GRU algorithm in different data collections are identified in this research for future reference. The simulation outcomes emphasize the influential combined CNN and GRU with the most remarkable improvements of 79.48% CRE, 83.54% validating CRE, 14.14% CP, 20.89% validating CP, 66.24% MAE, 63.59% validating MAE, 73.24% MSE, and 71.59% validating MSE benchmarks, compared with traditional GRU, LSTM, Bi-GRU, and Bi-LSTM methodologies. The additional CNN layers outperform the original GRU and the Bi-GRU algorithms in distinguishing the insulator's leakage current, which achieve the maximum improvements of 79.48%, 83.54%, 8.58%, 5.88%, 66.24%, 58.40%, 73.24%, and 71.59% for CRE, CP, MAE, and MSE in the training and validating operations, respectively. The developed hybrid CNN-GRU technology provides comprehensive information about the contamination degrees of insulator surfaces derived from the property of leakage currents. Therefore, this study completely fills the research gap by deploying the hybrid CNN and GRU algorithms in classifying the leakage current of 15 kV HDPE insulators.

Author Contributions: Methodology, W.-B.L.; Software, P.N.T.; Validation, Writing—review and editing, T.N.D.; Visualization, M.-Y.C. All authors have read and agreed to the published version of the manuscript.

Funding: This research received no external funding.

Institutional Review Board Statement: Not applicable.

Informed Consent Statement: Not applicable.

Data Availability Statement: Data available on request due to restrictions.

Conflicts of Interest: The authors declare no conflict of interest.

References

1. Ramirez, I.; Hernandez, R.; Montoya, G. Measurement of Leakage Current for Monitoring the Performance of Outdoor Insulators in Polluted Environments. *IEEE Electr. Insul. Mag.* **2012**, *28*, 29–34. [CrossRef]
2. Amarh, F.; Karady, G.G.; Sundararajan, R. Linear Stochastic Analysis of Polluted Insulator Leakage Current. *IEEE Trans. Power Deliv.* **2002**, *17*, 1063–1069. [CrossRef]
3. Salem, A.A.; Abd-Rahman, R.; Al-Gailani, S.A.; Salam, Z.; Kamarudin, M.S.; Zainuddin, H.; Yousof, M.F.M. Risk Assessment of Polluted Glass Insulator Using Leakage Current Index under Different Operating Conditions. *IEEE Access* **2020**, *8*, 175827–175839. [CrossRef]
4. Salem, A.A.; Abd-Rahman, R.; Al-Gailani, S.A.; Kamarudin, M.S.; Ahmad, H.; Salam, Z. The Leakage Current Components as a Diagnostic Tool to Estimate Contamination Level on High Voltage Insulators. *IEEE Access* **2020**, *8*, 92514–92528. [CrossRef]
5. Du, B.X.; Liu, Y.; Liu, H.J.; Yang, Y.J. Recurrent Plot Analysis of Leakage Current for Monitoring Outdoor Insulator Performance. *IEEE Trans. Dielectr. Electr. Insul.* **2009**, *16*, 139–146. [CrossRef]
6. Fierro-Chavez, J.L.; Ramirez-Vazquez, Z.; Montoya-Tena, G. On-Line Leakage Current Monitoring of 400 KV Insulator Strings in Polluted Areas. *IEE Proc.-Gener. Transm. Distrib.* **1996**, *143*, 560–564. [CrossRef]
7. Amin, M.; Amin, S.; Ali, M. Monitoring of Leakage Current for Composite Insulators and Electrical Devices. *Rev. Adv. Mater. Sci.* **2009**, *21*, 75–89.
8. Fontana, E.; Oliveira, S.C.; Cavalcanti, F.J.M.M.; Lima, R.B.; Martins-Filho, J.F.; Meneses-Pacheco, E. Novel Sensor System for Leakage Current Detection on Insulator Strings of Overhead Transmission Lines. *IEEE Trans. Power Deliv.* **2006**, *21*, 2064–2070. [CrossRef]
9. Suda, T. Frequency Characteristics of Leakage Current Waveforms of an Artificially Polluted Suspension Insulator. *IEEE Trans. Dielectr. Electr. Insul.* **2001**, *8*, 705–709. [CrossRef]
10. Chandrasekar, S.; Kalaivanan, C.; Cavallini, A.; Montanari, G.C. Investigations on Leakage Current and Phase Angle Characteristics of Porcelain and Polymeric Insulator under Contaminated Conditions. *IEEE Trans. Dielectr. Electr. Insul.* **2009**, *16*, 574–583. [CrossRef]
11. Douar, M.A.; Mekhaldi, A.; Bouzidi, M.C. Flashover Process and Frequency Analysis of the Leakage Current on Insulator Model under Non-Uniform Pollution Conditions. *IEEE Trans. Dielectr. Electr. Insul.* **2010**, *17*, 1284–1297. [CrossRef]
12. Montoya, G.; Ramirez, I.; Montoya, J.I. Correlation among ESDD, NSDD and Leakage Current in Distribution Insulators. *IEE Proc. Gener. Transm. Distrib.* **2004**, *151*, 334–340. [CrossRef]
13. Ahmadi-Joneidi, I.; Shayegani-Akmal, A.A.; Mohseni, H. Leakage Current Analysis of Polymeric Insulators under Uniform and Non-uniform Pollution Conditions. *IET Gener. Transm. Distrib.* **2017**, *11*, 2947–2957. [CrossRef]
14. Li, J.Y.; Sun, C.X.; Sebo, S.A. Humidity and Contamination Severity Impact on the Leakage Currents of Porcelain Insulators. *IET Gener. Transm. Distrib.* **2011**, *5*, 19–28. [CrossRef]
15. Li, J.; Sima, W.; Sun, C.; Sebo, S.A. Use of Leakage Currents of Insulators to Determine the Stage Characteristics of the Flashover Process and Contamination Level Prediction. *IEEE Trans. Dielectr. Electr. Insul.* **2010**, *17*, 490–501. [CrossRef]

16. Zhao, L.; Jiang, J.; Duan, S.; Fang, C.; Wang, J.; Wang, K.; Cao, P.; Zhou, J. The Prediction of Post Insulators Leakage Current from Environmental Data. In Proceedings of the 2011 International Conference on Electrical and Control Engineering, Yichang, China, 16–18 September 2011; pp. 5103–5106.
17. Jahromi, A.N.; El-Hag, A.H.; Jayaram, S.H.; Cherney, E.A.; Sanaye-Pasand, M.; Mohseni, H. A Neural Network Based Method for Leakage Current Prediction of Polymeric Insulators. *IEEE Trans. Power Deliv.* **2006**, *21*, 506–507. [CrossRef]
18. El-Hag, A.H.; Jahromi, A.N.; Sanaye-Pasand, M. Prediction of Leakage Current of Non-Ceramic Insulators in Early Aging Period. *Electr. Power Syst. Res.* **2008**, *78*, 1686–1692. [CrossRef]
19. Medeiros, A.; Sartori, A.; Stefenon, S.F.; Meyer, L.H.; Nied, A. Comparison of Artificial Intelligence Techniques to Failure Prediction in Contaminated Insulators Based on Leakage Current. *J. Intell. Fuzzy Syst.* **2021**, *42*, 3285–3298. [CrossRef]
20. Thanh, P.N.; Cho, M.-Y.; Da, T.N. Insulator Leakage Current Prediction Using Surface Spark Discharge Data and Particle Swarm Optimization Based Neural Network. *Electr. Power Syst. Res.* **2021**, *191*, 106888. [CrossRef]
21. de Santos, H.; Sanz-Bobi, M.Á. A Cumulative Pollution Index for the Estimation of the Leakage Current on Insulator Strings. *IEEE Trans. Power Deliv.* **2020**, *35*, 2438–2446. [CrossRef]
22. Jahromi, A.N.; El-Hag, A.H.; Cherney, E.A.; Jayaram, S.H.; Sanaye-Pasand, M.; Mohseni, H. Prediction of Leakage Current of Composite Insulators in Salt Fog Test Using Neural Network. In Proceedings of the CEIDP'05 2005 Annual Report Conference on Electrical Insulation and Dielectric Phenomena, Nashville, TN, USA, 16–19 October 2005; pp. 309–312.
23. Pinotti, M.A.; Meyer, L.H. Mathematical Model for Prediction of the Leakage Current on Distribution Insulators of 25 KV Class. In Proceedings of the 2017 IEEE Electrical Insulation Conference (EIC), Baltimore, MD, USA, 11–14 June 2017; pp. 256–260.
24. Kordkheili, H.H.; Abravesh, H.; Tabasi, M.; Dakhem, M.; Abravesh, M.M. Determining the Probability of Flashover Occurrence in Composite Insulators by Using Leakage Current Harmonic Components. *IEEE Trans. Dielectr. Electr. Insul.* **2010**, *17*, 502–512. [CrossRef]
25. Sidthik, A.S.; Kalaivani, L.; Iruthayarajan, M.W. Evaluation and Prediction of Contamination Level in Coastal Region Insulators Based on Leakage Current Characteristics. In Proceedings of the 2013 International Conference on Circuits, Power and Computing Technologies (ICCPCT), Nagercoil, India, 20–21 March 2013; pp. 132–137.
26. Salem, A.A.; Lau, K.Y.; Abdul-Malek, Z.; Mohammed, N.; Al-Shaalan, A.M.; Al-Shamma'a, A.A.; Farh, H.M.H. Polymeric Insulator Conditions Estimation by Using Leakage Current Characteristics Based on Simulation and Experimental Investigation. *Polymers* **2022**, *14*, 737. [CrossRef] [PubMed]
27. Volat, C.; Meghnefi, F.; Farzaneh, M.; Ezzaidi, H. Monitoring Leakage Current of Ice-Covered Station Post Insulators Using Artificial Neural Networks. *IEEE Trans. Dielectr. Electr. Insul.* **2010**, *17*, 443–450. [CrossRef]
28. Ghiasi, Z.; Faghihi, F.; Shayegani-Akmal, A.A. Artificial Neural Network Approach for Prediction of Leakage Current of Polymeric Insulator under Non-Uniform Fan-Shaped Contamination. *Electr. Power Syst. Res.* **2022**, *209*, 107920. [CrossRef]
29. Ali, H. Leakage Current Prediction for High Voltage Insulators Flashover Based on Extreme Value Theory. In Proceedings of the 2016 IEEE International Symposium on Computer, Consumer and Control, IS3C 2016, Xi'an, China, 4–6 July 2016; pp. 870–873.
30. Li, J.; Sun, C.; Sima, W.; Yang, Q.; Hu, J. Contamination Level Prediction of Insulators Based on the Characteristics of Leakage Current. *IEEE Trans. Power Deliv.* **2010**, *25*, 417–424. [CrossRef]
31. Vosloo, W.L.; Holtzhausen, J.P. The Prediction of Insulator Leakage Currents from Environmental Data. In Proceedings of the IEEE AFRICON. 6th Africon Conference in Africa, George, South Africa, 2–4 October 2002; Volume 2, pp. 603–608.
32. Al Khafaf, N.; El-Hag, A. Bayesian Regularization of Neural Network to Predict Leakage Current in a Salt Fog Environment. *IEEE Trans. Dielectr. Electr. Insul.* **2018**, *25*, 686–693. [CrossRef]
33. Wang, J.; Xi, Y.; Fang, C.; Cai, L.; Wang, J.; Fan, Y. Leakage Current Response Mechanism of Insulator String with Ambient Humidity on Days Without Rain. *IEEE Access* **2019**, *7*, 55229–55236. [CrossRef]
34. Deb, S.; Choudhury, N.R.; Ghosh, R.; Chatterjee, B.; Dalai, S. Short Time Modified Hilbert Transform-Aided Sparse Representation for Sensing of Overhead Line Insulator Contamination. *IEEE Sens. J.* **2018**, *18*, 8125–8132. [CrossRef]
35. Abouzeid, A.K.; El-Hag, A.; Assaleh, K. Equivalent Salt Deposit Density Prediction of Silicone Rubber Insulators under Simulated Pollution Conditions. *Electr. Power Compon. Syst.* **2018**, *46*, 1123–1133. [CrossRef]
36. Sun, J.; Zhang, H.; Li, Q.; Liu, H.; Lu, X.; Hou, K. Contamination Degree Prediction of Insulator Surface Based on Exploratory Factor Analysis-least Square Support Vector Machine Combined Model. *High Volt.* **2021**, *6*, 264–277. [CrossRef]
37. Yeh, C.-T.; Thanh, P.N.; Cho, M.-Y. Real-Time Leakage Current Classification of 15 kV and 25 kV Distribution Insulators Based on Bidirectional Long Short-Term Memory Networks with Deep Learning Machine. *IEEE Access* **2022**, *10*, 7128–7140. [CrossRef]
38. Nguyen, T.-P.; Yeh, C.-T.; Cho, M.-Y.; Chang, C.-L.; Chen, M.-J. Convolutional Neural Network Bidirectional Long Short-Term Memory to Online Classify the Distribution Insulator Leakage Currents. *Electr. Power Syst. Res.* **2022**, *208*, 107923. [CrossRef]
39. Benesty, J.; Chen, J.; Huang, Y.; Cohen, I. Pearson Correlation Coefficient. In *Noise Reduction in Speech Processing*; Springer: Berlin/Heidelberg, Germany, 2009; pp. 1–4.
40. Adler, J.; Parmryd, I. Quantifying Colocalization by Correlation: The Pearson Correlation Coefficient Is Superior to the Mander's Overlap Coefficient. *Cytom. Part A* **2010**, *77*, 733–742. [CrossRef]
41. Feurer, M.; Hutter, F. Hyperparameter Optimization. In *Automated Machine Learning*; Springer: Cham, Switzerland, 2019; pp. 3–33.
42. Refaeilzadeh, P.; Tang, L.; Liu, H. Cross-Validation. *Encycl. Database Syst.* **2009**, *5*, 532–538.
43. Rodriguez, J.D.; Perez, A.; Lozano, J.A. Sensitivity Analysis of K-Fold Cross Validation in Prediction Error Estimation. *IEEE Trans. Pattern Anal. Mach. Intell.* **2009**, *32*, 569–575. [CrossRef]

44. Hochreiter, S.; Schmidhuber, J. Long Short-Term Memory. *Neural Comput.* **1997**, *9*, 1735–1780. [CrossRef]
45. Graves, A. Long Short-Term Memory. In *Supervised Sequence Labelling with Recurrent Neural Networks*; Springer: Berlin/Heidelberg, Germany, 2012; pp. 37–45.
46. Chung, J.; Gulcehre, C.; Cho, K.; Bengio, Y. Empirical Evaluation of Gated Recurrent Neural Networks on Sequence Modeling. *arXiv* **2014**, arXiv:1412.3555.
47. Dey, R.; Salem, F.M. Gate-Variants of Gated Recurrent Unit (GRU) Neural Networks. In Proceedings of the 2017 IEEE 60th International Midwest Symposium on Circuits and Systems (MWSCAS), Medford, MA, USA, 6–9 August 2017; pp. 1597–1600.
48. Graves, A.; Schmidhuber, J. Framewise Phoneme Classification with Bidirectional LSTM and Other Neural Network Architectures. *Neural Netw.* **2005**, *18*, 602–610. [CrossRef]
49. Fukushima, K.; Miyake, S. Neocognitron: A Self-Organizing Neural Network Model for a Mechanism of Visual Pattern Recognition. In *Competition and Cooperation in Neural Nets*; Springer: Berlin/Heidelberg, Germany, 1982; pp. 267–285.
50. LeCun, Y.; Boser, B.; Denker, J.; Henderson, D.; Howard, R.; Hubbard, W.; Jackel, L. Handwritten Digit Recognition with a Back-Propagation Network. *Adv. Neural Inf. Process. Syst.* **1989**, *2*, 396–404.
51. Nair, V.; Hinton, G.E. Rectified Linear Units Improve Restricted Boltzmann Machines. In Proceedings of the 27th International Conference on Machine Learning (ICML-10), Haifa, Israel, 21–24 June 2010.
52. Boureau, Y.-L.; Ponce, J.; LeCun, Y. A Theoretical Analysis of Feature Pooling in Visual Recognition. In Proceedings of the 27th International Conference on Machine Learning (ICML-10), Haifa, Israel, 21–24 June 2010; pp. 111–118.
53. Wang, T.; Wu, D.J.; Coates, A.; Ng, A.Y. End-to-End Text Recognition with Convolutional Neural Networks. In Proceedings of the 21st International Conference on Pattern Recognition (ICPR2012), Tsukuba, Japan, 11–15 November 2012; pp. 3304–3308.
54. Bergstra, J.; Bardenet, R.; Bengio, Y.; Kégl, B. Algorithms for Hyper-Parameter Optimization. *Adv. Neural Inf. Process. Syst.* **2011**, *24*, 2546–2554.
55. Bergstra, J.; Bengio, Y. Random Search for Hyper-Parameter Optimization. *J. Mach. Learn. Res.* **2012**, *13*, 281–305.

energies

MDPI

Article

Regional Load Frequency Control of BP-PI Wind Power Generation Based on Particle Swarm Optimization

Jikai Sun [1], Mingrui Chen [2], Linghe Kong [1], Zhijian Hu [3] and Veerapandiyan Veerasamy [3,*]

1 College of Electrical Engineering, Qingdao University, Qingdao 266071, China
2 School of Artificial Intelligence and Automation, Huazhong University of Science and Technology, Wuhan 430074, China
3 School of Electrical and Electronic Engineering, Nanyang Technological University, Singapore 639798, Singapore
* Correspondence: veerapandiyan.v@ntu.edu.sg

Abstract: The large-scale integration of wind turbines (WTs) in renewable power generation induces power oscillations, leading to frequency aberration due to power unbalance. Hence, in this paper, a secondary frequency control strategy called load frequency control (LFC) for power systems with wind turbine participation is proposed. Specifically, a backpropagation (BP)-trained neural network-based PI control approach is adopted to optimize the conventional PI controller to achieve better adaptiveness. The proposed controller was developed to realize the timely adjustment of PI parameters during unforeseen changes in system operation, to ensure the mutual coordination among wind turbine control circuits. In the meantime, the improved particle swarm optimization (IPSO) algorithm is utilized to adjust the initial neuron weights of the neural network, which can effectively improve the convergence of optimization. The simulation results demonstrate that the proposed IPSO-BP-PI controller performed evidently better than the conventional PI controller in the case of random load disturbance, with a significant reduction to near 10 s in regulation time and a final stable error of less than 10^{-3} for load frequency. Additionally, compared with the conventional PI controller counterpart, the frequency adjustment rate of the IPSO-BP-PI controller is significantly improved. Furthermore, it achieves higher control accuracy and robustness, demonstrating better integration of wind energy into traditional power systems.

Keywords: wind power generation; sudden load disturbance; load frequency control; BP neural network; particle swarm optimization algorithm

Citation: Sun, J.; Chen, M.; Kong, L.; Hu, Z.; Veerasamy, V. Regional Load Frequency Control of BP-PI Wind Power Generation Based on Particle Swarm Optimization. *Energies* **2023**, *16*, 2015. https://doi.org/10.3390/en16042015

Academic Editor: Adrian Ilinca

Received: 25 January 2023
Revised: 13 February 2023
Accepted: 15 February 2023
Published: 17 February 2023

1. Introduction

With the increasing energy crisis and environmental pollution of today's society, the development of renewable energy resources is gradually gaining attention all over the world. Wind energy is gradually recognized as one of the most essential and promising energy sources due to its advantages of clean environmental protection and high feasibility [1,2]. While wind power penetration gradually increases, its volatility and unpredictability bring great challenges to the stable operation of the wind power system. Compared with traditional thermal generating units, the method of using wind turbine units to provide electric power can often quickly respond to the change in load frequency. In addition, it is more suitable to achieve the frequency regulation of the power system. Therefore, it is crucial to explore an advanced control strategy to address the negative effects of load frequency fluctuation.

In order to realize stable load frequency control (LFC) for power systems with wind turbine integration and effectively improve the frequency regulation capability of wind power regions, researchers have carried out extensive research in related fields and have achieved a series of advanced results [3–9]. In [3], the authors proposed a robust control strategy based on equivalent input disturbance (EID) that uses equivalent input disturbance to compensate the imbalance between power generation and load. This improves the anti-disturbance performance of the power system in wind farms. However, at present, robust control cannot be accurately guaranteed to work in the optimal state for a long time [4]. Considering the existence of multiple energy sources serving local areas, the whale optimization algorithm (WOA) is used to optimize fuzzy integral PI controllers (FIPIs); then, the LFC problem in some regions and all regions in the interconnected power system of heat–water–wind areas are analyzed under the condition of sudden disturbance, which proves the superiority of the control method [5]. Additionally, an H_2/H_∞ load frequency robust multi-objective TS fuzzy control method is applied in multi-area interconnected power systems, and a distributed compensation scheme is used to design the entire control system. Simulation data results show the robustness and effectiveness of this method when dealing with load disturbance, model uncertainty, transmission delay, and model nonlinear influence [4]. In [6], an ANFIS controller with artificial neural network was designed considering the strength of the nonlinear characteristics of doubly fed induction (DFIG) wind turbines and was successfully applied to the wind power generation system in two regions. The simulation results show that the controller was helpful in reducing overkill and shortening the regulation time of the nonlinear wind power generation system. The study [7] designed a load frequency controller based on PSO-MPC, which not only solved the randomness of wind turbine generation but also reduced the complexity of traditional MPC calculation, with fast response speed and high stability. However, this control method requires high precision of the model, and its continuous stability cannot be guaranteed. To achieve better system performance, the sliding mode control (SMC) strategy was used to analyze the frequency deviation and tie line power change of the interconnected power system in two regions under the disturbance of a fixed load, and a comparison with the traditional integral controller was made. The results show that the control method improved the response of the main loop to a great extent and that the frequency control effect was evident in case of overshoot and undershoot. Moreover, it can be widely used in situations where wind and conventional power generation work together [8]. The study [9] applied an optimal fuzzy PID sagging controller with adaptive and self-tuning functions to the structure of wind turbines, which effectively compensated for the decrease in the total inertia of the power system with the participation of the wind farm. At the same time, the artificial bee colony algorithm was used to optimize the membership function of input and output signals based on the multi-objective function. The simulation results show that the control method is reasonable, with short stabilizing time and high frequency control precision. In the paper [10], the researchers presented a data-driven and model-free frequency control method based on deep reinforcement learning (DRL) in the continuous-motion domain, which greatly differs from the traditional model-based frequency control method. It is more adaptable to the dynamics of the unmodeled system and has obvious advantages in solving the problem of renewable power supply. Recently, some researchers investigated an asynchronous tracking control method for amplitude signals and proved that the stochastically stable ability is significant in tracking error systems [11]. The distributed LFC method uses a modular, distributed architecture, where each region has an independent control system and is under the unified control of a central control system, so it is highly flexible and coordinated. The interconnection of signals from multiple regions solves, to a certain extent, the problem of the failure of part of a single system leading to the collapse of the frequency control of the entire power system. However, this control method has high technical requirements and a large communication volume, which also brings some problems, such as data security and confidentiality [12–14]. Recently, designers proposed a static/dynamic event-triggered and self-triggered discrete gain scheduled controller with designable parametric minimal interevent time (MIET) to effectively achieve the semi-global stabilization of a linear system with input constraints, which can save communication resources and avoid the monitoring of all states, which gives us an idea of how to solve the above problem [15].

LFC, as one of the important means to maintain the balance of power systems, plays an indispensable role in the stable operation of power systems. At present, LFC still adopts PI control to realize the normal operation of AGC, that is, traditional PI control is used to adjust the wind turbine within the specified output adjustment range, track the instructions issued by the power dispatching and trading institutions, and adjust the power output in real time according to a certain adjustment rate, so as to meet the frequency adjustment of the power system. Although traditional PI control can meet the basic requirements of frequency regulation in wind power regions to a certain extent when the power system is subjected to random disturbance, the load frequency can fluctuate. Thus, the balance between the power generation and consumption of the power system can be easily broken, which has negative effects on the stable operation of the whole power system. Therefore, the adaptability of frequency control of wind power generation systems needs to be further improved [16].

In recent years, modern artificial intelligence technology has been broadly utilized in the field of LFC. Mainstream expert systems [17], artificial neural networks [18–21], and fuzzy logic [22–24] have gradually replaced the classical control approaches and continuously penetrated every link of electric automation control, providing guarantees for the safe, secure, and stable operation of power systems. In this paper, an improved particle swarm optimization algorithm is proposed to optimize the initial neuron weights of the BP neural network (BPNN), and it is used to adjust PI control parameters in real time. This research is very adaptive compared with some existing secondary frequency regulation methods. The fixed gain parameter characteristic of the conventional method is altered in order to facilitate the appropriate frequency response for different load requirements [25]. At the same time, the method is less informative and more stable and has good application prospects to effectively improve the efficiency of wind energy utilization. The main innovations of this work lie in two aspects:

(1) The BP neural network algorithm has the defects of slow convergence speed and local minimization, which are mainly due to the random selection of initial weights [26]. In this paper, the optimal initial neuron connection weights of the BP-PI controller are determined by combining IPSO with its fast convergence speed and global optimization features. Thus, the individual parameters of the PI controller are continuously adjusted by the improved BPNN algorithm, which can achieve better dynamic performance.
(2) Coordinating modern artificial intelligence control with traditional PI control can effectively improve the efficiency and accuracy of an algorithm [27]. The proposed strategy applied to the wind turbine can effectively increase the anti-interference ability of the wind power region, enhance the stability of the power system, and thus have a promising development prospect in the field of new energy power generation.

The rest of this paper is structured as follows: Section 2 describes the modeling process of the involvement of the wind turbine in LFC. Section 3 illustrates the improved particle swarm algorithm for optimizing BP neural networks and the application of the PI controller. Section 4 uses simulation results to verify the capability of the IPSO-BP-PI controller. Moreover, Section 5 presents a concise statement of the final conclusions.

2. System Dynamics

In this paper, a doubly fed induction generator (DFIG) is used as the generating output of a wind turbine. In the local area, electrical energy is generated by the DFIG, and its detailed structure is shown in Figure 1. The simplified mathematical model of the wind turbine model is discussed in the below, and the detailed parameters are given in Table 1. The main source of these equations is [14], and we made appropriate adjustments to the form and dimensionality of some equations.

$$\dot{i}_{qr} = -\left(\frac{1}{T_1}\right)i_{qr} + \left(\frac{X_2}{T_1}\right)V_{qr} \tag{1}$$

$$\dot{w}_r = -\left(\frac{X_3}{2H_t}\right)i_{qr} + \left(\frac{1}{2H_t}\right)T_m \tag{2}$$

$$P_e = w_r X_3 i_{qr} \tag{3}$$

$$T_e = i_{qs} = -\frac{L_m}{L_s} i_{qr} \tag{4}$$

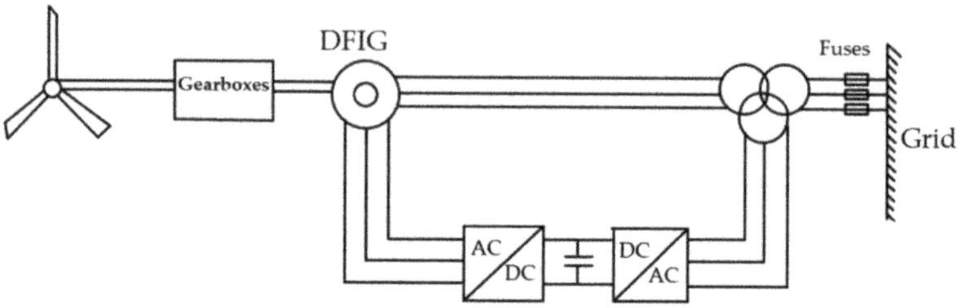

Figure 1. Structure of a doubly fed wind turbine.

Table 1. Parameters of power system.

Parameter	Physical Significance
f	Nominal system frequency of the power system
P_{mech}	Mechanical power of gas turbine
P^L	Sudden load disturbance
P_v	Steam valve position
P_e	Supplementary control action
M^a	Angular momentum
T_G	Speed governor time constant
T_{CH}	Changing time constant (prime mover)
P_{ref}	Reference setpoint
D	Equivalent damping coefficient of generator
β_0	Frequency bias constant
y	System measurement output

By linearizing the Taylor series, we can reduce (3) to

$$P_e = w_{opt} X_3 i_{qr} \tag{5}$$

where w_{opt} is the rotational speed operating point of the wind turbine, T_e is the electric revolution, T_m is the mechanical torque, i_{qr} is the q-axis component of the rotor current, i_{qs} is the q-axis component of the stator current, V_{qr} is the q-axis component of the rotor voltage, H_t is the equivalent inertia constant of the wind turbine, and w_r is the rotational speed of the wind turbine.

$$X_2 = \frac{1}{R_r},\ X_3 = \frac{L_m}{L_s},\ T_1 = \frac{L_0}{w_s R_r}$$

$$L_0 = L_r + \frac{L_m^2}{L_S},\ L_s = L_{ls} + L_m,\ L_r = L_{lr} + L_m \tag{6}$$

where w_s is the synchronous speed of the wind turbine, L_m is the magnetized inductance, L_{lr} is the rotor leakage inductance, L_{ls} is the stator leakage inductance, L_r is the rotor self-inductance, L_s is the stator self-inductance, and R_r is the rotor resistance.

The frequency deviation of the wind power region can be described by the following dynamic equation:

$$\Delta\dot{f} = -\frac{D}{M^a}\Delta f + \frac{1}{M^a}\Delta P_m + \frac{1}{2M^a}\Delta P_e - \frac{1}{M^a}\Delta P^L \tag{7}$$

The turbine dynamic in the wind power region is depicted as

$$\Delta\dot{P_m} = -\frac{1}{T_{CH}}\Delta P_m + \frac{1}{T_{CH}}\Delta P_v \tag{8}$$

The governor dynamic of the wind power region can be given as

$$\Delta\dot{P_v} = -\frac{1}{T_G}\Delta f - \frac{1}{T_G}\Delta P_v + \frac{1}{T_G}\Delta P_{ref} \tag{9}$$

The area control error of the region is given as

$$ACE = \beta_0 \Delta f \tag{10}$$

According to the above expressions, the frequency response of the system under load disturbance $z = \Delta P^L$ can be described in the form of the following state space equation [14]:

$$\dot{x} = Ax + Bu + Ez \tag{11}$$

$$y = Cx \tag{12}$$

$$A = \begin{bmatrix} -\frac{D}{M^a} & \frac{1}{M^a} & 0 & -\frac{X_3 W_{opt}}{M^a} \\ 0 & -\frac{1}{T_{CH}} & \frac{1}{T_{CH}} & 0 \\ -\frac{1}{T_G} & 0 & -\frac{1}{T_G} & 0 \\ 0 & 0 & 0 & -\frac{1}{T_1} \end{bmatrix}, \; B = \begin{bmatrix} 0 & 0 \\ 0 & 0 \\ \frac{1}{T_G} & 0 \\ 0 & \frac{X_2}{T_1} \end{bmatrix} \tag{13}$$

$$E = \begin{bmatrix} -\frac{1}{M^a} \\ 0 \\ 0 \\ 0 \end{bmatrix}, \; C = \begin{bmatrix} \beta_0 & 0 & 0 & 0 \end{bmatrix} \tag{14}$$

$$x = \begin{bmatrix} \Delta f \\ \Delta P_m \\ \Delta P_v \\ \Delta i_{qr} \end{bmatrix}, \; u = \begin{bmatrix} \Delta P_{ref} \\ \Delta V_{qr} \end{bmatrix} \tag{15}$$

where x represents the state of the system, u represents the control input, and y represents the output of the system.

3. Frame and Algorithm of Wind Turbine

3.1. Particle Swarm Optimization Analysis

3.1.1. Basic Particle Swarm Optimization Algorithm

Particle swarm optimization is widely used as a classical swarm intelligent optimization algorithm. The algorithm simulates the predation behavior of a flock of birds. Each individual in the flock is regarded as a particle with velocity and position attributes. Each particle searches for the optimal solution of the individual in the search space and shares the optimal solution of the individual with the group to obtain the global optimal solution of the current state. At the same time, each particle also adjusts its flight speed and direction according to the optimal solution information of the population and eventually forms a new population. Compared with other random search algorithms, the PSO algorithm does not have too many adjustable parameters. It is simple and easy to implement and has high precision and fast convergence, which can achieve good application effects in LFC for power system fields [28,29].

In the D-dimensional search space, a population contains N particles, and the position of the ith particle is expressed as $X_i = (x_{i1}, x_{i2} \dots x_{iD})$ in the D-dimensional space, with velocity $V_i = (v_{i1}, v_{i2} \dots v_{iD})$, $i = 1, 2, \dots N$. At the same time, the iteration formula is used to update the velocity and position of the particles, and the individual and population optimal values of the next generation of particles are obtained.

$$V_{id}(k+1) = wV_{id}(k) + c_1r_1(P_{id}(k) - X_{id}(k)) + c_2r_2(P_{gd}(k) - X_{id}(k) \quad (16)$$

$$X_{id}(k+1) = X_{id}(k) + V_{id}(k+1) \quad (17)$$

where w is the inertia factor, r_1 and r_2 are random numbers in [0, 1], c_1 and c_2 are learning factors, and P_i and P_g are the individual optimal position and the global optimal position of the iteration up to the present.

3.1.2. Improvement in Particle Swarm Optimization Algorithm

The traditional PSO algorithm still has the risk of non-convergence, so this paper adopts an improved strategy for the PSO algorithm, that is, introducing the iterative process of asynchronous learning factors. By improving the learning factors, the global search ability in the early stage is balanced with the local search ability in the late stage. We select larger c_1 with smaller c_2 in the early stage, enhance the global search ability, and avoid falling into the local optimum. In the later stage, the opposite happens, i.e., we enhance the local search ability and accurately obtain the global optimal solution.

$$c_1(t+1) = c_{1max} - (c_{1max} - c_{1\min})\frac{t}{t_m} \quad (18)$$

$$c_2(t+1) = c_{2max} - (c_{2max} - c_{2\min})\frac{t}{t_m} \quad (19)$$

where t is the current iteration number, t_m is the maximum iteration number, c_{1max} and c_{2max} are the maximum learning factors, and $c_{1\min}$ and $c_{2\min}$ are the minimum learning factors.

3.2. Principle and Framework of BP-PI Control Algorithm

Traditional PI control is one of the most widely used approaches in power industries, and refers to the adjustment of the controlled object by incorporating two fixed parameters through proportion. This paper selects a self-tuning BP-PI controller, which is possible with the participation of an artificial intelligence control algorithm. The controller constantly adjusts the weight of the neural network to achieve the optimal combination of PI gain output by the neural network, which is shown in Figure 2. In the discrete state, the mathematical formulation of PI control is described by

$$u(k) = K_P \times e(k) + K_I \times \sum e(k) \quad (20)$$

where K_P and K_I are the proportion and integration coefficients and $e(k)$ represents the deviation of system state input and output at time k.

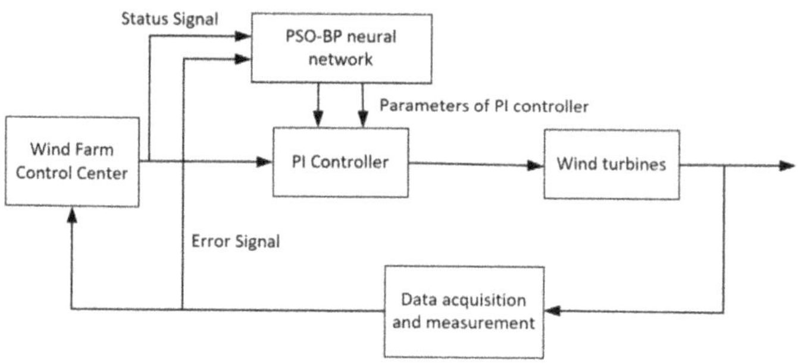

Figure 2. IPSO-BP-PI controller of the power system.

For LFC with wind turbine participation, since the controller achieves the control effect in the mode of multi-input and multi-output, improved PI control based on the above state space equation is written as

$$u_1(k) = Q_A \times e(k) + R_A \times \sum e(k) \tag{21}$$

$$u_2(k) = Q_B \times e(k) + R_B \times \sum e(k) \tag{22}$$

Q_A, Q_B, R_A, and R_B are expressed as

$$\begin{aligned} Q_A &= \begin{bmatrix} K_{Pa1} & K_{Pa2} & K_{Pa3} & K_{Pa4} \end{bmatrix},\ Q_B = \begin{bmatrix} K_{Pb1} & K_{Pb2} & K_{Pb3} & K_{Pb4} \end{bmatrix}, \\ R_A &= \begin{bmatrix} K_{Ia1} & K_{Ia2} & K_{Ia3} & K_{Ia4} \end{bmatrix},\ R_B = \begin{bmatrix} K_{Ib1} & K_{Ib2} & K_{Ib3} & K_{Ib4} \end{bmatrix} \end{aligned} \tag{23}$$

where Q_A and Q_B represent the proportional coefficient matrices of PI control, R_A and R_B represent the integral coefficient matrices of PI control, $u_1(k)$ represents the control quantity of valve position deviation quantity ΔP_{ref} at moment k, $u_2(k)$ represents the control quantity of voltage ΔV_{qr} at moment k, and $e(k)$ is the deviation of input and output of each state quantity in the system at time k.

The BP-trained neural network (BPNN) is a multilayer feedforward network trained by an error backpropagation algorithm. Its main characteristics are the forward propagation of signal and the backpropagation of error. Its powerful nonlinear mapping capability enables the BPNN to approximate any nonlinear continuous function with arbitrary precision. At the same time, BPNN has the functions of self-adaptation and self-learning, and the learning content is memorized in the network weight. In addition, the weight of the network is adjusted by means of backpropagation; finally, the network error sum of squares is minimized. In this paper, a three-layer BPNN is used to constantly adjust the weight; finally, the optimal PI parameter matrix combination is obtained. The network structure of the BPNN is shown in Figure 3.

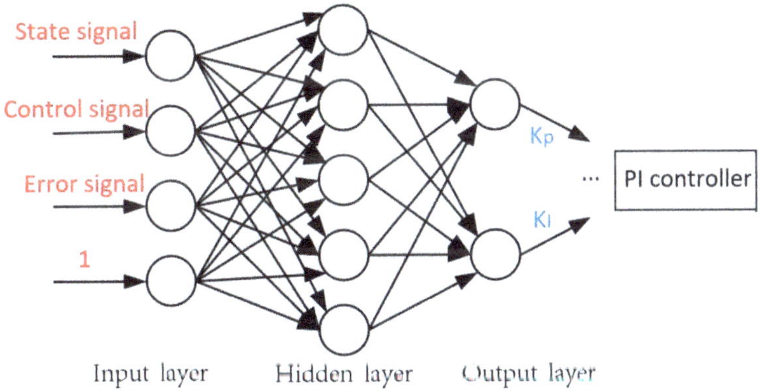

Figure 3. Structure of BP neural network.

A group of state quantity x and control quantity u, as well as the deviation quantity error and adjustment value 1 of input and output, are selected each time as the input of the BPNN. The structure of the BPNN adopts a 4-5-2 structure. The neural network outputs a group of parameter values of K_p and K_i by learning each time, and a total of 8 groups of values of K_p and K_i are output. The corresponding control effect can be obtained by applying the values of K_p and K_i to the PI controller.

We set the input variable of the BPNN as x_j; then, the input–output relationship of the hidden layer can be expressed as

$$net_i^{(2)}(k) = \sum_{j=1}^{M} w_{ij}^{(2)}(k) x_j^{(1)} \tag{24}$$

$$O_i^{(2)}(k) = f_1\left(net_i^{(2)}(k) - \theta_i\right) \tag{25}$$

where $net_i^{(2)}(k)$ denotes the node input of the hidden layer of the neural network, θ_i is the threshold value of the neuron of the hidden layer, $w_{ij}^{(2)}(k)$ signifies the weight of the neuron of the input layer and the neuron of the hidden layer, and $O_i^{(2)}(k)$ is the node output of the hidden layer. The excitation function of the hidden layer is

$$f_1(x) = \frac{e^x - e^{-x}}{e^x + e^{-x}} \tag{26}$$

The relationship of the output layer of the BPNN is given by

$$net_l^{(3)}(k) = \sum_{j=1}^{P} w_{li}^{(3)}(k) O_i^{(2)}(k) \tag{27}$$

$$O_l^{(3)}(k) = f_2\left(net_l^{(3)}(k) - \theta_l\right) \tag{28}$$

where $w_{li}^{(3)}(k)$ is the weight of the hidden layer neuron and the output layer neuron, θ_l is the threshold value of the output layer neuron, and $O_l^{(3)}(k)$ represents the output of the output layer. The excitation function of the output layer selects the non-negative sigmoid function, whose expression form is

$$f_2\left(net_l^{(3)}(k)\right) = \frac{e^x}{e^x + e^{-x}} \tag{29}$$

In the BP neural network backpropagation process, we adjust the connection weights by setting the backpropagation of the error function. The difference between the actual measured output state quantity and the expected output state quantity of the control system is defined as the calculation error of backpropagation, and the performance index function is obtained as follows:

$$e(k) = yout(k) - rin(k)$$
$$E(k) = \frac{1}{2}e^2(k) \tag{30}$$

The BP neural network usually adopts the gradient descent method to modify neural network parameters, that is, to adjust connection weights according to the direction of the negative gradient change of performance index function $E(k)$. The momentum term is introduced in the backpropagation process, and a value proportional to the previous weight change is added to each weight change. The weight updating processes of the output layer and the hidden layer can be described as

$$\Delta w_{li}^{(3)}(k+1) = -\beta \frac{\partial E(k)}{\partial w_{li}^{(3)}(k)} + \alpha \Delta w_{li}^{(3)}(k) \tag{31}$$

$$\Delta w_{ij}^{(2)}(k+1) = -\beta \frac{\partial E(k)}{\partial w_{ij}^{(2)}(k)} + \alpha \Delta w_{ij}^{(2)}(k) \tag{32}$$

where $\Delta w_{li}^{(3)}(k+1)$ and $\Delta w_{ij}^{(2)}(k+1)$ represent the weight increments of output layer and hidden layer at moment $k + 1$, $\Delta w_{li}^{(3)}(k)$ and $\Delta w_{ij}^{(2)}(k)$ stand for the weight increments of output layer and hidden layer at moment $k + 1$, β is the learning rate (in the range of $0 \sim 1$), and α is the momentum factor (in the range of $0 \sim 1$).

The process of signal forward transmission and error backpropagation of the BPNN is the training process of the network. This process makes the actual output of the network gradually approximate to the expected output by adjusting the connection weights of the hidden layer and the output layer inside the network. In general, the typical BPNN needs to be trained on multiple sets of data to achieve the desired effect.

3.3. Specific Implementation Process of IPSO-BPNN-PI-Based Secondary Frequency Control

We use the IPSO-BP-PI control strategy to achieve effective control in LFC, as shown in Figure 4. The specific implementation process of IPSO-BPNN-PI-based LFC can be divided into the following 6 steps.

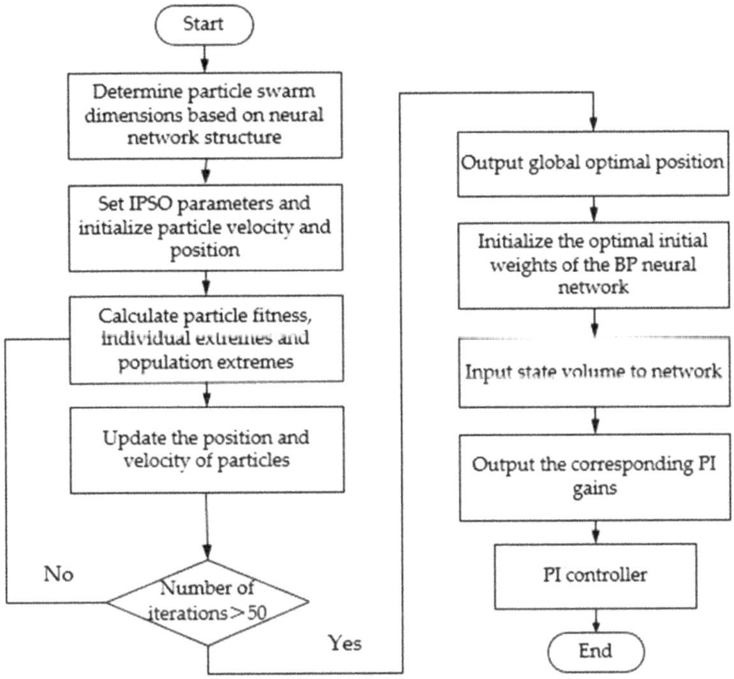

Figure 4. IPSO algorithm for optimizing neural network process.

Step 1: The dimension of the particle swarm is determined; the velocity and position of particles are initialized; and learning factors c_1 and c_2, as well as initial individual optimal value P_{0best} and global optimal value G_{0best}, and other related parameters are set.

Step 2: The adaptation value after particle position update and the following fitness function are calculated:

$$fitness = \frac{1}{n}\sum_{k=1}^{n} e^2(k) \tag{33}$$

Step 3: The above improved PSO is adopted to iteratively update the individual optimal value and global optimal value of the particle, and the fitness of the new particle is compared with the fitness of the previous particle. If the number of iterations reaches the maximum set number of iterations, the iteration is ended, and the current optimal solution is output.

Step 4: The optimal solution of the PSO algorithm is used to initialize the optimal initial weight of the BPNN. Then, the network obtains the optimal PI controller parameter combination through the process of signal forward transmission and error reverse transmission.

Step 5: The adjustable parameters of the neural network output are used to adapt the gains of the PI controller in order to reach valve position ΔP_{ref} and the voltage adjustment of ΔV_{qr}.

Step 6: Control quantities ΔP_{ref} and ΔV_{qr} jointly act on the wind turbine and finally achieve the control of the load frequency output of the power system.

4. Simulation and Discussion

An LFC system involving a wind turbine in a certain region was set up. We here set the base simulation sampling interval to 0.3 s. Assuming that there is no power flow switching between the regions, the actual nominal parameters used in the power system simulation are shown below in Table 2 [7,14].

Table 2. Model parameters of power system.

$D = 2.75$ pu/Hz	$w_{opt} = 1.18$ rad/s
$M^a = 3.15$ N·m	$R_r = 0.00552$ pu
$T_{CH} = 8.0$	$R_s = 0.00491$ pu
$T_G = 12.5$	$L_m = 3.62$ H
$\beta_0 = 1$	$L_r = 0.17$ H

The number of iterations of the control system was set to 50. After 50 iterations, the fitness changes of the optimal individuals of the particle swarm are shown in Figure 5. After 50 iterations, the fitness of the particle swarm tends to be stable, and changes are no longer observed.

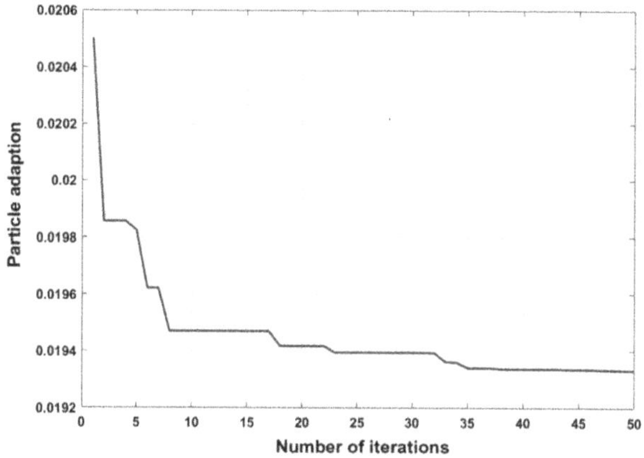

Figure 5. Convergence curve of IPSO.

Case study one: Frequency adjustment in the presence of initial frequency errors

In order to verify the dynamic performance of the system, a simulation test was carried out with an IPSO-BP-PI controller, and the results were evaluated within 60 s to obtain the effect of the change in the frequency deviation of the system output. Meanwhile, we compared this control strategy with conventional PI control and BP-PI control, whose results show that the IPSO-BP-PI control strategy had the best adjustment effect. In the simulation, we set the initial frequency error to 0.03 Hz; the dynamic response curves of the frequency deviation are shown in Figures 6 and 7.

Case study two: Adjustment process with application of random load disturbance

Figures 8–10 show the control effects of Δf, ΔP_m, and ΔP_v, respectively, following the application of random disturbance ΔP^L at the initial moment. On the whole, the frequency deviation, the variation range of mechanical power deviation, and the governor output power deviation were relatively stable from a numerical point of view, and all converged to 0 within a short period of time after a period of fluctuation. Furthermore, to verify the proposed scheme, we used MATLAB to test the maximum overshoot, minimum overshoot, adjustment time, and the magnitude of the stability error, which are given in Table 3. The measurements validated the effectiveness of the IPSO-BP-PI control method.

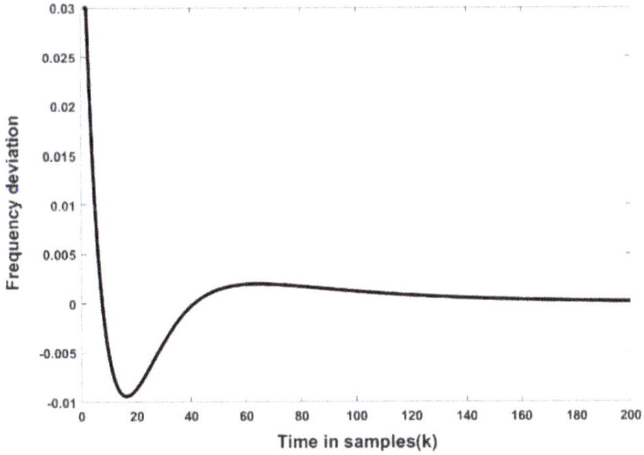

Figure 6. Dynamic response of frequency using IPSO-BP-PI strategy with DFIG-based wind turbine.

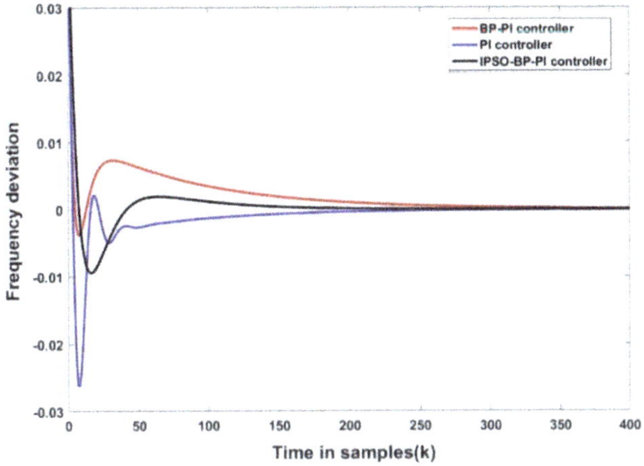

Figure 7. Dynamic responses of frequency using three strategies with DFIG-based wind turbine.

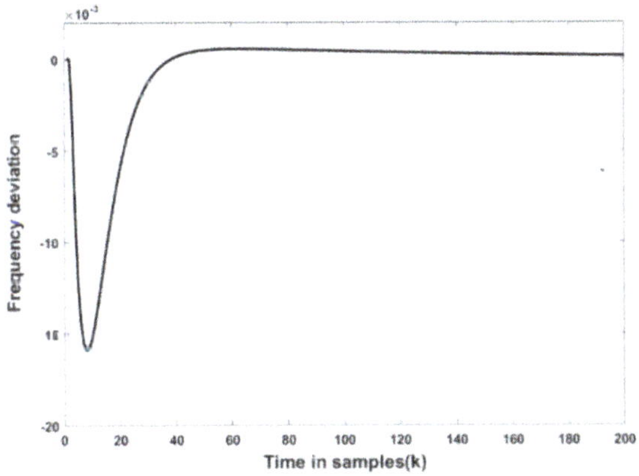

Figure 8. Frequency dynamic curve in the presence of load disturbance using IPSO-BP-PI technique.

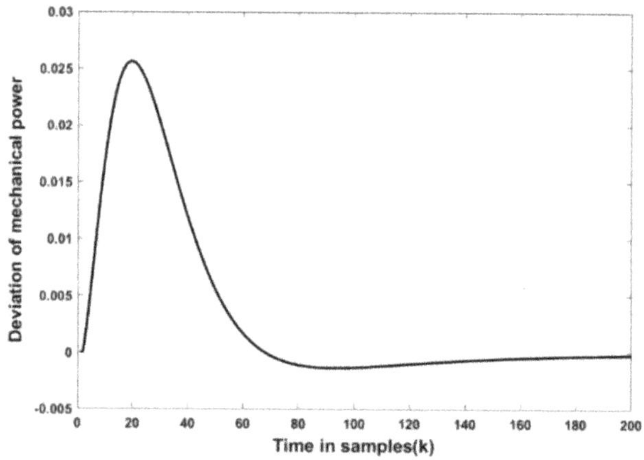

Figure 9. Dynamic response of mechanical power using IPSO-BP-PI technique.

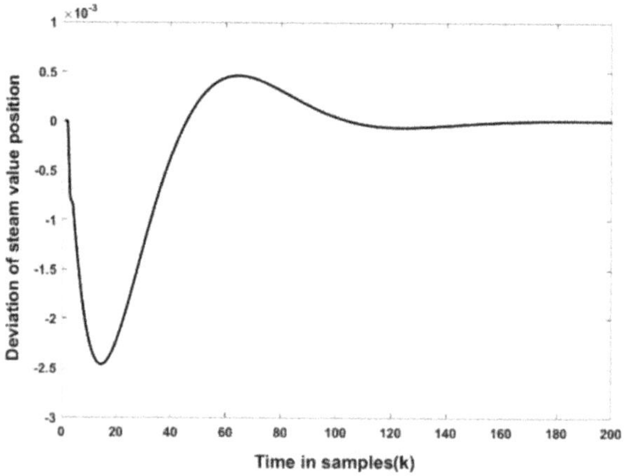

Figure 10. Dynamic response of steam value position using IPSO-BP-PI technique.

Table 3. Specific values for random load demand.

Parameter	Max. Value	Min. Value	Stable Time	Stable Error
Δf	0.00086	−0.01596	15.81	−0.00024
ΔP_m	0.02611	−0.00137	33.48	0.00012
ΔP_v	0.00047	−0.00248	37.02	0.00007

Case study three: Frequency tuning with perturbations applied at subsequent intervals

Here, we simulated the frequency adjustment more realistically using the IPSO-BP-PI controller when the actual area is subjected to load disturbance. Power systems are always subjected to varying degrees of load demand changes during real-time operation. To clearly demonstrate the frequency tuning capability of the IPSO-BP-PI controller, we simulated the frequency response when random perturbations $\Delta P_L = 0.1 \times (rand(1) - 0.5)$ were applied every other 60 s. The frequency response effects with PI control, BP-PI control, and IPSO-BP-PI control strategies were respectively compared. The frequency responses in these cases using different techniques are shown in Figure 11. With the participation of the doubly fed wind turbine, the three controllers all converged to zero and had a certain effect on LFC, ignoring the information exchange between the region and the outside region. In comparison, the IPSO-BP-PI controller showed shorter stabilizing time and smaller overshoot, which is ideal for regulating different load changes. Meanwhile, the method effectively improved the frequency response of the system and avoided redundant oscillation links, as clearly shown in Table 4. The superior effect of the IPSO-BP-PI controller on the LFC of local random disturbance was also verified.

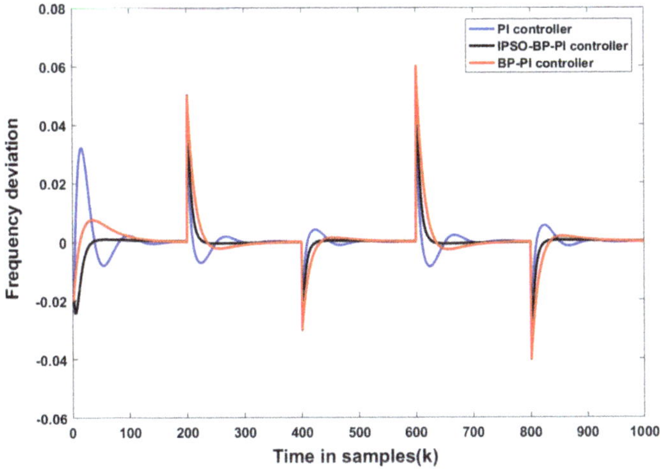

Figure 11. Frequency regulation curves with random perturbation using three techniques.

Table 4. Stable time of the system.

	Stable Time 1	Stable Time 2	Stable Time 3	Stable Time 4	Stable Time 5
PI	46.86	30.39	28.47	38.25	35.17
BP-PI	36.57	25.72	23.62	33.51	31.64
IPSO-BP-PI	15.85	10.67	9.91	11.76	10.26

5. Conclusions

The requirements of grid-connected wind power rules around the world make it easy to conclude that system frequency regulation involving wind power will likely become an inevitable requirement for future new energy development [30,31]. In this paper, an IPSO-BP-PI frequency control method for wind power systems is proposed to solve the problems of large load disturbance and high uncertainty in wind power systems. The improved PSO algorithm is used to optimize the initial weight of the BPNN, which reduces and stabilizes particle fitness around 0.194 and increases the stability of the neural network. Using the adaptive ability of the optimized BPNN to adjust the gain parameters of the PI controller in real time overcomes the problem of the traditional PI controller used in time-varying systems. The optimization scheme was applied to regional LFC involving wind turbines. The simulation test results show that the IPSO-BP-PI control strategy had better instantaneous response effect, which effectively reduced the stable error of the frequency to less than 10^{-3}. Additionally, the IPSO-BP-PI control approach improves the accuracy and significantly shortens adjustment times for frequency stabilization, which has a remarkable effect on solving the load disturbance caused by uncertain factors compared with traditional PI control and BP-PI control. In the future, we aim to investigate LFC for power systems with the integration of multiple renewable energy resources in the presence of unforeseen disturbances.

Author Contributions: Conceptualization, J.S.; methodology, J.S. and Z.H.; software, J.S.; validation, M.C. and L.K.; formal analysis, Z.H.; investigation, M.C. and L.K.; writing—original draft preparation, J.S.; writing—review and editing, Z.H. and V.V.; supervision, V.V. All authors have read and agreed to the published version of the manuscript.

Funding: This research received no external funding.

Institutional Review Board Statement: Not applicable.

Informed Consent Statement: Not applicable.

Data Availability Statement: Not applicable.

Conflicts of Interest: The authors declare no conflict of interest.

References

1. Sadorsky, P. Wind energy for sustainable development: Driving factors and future outlook. *J. Clean. Prod.* **2021**, *289*, 125779. [CrossRef]
2. Veerasamy, V.; Wahab, N.I.A.; Ramachandran, R.; Othman, M.L.; Hizam, H.; Irudayaraj, A.X.R.; Guerrero, J.M.; Kumar, J.S. A Hankel Matrix Based Reduced Order Model for Stability Analysis of Hybrid Power System Using PSO-GSA Optimized Cascade PI-PD Controller for Automatic Load Frequency Control. *IEEE Access* **2020**, *8*, 71422–71446. [CrossRef]
3. Liu, F.; Ma, J. Equivalent input disturbance-based robust LFC strategy for power system with wind farms. *IET Gener. Transm. Distrib.* **2018**, *12*, 4582–4588. [CrossRef]
4. Mohseni, N.; Bayati, N. Robust Multi-Objective H2/H∞ Load Frequency Control of Multi-Area Interconnected Power Systems Using TS Fuzzy Modeling by Considering Delay and Uncertainty. *Energies* **2022**, *15*, 5525. [CrossRef]
5. Debnath, M. Whale Optimization Algorithm Tuned Fuzzy Integrated PI Controller for LFC Problem in Thermal-hydro-wind Interconnected System. In *Applications of Computing, Automation and Wireless Systems in Electrical Engineering: Proceedings of MARC 2018*; Springer: Singapore, 2019; pp. 67–77.
6. Payal, R.; Kumar, D. Load Frequency Control of Wind Integrated Power System Using Intelligent Control Techniques. *Int. J. Adv. Eng. Manag. (IJAEM)* **2021**, *3*, 1339–1353.
7. Fan, W.; Hu, Z.; Veerasamy, V. PSO-Based Model Predictive Control for Load Frequency Regulation with Wind Turbines. *Energies* **2022**, *15*, 8219. [CrossRef]
8. Abhayadev, S.; Kumar, P.R. Effect of DFIG Wind Turbines on LFC in a Sliding Mode Controlled Power System. In Proceedings of the 6th International Conference for Convergence in Technology (I2CT), Maharashtra, India, 2–4 April 2021; pp. 1–6. [CrossRef]
9. Abazari, A.; Monsef, H.; Wu, B. Load frequency control by de-loaded wind farm using the optimal fuzzy-based PID droop controller. *IET Renew. Power Gener.* **2018**, *13*, 180–190. [CrossRef]
10. Yan, Z.; Xu, Y. Data-Driven Load Frequency Control for Stochastic Power Systems: A Deep Reinforcement Learning Method with Continuous Action Search. *IEEE Trans. Power Syst.* **2018**, *34*, 1653–1656. [CrossRef]
11. Xu, Z.; Wang, D.; Yi, G.; Hu, Z. Asynchronous Tracking Control of Amplitude Signals in Vibratory Gyroscopes with Partially Unknown Mode Information. *IEEE Trans. Ind. Electron.* **2022**, 1–10. [CrossRef]
12. Hu, Z.; Liu, S.; Luo, W.; Wu, L. Intrusion-Detector-Dependent Distributed Economic Model Predictive Control for Load Frequency Regulation with PEVs Under Cyber Attacks. *IEEE Trans. Circuits Syst. I Regul. Pap.* **2021**, *68*, 3857–3868. [CrossRef]
13. Hu, Z.; Liu, S.; Wu, L. Credibility-based distributed frequency estimation for plug-in electric vehicles participating in load frequency control. *Int. J. Electr. Power Energy Syst.* **2021**, *130*, 106997. [CrossRef]
14. Ma, M.; Liu, X.; Zhang, C. LFC for multi-area interconnected power system concerning wind turbines based on DMPC. *IET Gener. Transm. Distrib.* **2017**, *11*, 2689–2696. [CrossRef]
15. Zhang, K.; Zhou, B.; Duan, G. Event-Triggered and Self-Triggered Control of Discrete-Time Systems with Input Constraints. *IEEE Trans. Syst. Man Cybern. Syst.* **2020**, *52*, 1948–1957. [CrossRef]
16. Mohamed, T.H.; Alamin, M.A.M.; Hassan, A.M. A novel adaptive load frequency control in single and interconnected power systems. *Ain Shams Eng. J.* **2020**, *12*, 1763–1773. [CrossRef]
17. Navarrete, E.C.; Trejo Perea, M.; Jáuregui Correa, J.C.; Carrillo Serrano, R.V.; Moreno, G.J.R. Expert Control Systems Implemented in a Pitch Control of Wind Turbine: A Review. *IEEE Access* **2019**, *7*, 13241–13259. [CrossRef]
18. Yu, T.; Liu, J.; Zeng, Y.; Zhang, X.; Zeng, Q.; Wu, L. Stability Analysis of Genetic Regulatory Networks with Switching Parameters and Time Delays. *IEEE Trans. Neural Netw. Learn. Syst.* **2017**, *29*, 3047–3058. [CrossRef] [PubMed]
19. Prasad, S.; Ansari, M.R. Frequency regulation using neural network observer based controller in power system. *Control Eng. Pract.* **2020**, *102*, 104571. [CrossRef]
20. Yu, T.; Zhang, X.; Wang, X. Global Exponential Stability Analysis for A Class of Coupled Cyclic Genetic Regulatory Networks with Multiple Time-Varying Delays. *IFAC-PapersOnLine* **2021**, *54*, 23–28. [CrossRef]
21. Wang, Y.; Wang, T.; Yang, X.; Yang, J. Gradient Descent-Barzilai Borwein-Based Neural Network Tracking Control for Non-linear Systems with Unknown Dynamics. *IEEE Trans. Neural Netw. Learn. Syst.* **2023**, *34*, 305–315. [CrossRef] [PubMed]

22. Hu, Z.; Liu, S.; Luo, W.; Wu, L. Resilient Distributed Fuzzy Load Frequency Regulation for Power Systems Under Cross-Layer Random Denial-of-Service Attacks. *IEEE Trans. Cybern.* **2020**, *52*, 2396–2406. [CrossRef] [PubMed]
23. Sahu, P.C.; Mishra, S.; Prusty, R.C.; Panda, S. Improved-salp swarm optimized type-II fuzzy controller in load frequency control of multi area islanded AC microgrid. *Sustain. Energy Grids Netw.* **2018**, *16*, 380–392. [CrossRef]
24. Rahman, J.; Tafticht, T.; Doumbia, M.; Mutombo, N. Dynamic Stability of Wind Power Flow and Network Frequency for a High Penetration Wind-Based Energy Storage System Using Fuzzy Logic Controller. *Energies* **2021**, *14*, 4111. [CrossRef]
25. Mokhtar, M.; Marei, M.I.; Sameh, M.A.; Attia, M.A. An Adaptive Load Frequency Control for Power Systems with Renewable Energy Sources. *Energies* **2022**, *15*, 573. [CrossRef]
26. Li, J.; Cheng, J.-H.; Shi, J.-Y.; Huang, F. Brief Introduction of Back Propagation (BP) Neural Network Algorithm and its Improvement. In *Advances in Computer Science and Information Engineering*; Springer: Berlin/Heidelberg, Germany, 2012; pp. 553–558. [CrossRef]
27. Veerasamy, V.; Wahab, N.I.A.; Ramachandran, R.; Othman, M.L.; Hizam, H.; Kumar, J.S.; Irudayaraj, A.X.R. Design of single- and multi-loop self-adaptive PID controller using heuristic based recurrent neural network for ALFC of hybrid power system. *Expert Syst. Appl.* **2021**, *192*, 116402. [CrossRef]
28. Panda, D.K.; Das, S.; Townley, S. Toward a More Renewable Energy-Based LFC Under Random Packet Transmissions and Delays with Stochastic Generation and Demand. *IEEE Trans. Autom. Sci. Eng.* **2020**, *19*, 1217–1232. [CrossRef]
29. Shouran, M.; Anayi, F.; Packianather, M.; Habil, M. Different Fuzzy Control Configurations Tuned by the Bees Algorithm for LFC of Two-Area Power System. *Energies* **2022**, *15*, 657. [CrossRef]
30. Luo, W.; Wang, F.; Li, Q.J.; Xie, H. Research on Development of Wind Power Grid Integration in China. *Adv. Mater. Res.* **2012**, *608-609*, 560–563. [CrossRef]
31. Holjevac, N.; Zidar, M.; Krpan, M.; Kuzle, I. Optimizing the Grid Connection Scheme of the Wind Power Plant. In Proceedings of the 14th Conference on Sustainable Development of Energy, Water and Environment Systems (SDEWES), Dubrovnik, Croatia, 1–6 October 2019.

energies

MDPI

Article

A Hybrid Grey Wolf Assisted-Sparrow Search Algorithm for Frequency Control of RE Integrated System

Bashar Abbas Fadheel [1,2,*], Noor Izzri Abdul Wahab [1,*], Ali Jafer Mahdi [2,3], Manoharan Premkumar [4], Mohd Amran Bin Mohd Radzi [1], Azura Binti Che Soh [1], Veerapandiyan Veerasamy [5] and Andrew Xavier Raj Irudayaraj [1]

1 Advanced Lightning, Power, and Energy Research (ALPER), Department of Electrical and Electronics Engineering, Faculty of Engineering, Universiti Putra Malaysia (UPM), Serdang 43400, Malaysia
2 Department of Biomedical Engineering, University of Kerbala, Karbala 56001, Iraq
3 Department of Scientific Affairs, Al-Zahraa University for Women, Karbala 56001, Iraq
4 Department of Electrical and Electronic Engineering, Dayananda Sagar College of Engineering, Bengaluru 560078, India
5 School of Electrical and Electronic Engineering, Nanyang Technological University, Singapore 639798, Singapore
* Correspondence: gs58252@student.upm.edu.my (B.A.F.); izzri@upm.edu.my (N.I.A.W.); Tel.: +964-77-2656-1977 (B.A.F.); +60-13-229-1140 (N.I.A.W.)

Abstract: Nowadays, renewable energy (RE) sources are heavily integrated into the power system due to the deregulation of the energy market along with environmental and economic benefits. The intermittent nature of RE and the stochastic behavior of loads create frequency aberrations in interconnected hybrid power systems (HPS). This paper attempts to develop an optimization technique to tune the controller optimally to regulate frequency. A hybrid Sparrow Search Algorithm-Grey Wolf Optimizer (SSAGWO) is proposed to optimize the gain values of the proportional integral derivative controller. The proposed algorithm helps to improve the original algorithms' exploration and exploitation. The optimization technique is coded in MATLAB and applied for frequency regulation of a two-area HPS developed in Simulink. The efficacy of the proffered hybrid SSAGWO is first assessed on standard benchmark functions and then applied to the frequency control of the HPS model. The results obtained from the multi-area multi-source HPS demonstrate that the proposed hybrid SSAGWO optimized PID controller performs significantly by 53%, 60%, 20%, and 70% in terms of settling time, peak undershoot, control effort, and steady-state error values, respectively, than other state-of-the-art algorithms presented in the literature. The robustness of the proffered method is also evaluated under the random varying load, variation of HPS system parameters, and weather intermittency of RE resources in real-time conditions. Furthermore, the controller's efficacy was also demonstrated by performing a sensitivity analysis of the proposed system with variations of 75% and 125% in the inertia constant and system loading, respectively, from the nominal values. The results show that the proposed technique damped out the transient oscillations with minimum settling time. Moreover, the stability of the system is analyzed in the frequency domain using Bode analysis.

Keywords: automatic load frequency control; renewable energy resources; hybrid sparrow search algorithm-grey wolf optimization; hybrid power system; stability

Citation: Fadheel, B.A.; Wahab, N.I.A.; Mahdi, A.J.; Premkumar, M.; Radzi, M.A.B.M.; Soh, A.B.C.; Veerasamy, V.; Irudayaraj, A.X.R. A Hybrid Grey Wolf Assisted-Sparrow Search Algorithm for Frequency Control of RE Integrated System. *Energies* **2023**, *16*, 1177. https://doi.org/10.3390/en16031177

Academic Editor: Abdelali El Aroudi

Received: 28 November 2022
Revised: 10 January 2023
Accepted: 17 January 2023
Published: 20 January 2023

1. Introduction

Currently, rising energy demand and the depletion of fossil fuel resources have necessitated the extensive use of Renewable Energy (RE), which is considered clean and cost-effective, and likely to be connected to interconnected power systems for distributed utilization [1]. The integration of RE such as solar and wind leads to an uncertain variation in the generation of power due to the intermittent nature of RE, which introduces frequency oscillation that leads to multiple protection relays trip in the power system [2]. Automatic Load Frequency Control (ALFC) has proven very effective in regulating the system frequency by adjusting the generation system. Hence, the overall generation matches the load demand to successfully operate the interconnected power networks. To achieve this, a robust control technique is vital in each area of the interconnected power network to maintain the nominal system frequency within the prescribed limits of ±0.2 Hz and planned tie-line power exchange [3]. To study this, the system frequency and tie-line power exchange parameters are considered as Area Control Error (ACE), which serves as input to the controller [4].

For ALFC applications, researchers have developed numerous controllers in the past. A conventional Proportional-Integral (PI) controller is used to enhance the frequency stability of the power system using an electric vehicle as an energy storage unit [5]. In reference [6], a four-area multi-source thermal system is controlled using linear quadrat/linear Gaussian control. Usually, the conventional Proportional-Integral-Derivative (PID) controllers are the best option due to their extra degree of freedom, affordability, and simplicity when compared to more robust controllers such as the Sliding Mode Control (SMC) [7], the Model Predictive Controller (MPC) [8], and H2/H∞ controllers [9]. In reference [10], a hybrid fuzzy logic controller with sliding mode was implemented for a microgrid system to evaluate the penetration of Renewable Energy Resources (RERs) connected with a power system. Another hybrid controller based on PID with fuzzy logic technique has been evaluated for a microgrid system [11]. In this paper, the effect of designing high performance power converters has been demonstrated to achieve a stable power system. All these controllers are challenging to implement in real-time and need advanced mathematics to derive the control law [12]. On the other hand, tuning the parameters of the PID controller is critical for a large dynamic HPS model. Instability in the system may result from the improper tuning of the PID parameters, which displays poor dynamic response in the frequency regulation of the system. To resolve this, several papers have been published in the literature to optimize the gain values of the PID controller using Artificial Neural Network (ANN) [13] and Fuzzy Logic control (FLC) [14]. Despite their benefits, their application in the actual ALFC system is constrained by the computational time required to choose a rule foundation fuzzy logic and a big training database for ANN [15]. However, the selection has no precise mathematical meaning, which might occasionally result in poor ALFC performance.

On the other hand, numerous heuristic approaches are used to optimize the parameters of the PID controller for ALFC, as illustrated in Table 1. PID or other modified forms of PID controllers for ALFC applications, the system model considered, the recent optimization algorithms used, and the limitations of the control method are briefly depicted in Table 1.

Table 1. Review of optimization algorithms used to study the ALFC.

Ref. No.	Controller Type	Optimization	Model Description	Limitations
[16]	PID	GWO	Two-area thermal system	GRC, GDB, and BD nonlinearities in thermal systems and RER have not been considered in this work. The most significant limitation of this optimization technique is trapping in local optimal points
[17]	PFMPID	GOA	Three area multi source power system comprises thermal, hydro, wind, diesel, and RFB with GDB and GRC	BD has not been considered in this study. Moreover, the GOA optimization search process begins with a population or flock of grasshoppers whose locations are comparable to design vectors which leads to poor exploration

Table 1. *Cont.*

Ref. No.	Controller Type	Optimization	Model Description	Limitations
[18]	PID	CTO	Two-area multi-source with thermal, hydro, and gas units in each area	Non-conventional power resources have not been considered in this study. The CTO algorithm has the drawback of a greater number of parameters to be initialized
[19]	PID	GWO	Single area multi-source power system comprises thermal, hydro, and gas units	RER has not been considered, and the limitation of GWO is that poor exploration leads to trapping in local optimal points
[15]	PID	Hybrid PSO-GSA	Considering a two-area thermal system with GRC and GDB	For a highly nonlinear and big dimensional system such as ALFC, PSO exhibits poor exploration and takes longer to find global minima. The PSO-GSA approach overcomes this, although the study considered only a conventional source
[20]	PID	MPA	Two-area multi-source systems were taken into consideration, with WTPG, STPP, BES, and thermal plants in area 2 as well as STPP, PV, SMES, and thermal power plants combining GDB and GRC in area 1, and a reheat generator with wind and PV RERs in area 2. GRC has been considered in the two-area	MPA's main drawback is a slow convergence rate with poor exploration ability
[21]	PID	HIO	Two-area reheat turbine power plant with gas and hydro units in each area	RER has not been considered in this study
[22]	PID	DE	Two-area multi-source with hydro, thermal, and wind power plants in each area	Nonlinearities are not considered. The limitation of DE is inapplicable to solving many complex real-world problems in continuous domains
[23]	PIDA	MPA	Two-area non-reheat thermal system	This study has not considered RER, GRC, BD, or GDB
[24]	PID	WOA	Two-area reheat thermal system	Nonlinearities are not considered. The limitation of WOA is that whales are drawn to the coefficient vector during the later phases of WOA iteration convergence, and as a result, the whole whale population quickly enters the local optimum for the high-dimensional optimization problem
[25]	PID	SSO	Two-area thermal system with GRC and GDB considered with the wind power plant in both the areas	The drawback of SSO is that the update rule fails when one of the dimensions has a lower bound other than zero

The optimization algorithm listed in Table 1 is an individual metaheuristic algorithm. Individual algorithms are advantageous due to their minimum requirements for function evaluation and ease of usage throughout the whole optimization process. However, the probability of local optima stagnation is very high, and individual meta-heuristic algorithms have difficulty in balancing between exploration and exploitation in order to find the global optima [26]. To cater to this issue, various modifications are performed in meta-heuristic algorithms to enhance their performance. One of the most recent and popular methods is to apply chaos theory in meta-heuristic algorithms to improve the global convergence speed and exploration/exploitation capabilities of optimization algorithms, resulting in more diversity in solutions. Another way is by hybridizing two algorithms.

Hybrid algorithms have recently been a popular topic applied in several domains [27–30]. Combining the qualities and features of two different algorithms would provide an optimization tool that is more dependable and potent for solving complex problems. It is recognized that combining evolutionary algorithms can increase their performances with problem-dependent local searches [31]. In recent years, numerous swarm intelligence optimization algorithms have been proposed and investigated in the literature, such as the Ant Colony Optimization (ACO) [32], Manta-Ray Foraging Optimizer (MRFO) [33], Salp Swarm Optimization (SSO) [34], SSA [35], and GWO [36]. Among them, SSA is a new optimization intelligence algorithm proposed in 2020 [35] that mimics the foraging behavior of sparrows to tackle particular optimization problems. SSA is distinct from other intelligent optimization algorithms by its high search accuracy, quick convergence time, superior stability, and robustness [36]. However, in the later stages of convergence, SSA is prone to local optima due to poor exploitation [37]. These issues directly impact SSA's optimization effect, resulting in the inability to discover the optimal global solution.

Another popular swarm intelligence algorithm is GWO, developed by Mirjalili et al. [36], which has received a great reception in the optimization field. This algorithm simulates the natural hunting and dominance behavior of grey wolves. GWO has received great attention because of its simplicity and ease of implementation, and it has been used to tackle many real-time optimization problems. However, when handling complex, high-dimensional, and unimodal problems, GWO can be stuck in local optima due to its poor exploration [37]. Hence, to improve the convergence speed and avoid the optima problems, GWO can be hybridized with a good exploration ability algorithm.

Therefore, based on the advantages and disadvantages of SSA and GWO, the hybrid method is the possible solution to eradicate the limitations and combine the advantages of the two algorithms. Hence, this paper attempts to hybridize SSA with GWO to balance the exploration and exploitation phases, improving the controller performance in terms of steady-state and dynamic responses for frequency and power flow variation in interconnected power systems. The main contributions of this research study are summarized as follows:

1. A hybrid SSAGWO algorithm is proposed to improve the SSA algorithm exploitation ability, and the algorithm is tested using various classical benchmark functions to prove its effectiveness against other algorithms.
2. Auto-tuning of the PID controller parameters for ALFC of an RER-integrated HPS network is implemented using various optimization algorithms to verify the robustness of the proposed algorithm.
3. The proposed system is tested using the data of an actual solar power plant, emulated for extreme operating conditions.
4. A stability analysis is conducted to prove the efficacy and robustness of the proposed technique.

The paper is organized as follows: Section 2 presents the mathematical modeling of the two-area power system considered in this study. Next, Section 3 discusses the control strategy for load frequency control of the two-area power system. Section 4 discusses in detail the formulation procedure of the proposed hybrid SSAGWO algorithm. The results and critical analysis of the obtained results are discussed in Section 5, and Section 6 concludes the paper.

2. System Model

The study proposed in this paper consists of a two-area reheat steam power turbine, Photovoltaic (PV), and Wind Turbine Power Generator (WTPG) conversion, as shown in Figure 1. The two-area power system consists of a governor, reheat power thermal turbine generators with 0.05% of Governor Dead Band (GDB) and 10% of Generator Rate Constraints (GRC), and Boiler Dynamics (BD). In this research, two symmetrical power generator areas have been proposed with the parameters shown in Appendix A [15,38]. The BD has been considered for the realistic two-area reheat power systems. The thermal power plant uses the boiler dynamics arrangement to generate steam under high pressure. The long-term process of the steam flow on the boiler drum is considered in this model as combustion control. The block diagram configuration of the BD is shown in Figure 2. Oil-fired boiler type parameters [39] are listed in Appendix A.

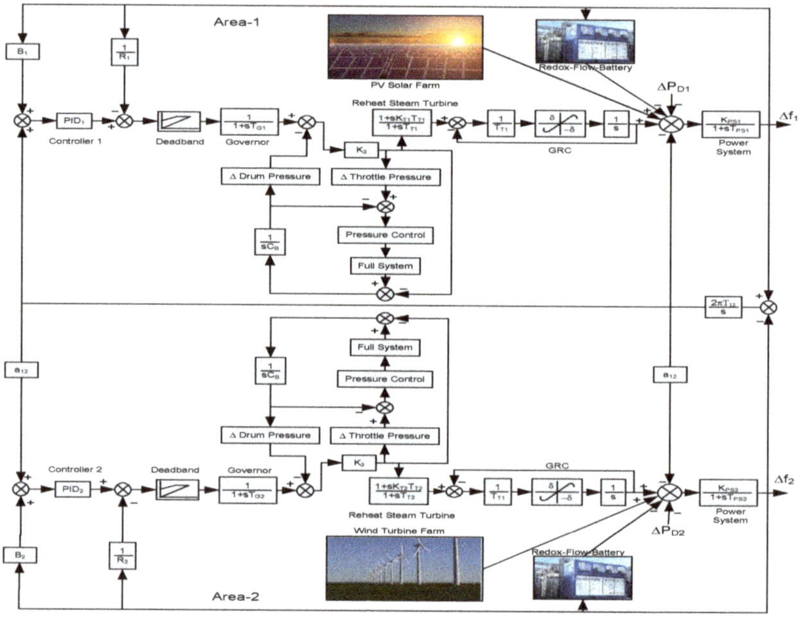

Figure 1. Transfer function model of the two-area power system.

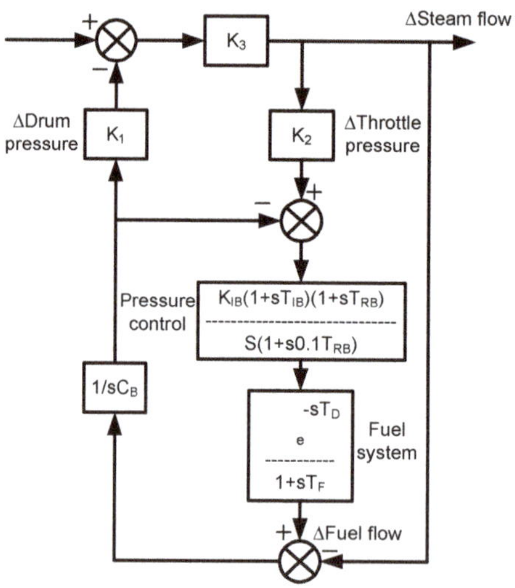

Figure 2. Block diagram of the boiler dynamics configuration system.

2.1. SPV Model

SPV cells comprising semiconductor materials can convert photon energy directly into electrical energy. Power loss is also modeled because of the boundary and external contact, represented by series resistors, and a leakage current, represented by parallel resistance. SPV power generation is intermittent and dependent on sun irradiation and temperature; thus, a random power source can represent SPV behavior [40]. The block diagram of the SPV is shown in Figure 3. The linearized SPV model, which is considered for the ALFC model, is given by [40]:

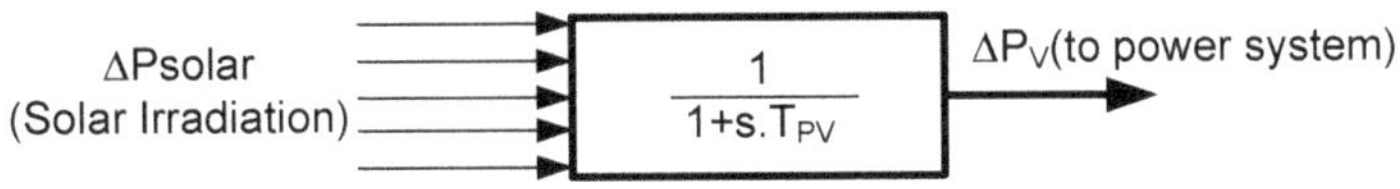

Figure 3. Block diagram of PV dynamics system.

Where ΔT_{PV} is the time constant of the SPV, ΔP_{Solar} is the change in solar irradiation power, and ΔP_{PV} denotes the change in power generated from solar to the power system.

2.2. WTPG Model

The mathematical model implementation for the wind power plant comprises a hydraulic pitch actuator, data fit pitch response, and blade characteristics, as shown in Figure 4. The pitch angle control mechanism keeps the pitch angle at the desired value based on wind speed. Thus, wind turbine production may be controlled by adjusting the pitch angle regardless of wind speed. The wind's mathematical modeling transfer function model can be given in [41]:

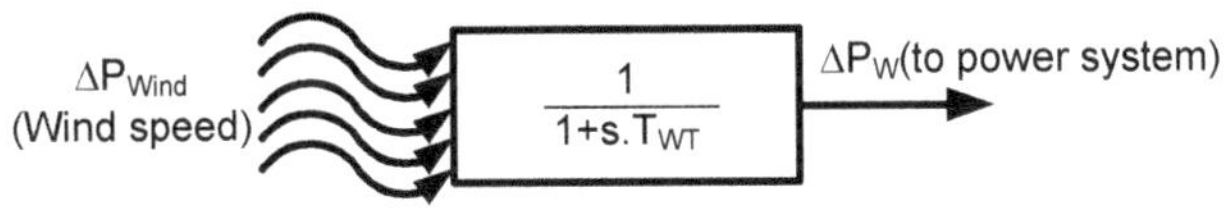

Figure 4. Block diagram of Wind turbine dynamics system.

Where ΔT_{WT} is the time constant of the wind turbine, ΔP_{Wind} is the change in wind speed, and ΔP_W is the change in power generated by the power system.

2.3. RFB Structure

RFB is a type of electrochemical energy storage compensation device that can convert energy from its electrical form to its chemical form and vice versa through electrochemical reaction processes [42]. RFB is an efficient method for reducing frequency deviations and tie-line power and is considered a fast active power compensation energy device. The RFB block diagram model integrated with areas 1 and 2 is shown in Figure 5 [41]:

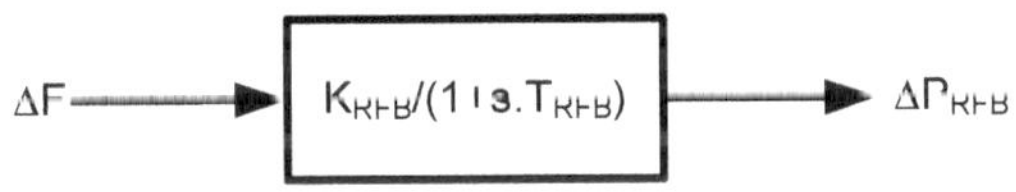

Figure 5. Block diagram of Redox-Flow-Battery (RFB) model.

Where T_{RFB} and K_{RFB} are the time and gain constant of the RFB. The main objective of the RFB is to charge and store energy from the power system as a very small load during a normally operating system, and it delivers the stored energy to the interconnected power system during any sudden change in load to reduce frequency fluctuations.

3. Control Strategy

The closed loop of the two-area power system can be simplified as shown in Figure 6:

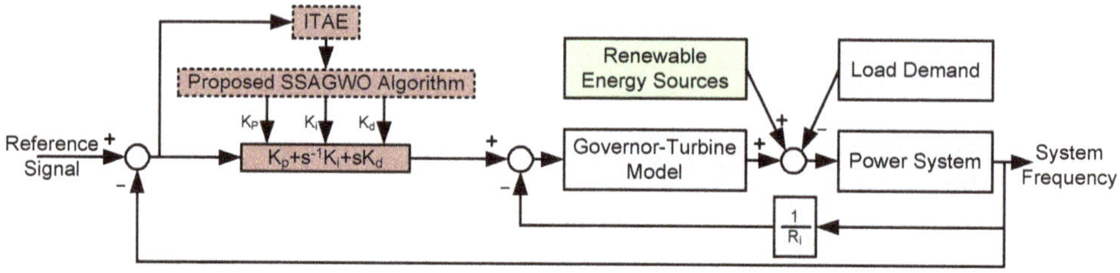

Figure 6. Close loop control system tuning.

Where ΔP_{tie} is the power deviation of the tie line of the power system. The transfer function of the PID controller is given by:

$$C(s) = k_p + \frac{k_i}{s} + sk_d \tag{1}$$

where k_p is the proportional gain, k_i is the integral gain, and k_d is the derivative gain. The PID controller output of the power system is as follows:

$$u_{1,2} = k_{p1}ACE_{1,2} + k_i\int_0^t ACE_{1,2}dt + \frac{dACE_{1,2}}{dt} \tag{2}$$

The gain of the PID controller in this study is designed to minimize the steady-state error guides, such as the performance index *ITAE* as given below:

$$J = ITAE = \int_0^t t|ACE_i|dt = \int_0^t t\{|\Delta f_i + \Delta P_{tie,i-j}|\}dt \tag{3}$$

where i and j are the area numbers from 1,2,3, 4, ... n with $i \neq j$.

This study aims to minimize J by tuning the values of PID controller parameters by using techniques such as the Z–N technique and intelligence optimization algorithms techniques such as PSO, GWO, SSA, and the proposed new hybrid SSAGWO.

4. Proposed Hybrid SSA-GWO

In this section, an effort was made to develop a new hybrid metaheuristic algorithm that combines the strength of swarm-inspired algorithms such as SSA with GWO to tune the values of the PID controller for a multi-area power system.

4.1. Grey Wolf Optimization

GWO is a well-known swarm intelligence optimization technique. The algorithm searches and optimizes by simulating the process of tracking, enclosing, and hunting, which was inspired by the predation behavior of the grey wolf population. There are various types of members in a wolf pack depending on the level of dominance, such as α, β, δ, and ω. The best three solutions are α, β, and δ, who leads the pack ω. The grey wolves hunt in a pack and collaborate to catch prey. The technique is to chase the prey and encircle it [43]. The mathematical model of the encircling behavior of grey wolves is in the following equations:

$$\vec{X}(t+1) = \frac{\vec{X}_1 + \vec{X}_2 + \vec{X}_3}{3} \tag{4}$$

$$\vec{X}_1 = \vec{X}_\alpha(t) - \vec{A}\cdot\vec{D}_\alpha, \vec{X}_2 = \vec{X}_\beta(t) - \vec{A}\cdot\vec{D}_\beta, \text{ and } \vec{X}_3 = \vec{X}_\delta(t) - \vec{A}\cdot\vec{D}_\delta \tag{5}$$

$$\vec{D}_\alpha = \left|\vec{C}_1\cdot\vec{X}_\alpha - \vec{X}\right|, \vec{D}_\beta = \left|\vec{C}_2\cdot\vec{X}_\beta - \vec{X}\right|, \text{ and } \vec{D}_\delta = \left|\vec{C}_3\cdot\vec{X}_\delta - \vec{X}\right| \tag{6}$$

$\vec{X}(t)$ represents the current position of the grey wolf; $\vec{A}$ is a coefficient vector; $\vec{D}$ represent the distance that varies according to the place of the prey $\vec{X}_p$; $\vec{a}$ is a factor that reduces from 2 to 0 linearly during optimization; $\vec{r}_1$ and $\vec{r}_2$ are random numbers between 0 and 1; t represents as the current iteration, and T is the maximum iteration [44].

To predict the prey, we suppose that the best solution discovered so far (the alpha wolf) is the prey's location because he is leading the pack and knows the prey's location, while the second-best solution is for β wolf, and the third-best solution is δ wolf in the current iteration [45]. A is a random number in the gap $[-2a, 2a]$. The wolves move to attack the prey when the value of $|A| < 1$. Exploration is the ability to search for prey, while exploitation is the ability to attack the prey. The arbitrary values of A are used to force the wolves to move away from the prey. Furthermore, when the value of $|A| > 1$, the wolves are forced to diverge from the prey (local minimum).

4.2. Sparrow Search Algorithm

The SSA is a swarm intelligence optimization technique. This technique mimics sparrow foraging behavior. It has three sorts of individual behavior: discoverer, follower, and investigator, and the sparrows update their locations according to their own rules. The discoverer searches for food and guides the rest of the community [35]. After discovering the discoverer's location, followers search for food around this location. The follower's location is updated according to Equation (7) [46,47]:

$$X_i^{t+1} = \left\{ \begin{array}{ll} Q \cdot \exp\left(\frac{X_{worst} - X_i^t}{i^2}\right), & i > \frac{n}{2} \\ X_{best}^{t+1} + \left| X_i^t - X_{best}^{t+1} \right| \cdot A \cdot L, & i \leq \frac{n}{2} \end{array} \right\} \quad (7)$$

where X_{best} represents the best individual position, which means the best current location; X_{worst} is the current worst global location; A is a $d \times d_{matrix}$, which contains each factor randomly assigned 1 or -1; n represents the number of sparrows; when $i \leq n/2$, it proposes the *ith* entrant is searching for food close to the best location, if $i > n/2$, it means the *ith* entrant needs to fly to another location for food.

Individuals from the population are chosen at random to be investigators. When predators attack, they send out signals that cause sparrows to flee to a safe location [46].

The mutation strategy in SSA directly affects convergence accuracy and speed. The SSA performs better in solving complex optimization problems but has the disadvantages of reduced population diversity and insufficient convergence accuracy. It could have a chance to fall into the local optimum, failing to achieve the optimum solution to the problems.

4.3. Proposed Optimization Algorithm

A new hybrid sparrow search algorithm based on the GWO, namely SSAGWO, is proposed in this paper. Its features allow the SSAGWO to avoid the local optimum while improving convergence speed and accuracy. The exploitation capability of the GWO is introduced into the sparrow search algorithm to improve its exploitation ability. The modification in the structure of SSA is improving the exploitation ability by using the behavior of exploration ability in GWO to compromise between exploitation and exploration. Improving the exploitation in SSA with the exploration effort should be achieved by hybridization between SSA and GWO. The variants of the position of SSA Equation (7) are mixed with the distance equation of GWO by new weight factor (θ) to keep the problem's final solutions near the optimal values. The modification-guided equations are rewritten as follows:

$$\vec{D}_\alpha = \left|\vec{C}_1 \cdot \vec{X}_\alpha - \theta\vec{X}\right|, \vec{D}_\beta = \left|\vec{C}_2 \cdot \vec{X}_\beta - \theta\vec{X}\right|, \text{ and } \vec{D}_\delta = \left|\vec{C}_3 \cdot \vec{X}_\delta - \theta\vec{X}\right| \quad (8)$$

The probability of shifting the position of all agents is calculated using Equation (9):

$$\Gamma[X(t+1)-X(t)] = \left| \frac{[X(t+1)-X(t)]}{\sqrt{[X(t+1)-X(t)]^2+1}} \right| \quad (9)$$

where Γ is the probability factor. If the result of $[X(t+1) - X(t)]$ is positive, the sparrows select the food, while if the result is negative, the sparrows move far away from predators. It is worth noting that the correct position is quickly achieved by selecting well the boundaries of the PID factors. The SSAGWO's specific steps are described in detail:

Step 1: Initialize the sparrow search population and its parameters (n is the total number of sparrows, T_{max} maximum iteration, d is the number of variables).
Step 2: While ($t < T_{max}$), rank the sparrows according to their fitness values by minimizing J in Equation (3). Find the current best value, which is the minimum fitness value, and the current worst value, which is the maximum fitness value.
Step 3: Update the sparrow location for the discoverer by using Equation (10).

$$X_i^{t+1} = \begin{Bmatrix} X_i^t \times \exp\left(\frac{-i}{\alpha \cdot T_{\max}}\right), & R_2 < STh \\ X_i^t + Q \cdot L, & R_2 \geq STh \end{Bmatrix} \quad (10)$$

With $X = [X_1, X_2, X_3, \ldots X_i \ldots X_n]^T, X_i = [X_{i,1}, X_{i,2}, X_{i,3} \ldots X_{i,d}]$.

Where t denotes the current iteration; X_i^t is the location of the ith sparrow at iteration t; α is a random number between 0 and 1; R_2 is the alarm value, which is $0 < R_2 \leq 1$; the safety threshold is STh, and it is valued $0.5 \leq STh < 1$; L is a $1 \times d_{matrix}$; d_{matrix} is the matrix including 1 in each factor; and Q is a random value with a mean of zero and a variance of one that follows the normal distribution. If $R_2 < STh$, it indicates that the foraging surroundings are safe, while $R_2 \geq STh$ denotes that some individuals already have faced dangerous animals and that all sparrows must flee to other safe locations as soon as possible.

Step 4: If the *ith* individuals at the current iteration are less or equal to half the sparrow population, then update the follower's position using Equation (7) and go to step 9. Except for that, run the GWO algorithm.
Step 5: Initialize the values of a, A, and C.
Step 6: Calculate the first-best value of the alpha wolf, the second-best value of the beta wolf, and third best value of the gamma wolf.
Step 7: Determine the distance between the wolves and prey using Equation (6). After that, calculate the value of the new position using Equations (4) and (5).
Step 8: Export the position of the best three wolves and exchange it with the current sparrows. The flowchart of the SSAGWO is shown in Figure 7.
Step 9: Update the follower's position using Equation (7). Then, update the investigator's location using Equation (11).

$$X_i^{t+1} = \begin{Bmatrix} X_{best}^t + \beta_e \left| X_i^t - X_{best}^t \right|, & f_i > f_b \\ X_{best}^{t+1} + K \frac{\left| X_i^t - X_{best}^{t+1} \right|}{(f_i - f_w) + \varepsilon}, & f_i = f_b \end{Bmatrix} \quad (11)$$

where β_e denotes the random step length control coefficient, which has a variance of 1 and a mean value of 0 and follows the normal distribution; K is a random number between -1 and 1; f_i is the fitness value of the *ith* individual; f_b is the current global best fitness; and f_w is the current global worst fitness.

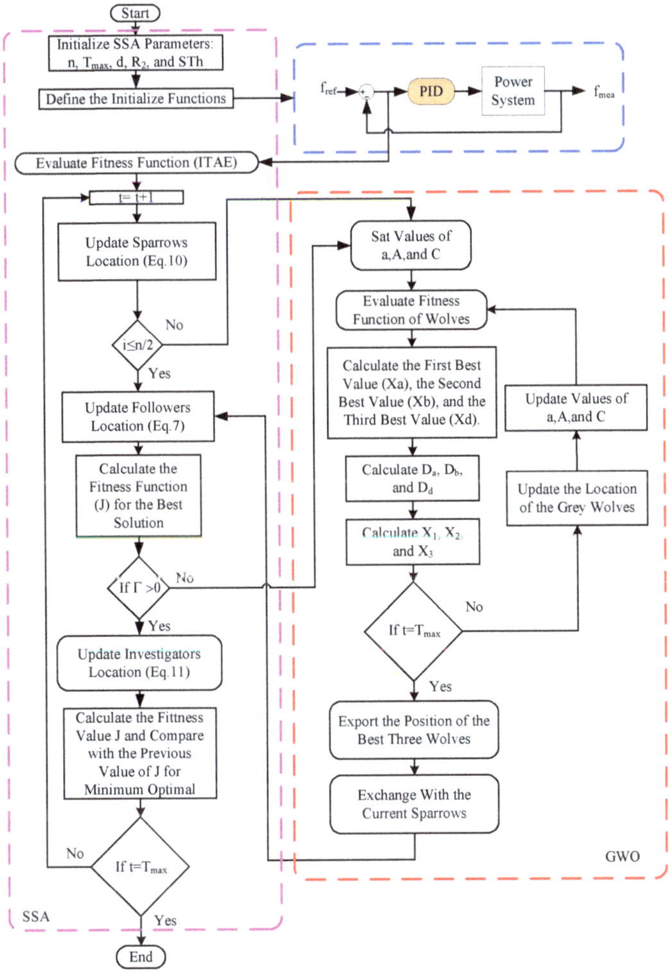

Figure 7. Flow chart of the proposed hybrid SSAGWO algorithm technique.

Step 10: By using Equation (9), if the probability factor is a positive value, calculate the fitness value using Equation (3) and compare it with the best fitness solution to obtain the minimum optimal value. However, if the probability factor is negative, go to step 5.

The parameters of the proposed SSAGWO algorithm technique are initialized at maximum iteration = 100, the number of search agents = 50, and the percentage of the total population size is selected as 0.2.

5. Simulation Results and Discussion

In this section, the proposed SSAGWO algorithm performance is first evaluated by comparing it with SSA and GWO algorithms in terms of statistical findings using five well-known benchmark functions from the literature. Furthermore, the proposed SSAGWO-tuned PID controller is analyzed for a two-area multi-source interconnected power system. The proposed SSAGWO optimized PID controller results are compared with SSA and GWO techniques. The obtained results for benchmark functions and various scenarios of the system model considered are discussed in detail in the following sections.

5.1. Validation of Benchmark Functions

Six classical benchmark functions with a wide range of characteristics are used to compare the proposed SSAGWO algorithm's performance to that of GWO and SSA algorithms. The description of the benchmark functions used to verify the performance of various hybrid SSAGWO algorithms is shown in Table 2. When referring to Table 2, the letter U indicates that the benchmark functions F1–F4 have a single global best and are unimodal. A function is said to be separable if and only if the letter S appears after the letter U in the notation. While if the letter N is written after the letter U, then the function is non-separable [3]. The exploitation ability of the optimization algorithm is investigated by the F1–F4 unimodal benchmark functions, as shown in Table 2, and they indicate that a robust local search capability is necessary for achieving good results. F5 and F6 are multimodal functions. These functions have multiple global bests and are used to investigate the optimization algorithm's exploration ability. The multimodal functions are denoted with the letter M as shown in Table 2. The multimodal functions are also classified into separable and non-separable functions.

Table 2. Benchmark functions.

Function Name		Type	Formula	Dimension (d)	Range
Sphere	F1	US	$f(x) = \sum_{i=1}^{d} x_i^2$	30	[−100, 100]
Schwefel 2.21	F2	US	$f(x) = \max_i\{\lvert x_i \rvert, 1 \le i \le d\}$	30	[−100, 100]
Schwefel 2.22	F3	UN	$f(x) = \sum_{i=1}^{d} \lvert x_i \rvert + \prod_{i=1}^{d} \lvert x_i \rvert$	30	[−10, 10]
Rosenbrock	F4	UN	$f(x) = \sum_{i=1}^{d-1} \left[100(x_{i+1} - x_i^2)^2 + (x_i - 1)^2\right]$	30	[−30, 30]
Rastrigin	F5	MS	$f(x) = 10d + \sum_{i=1}^{d} \left[x_i^d - 10\cos(2\pi x_i)\right]$	30	[−5.12, 5.12]
Ackley	F6	MN	$f(x) = -20\exp\left(-0.2\sqrt{\frac{1}{d}\sum_{i=1}^{d} x_i^2}\right) - \exp\left(\frac{1}{d}\sum_{i=1}^{d}\cos 2\pi x_i\right) + 20 + e$	30	[−32, 32]

Table 3 illustrates the statistical results of the optimization algorithm on the conventional benchmark functions. The results are compared based on the mean, minimum, and standard deviation of reaching the best values. The results were recorded for each algorithm 30 times running. Different benchmark functions are used to assess the algorithms GWO and SSA with SSAGWO by suggesting the algorithm population of 30 and 100 as a number of iterations. The results in Table 3 show that the SSAGWO has improved the exploration and exploitation ability compared with other optimization techniques. SSAGWO has mean, min, and standard deviation less than other algorithms.

Three cases are presented to analyze the time domain response of the HPS as follows. Case I: In this case, area 1 is integrated with an SPV source, and area 2 is integrated with a WPG source. A step load change of 50 MW and 35 MW of the rated power system occurs in area 1 and area 2, respectively. Figure 8 shows the frequency deviation of four optimization techniques for tuning PID controller parameters (PSO, GWO, SSA, and SSAGWO) and the Z–N technique. It is clear from the figure that the ST of the proposed SSAGWO-optimized PID controller is much faster than the other optimization-tuned PID controller and Z–N-tuned PID controller. Table 4 shows the results of the frequency deviation of area 1 and area 2 and the power deviation in the tie-line power. The ST of the proposed technique is improved by 75.06%, 74.29%, 71.60%, and 71.04% over Z–N, PSO, GWO, and SSA, respectively. Moreover, the RT of the frequency deviation of the proposed algorithm is less than other techniques by 85.18%, 78.54%, 73.65%, and 67.38%, respectively. In addition, the value of the undershoot of the frequency deviation is less than other techniques by 9.19%, 0.59%, 7.69%, and 0.59%, respectively. Furthermore, the results show that the steady-state error of the power system frequency when using the proposed SSAGWO is less by 85.06%, 78.49%, 73.60%, and 67.34%, respectively, than when using the other optimization techniques mentioned above. Furthermore, the steady-state values of the performance indices of the frequency deviation are improved by 30.43%, 70.45%, 39.50%, and 64.08%

for the ISE, ITSE, IAE, and ITAE, respectively, compared with the best performance index of the optimization techniques presented in Table 5. Moreover, the proposed algorithm reduces the controller efforts by 40.67%, 18.60%, 51.70%, and 9.85% for Z–N, PSO, GWO, and SSA, respectively.

Table 3. Benchmark functions statistics.

Function	Index	SSAGWO	SSA	GWO
F1	Best	2.2339×10^{-31}	2.7901×10^{-23}	42.0302
	SD	3.3466×10^{-17}	1.8513×10^{-19}	10.11
	Mean	1.0642×10^{-17}	6.6255×10^{-20}	19.9949
	Min	0	2.4457×10^{-47}	8.9247
F2	Best	5.1938×10^{-13}	8.0550×10^{-12}	22.7455
	SD	1.0443×10^{-10}	2.2659×10^{-10}	3.9943
	Mean	3.7025×10^{-11}	7.8765×10^{-11}	9.4191
	Min	1.6096×10^{-80}	4.3857×10^{-55}	5.3553
F3	Best	3.6567×10^{-13}	1.6810×10^{-11}	1.4005
	SD	4.9149×10^{-09}	6.1537×10^{-10}	0.3749
	Mean	1.7859×10^{-09}	3.9256×10^{-10}	1.3875
	Min	8.6574×10^{-58}	1.2831×10^{-24}	0.7203
F4	Best	2.8199×10^{-4}	2.7812×10^{-2}	548.2237
	SD	1.873×10^{-1}	4.97×10^{-2}	492.196
	Mean	1.113×10^{-1}	5.28×10^{-2}	963.3173
	Min	2×10^{-4}	9.3×10^{-3}	435.8798
F5	Best	0	0	76.6652
	SD	0	0	14.1344
	Mean	0	0	61.2244
	Min	0	0	36.401
F6	Best	8.8818×10^{-16}	8.8818×10^{-16}	2.4941
	SD	8.5631×10^{-11}	7.1236×10^{-10}	0.6564
	Mean	4.7806×10^{-11}	3.0156×10^{-10}	3.0160
	Min	8.8817×10^{-16}	8.8817×10^{-16}	1.3413

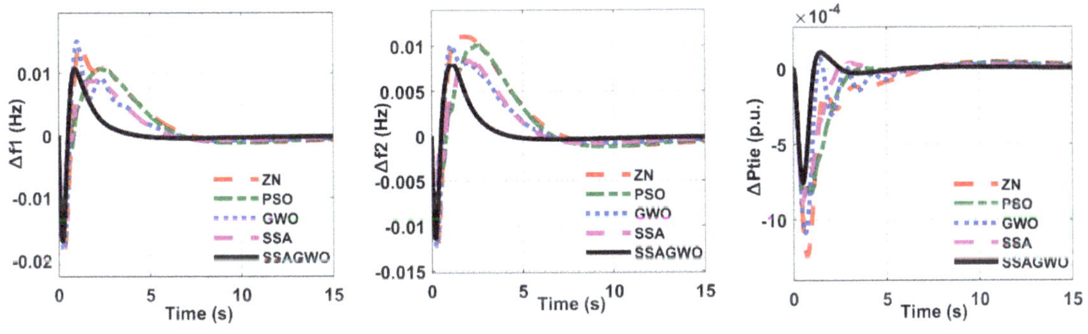

(a) (b) (c)

Figure 8. The frequency deviation responses of the optimization techniques and conventional Ziegler–Nichols method for tuning PID controller parameters for case I: (**a**) The frequency deviation of area 1; (**b**) The frequency deviation of area 2; (**c**) The power deviation of the tie line.

Table 4. Frequency and power deviation in the tie-line power of case I.

Optimization Technique	Controller Effort	ST (s)	RT (s)	Undershoot	Steady-State Error
			Δf_1		
Z–N	10.03×10^{-2}	15.5747	1.3000×10^{-3}	-1.85×10^{-2}	-2.416×10^{-4}
PSO	7.31×10^{-2}	15.1049	8.9766×10^{-4}	-1.69×10^{-2}	-1.678×10^{-4}
GWO	12.32×10^{-2}	13.6758	7.3107×10^{-4}	-1.82×10^{-2}	-1.367×10^{-4}
SSA	6.60×10^{-2}	13.4111	5.9055×10^{-4}	-1.69×10^{-2}	-1.105×10^{-4}
SSAGWO	5.95×10^{-2}	3.8834	1.9259×10^{-4}	-1.68×10^{-2}	-3.608×10^{-5}
			Δf_2		
Z–N	7.14×10^{-2}	16.7079	1.8×10^{-3}	-1.23×10^{-2}	-2.366×10^{-4}
PSO	7.14×10^{-2}	16.1989	1.3×10^{-3}	-1.14×10^{-2}	-1.643×10^{-4}
GWO	6.24×10^{-2}	15.1067	1.0×10^{-3}	-1.22×10^{-2}	-1.337×10^{-4}
SSA	6.02×10^{-2}	14.7612	8.2627×10^{-4}	-1.14×10^{-2}	-1.081×10^{-4}
SSAGWO	5.18×10^{-2}	9.9788	2.6857×10^{-4}	-1.13×10^{-2}	-3.521×10^{-5}
			ΔP_{tie}		
Z–N	6.6906×10^{-4}	13.6500	2.957×10^{-1}	-1.2×10^{-3}	10.035×10^{-6}
PSO	5.5430×10^{-4}	14.0710	4.0090	-0.8×10^{-3}	7.325×10^{-6}
GWO	3.4650×10^{-4}	11.5746	5.7×10^{-3}	-1.1×10^{-3}	6.509×10^{-6}
SSA	5.3688×10^{-4}	12.0433	2.87×10^{-2}	-0.8×10^{-3}	5.096×10^{-6}
SSAGWO	1.1460×10^{-4}	4.9285	2.0×10^{-3}	-0.7×10^{-3}	1.912×10^{-6}

Table 5. Steady-state indices in case I.

Performance Index	Δf_1				
	Z–N	PSO	GWO	SSA	SSAGWO
ITAE	22.81×10^{-2}	20.45×10^{-2}	14.77×10^{-2}	13.860×10^{-2}	4.978×10^{-2}
IAE	5.787×10^{-2}	4.785×10^{-2}	4.296×10^{-2}	3.625×10^{-2}	2.193×10^{-2}
ITSE	9.9×10^{-4}	8.1×10^{-4}	5×10^{-4}	4.4×10^{-4}	1.3×10^{-4}
ISE	5×10^{-4}	3.4×10^{-4}	3.4×10^{-4}	2.3×10^{-4}	1.6×10^{-4}
Performance Index	Δf_2				
	Z–N	PSO	GWO	SSA	SSAGWO
ITAE	22.22×10^{-2}	20.070×10^{-2}	14.4×10^{-2}	13.64×10^{-2}	4.845×10^{-2}
IAE	5.236×10^{-2}	4.425×10^{-2}	3.816×10^{-2}	3.322×10^{-2}	1.862×10^{-2}
ITSE	8.9×10^{-4}	7.5×10^{-4}	4.3×10^{-4}	4×10^{-4}	1×10^{-4}
ISE	3.8×10^{-4}	2.7×10^{-4}	2.3×10^{-4}	1.7×10^{-4}	0.997×10^{-4}

Case II: In this case, area 1 and area 2 are integrated with SPV resources. The step change in demand load applied in case II is similar to that of case I. Figure 9 shows the frequency deviation of the two areas and the power deviation in tie-line power for tuning PID parameters of the proposed SSAGWO compared with PSO, GWO, and SSA and the Z–N technique. Figure 9 demonstrates that the ST of the proposed SSAGWO-optimized PID controller is significantly less than that of the other optimization-tuned PID controllers and the Z–N-tuned PID controller. The results of the frequency deviation and tie-line power of case II are shown in Table 6. The ST of the proposed algorithm has been enhanced by 75.94%, 74.23%, 72.53%, and 70.83% compared with the ST of Z–N, PSO, GWO, and SSA, respectively. In addition, the RT of the frequency deviation of the proposed SSAGWO is less than other techniques by 85.63%, 80.60%, 75.24%, and 69.86%, respectively. Moreover, the peak undershoots value of the frequency deviation is less than other techniques by 8.15%, 8.15%, 7.65%, and 7.65%, respectively. Furthermore, the results shown in Table 6 depict that the steady-state error of the frequency deviation of the power system is less than other techniques by 85.93%, 80.54%, 75.20%, and 69.83%, respectively, when using the proposed SSAGWO to tune PID controller parameters.

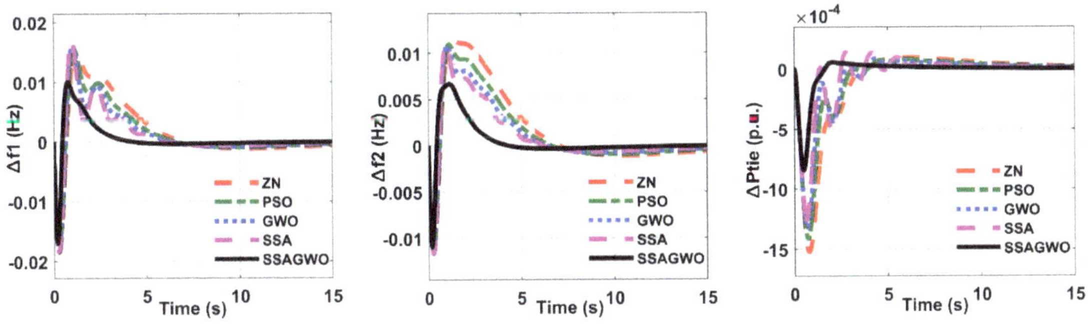

(**a**) (**b**) (**c**)

Figure 9. The frequency deviation responses of the optimization techniques and conventional Ziegler–Nichols method for tuning PID controller parameters for case II: (**a**) The frequency deviation of area 1; (**b**) The frequency deviation of area 2; (**c**) The power deviation of the tie line.

Table 6. Frequency and power deviation in the tie-line power of case II.

Optimization Technique	Controller Effort	ST (s)	RT (s)	Undershoot	Steady-State Error
			Δf_1		
Z–N	14.71×10^{-2}	15.5119	12.000×10^{-4}	-1.84×10^{-2}	-2.296×10^{-4}
PSO	17.09×10^{-2}	14.4838	8.8905×10^{-4}	-1.84×10^{-2}	-1.661×10^{-4}
GWO	19.04×10^{-2}	13.5912	6.9661×10^{-4}	-1.83×10^{-2}	-1.303×10^{-4}
SSA	20.49×10^{-2}	12.7962	5.7231×10^{-4}	-1.83×10^{-2}	-1.071×10^{-4}
SSAGWO	8.060×10^{-2}	3.73260	1.7247×10^{-4}	-1.69×10^{-2}	-3.231×10^{-5}
			Δf_2		
Z–N	8.96×10^{-2}	16.7318	17.000×10^{-4}	-1.16×10^{-2}	-2.264×10^{-4}
PSO	9.55×10^{-2}	15.9302	13.000×10^{-4}	-1.16×10^{-2}	-1.637×10^{-4}
GWO	10.32×10^{-2}	15.1767	9.8148×10^{-4}	-1.16×10^{-2}	-1.283×10^{-4}
SSA	11.09×10^{-2}	14.4961	8.0590×10^{-4}	-1.16×10^{-2}	-1.054×10^{-4}
SSAGWO	5.19×10^{-2}	9.69000	2.4235×10^{-4}	-1.09×10^{-2}	-3.178×10^{-5}
			ΔP_{tie}		
Z–N	6.8776×10^{-4}	12.0385	6.85×10^{-2}	-1.5×10^{-3}	6.506×10^{-6}
PSO	6.9817×10^{-4}	11.2029	81.85×10^{-2}	-1.3×10^{-3}	5.092×10^{-6}
GWO	8.2955×10^{-4}	10.5142	7.3×10^{-3}	-1.3×10^{-3}	4.183×10^{-6}
SSA	17.000×10^{-4}	9.7186	10.9×10^{-3}	1.2×10^{-3}	3.547×10^{-6}
SSAGWO	6.1746×10^{-4}	6.4993	3.6×10^{-3}	-0.8×10^{-3}	1.167×10^{-6}

The steady-state values of the ISE, ITSE, IAE, and ITAE all improve when using the proposed SSAGWO upon the best performance index of the optimization strategies shown in Table 7 by 57.57%, 74.42%, 46.93%, and 62.58%, respectively. Moreover, the controller effort was reduced by 45.20%, 52.83%, 57.66%, and 60.66%, respectively, when using SSAGWO compared with other optimization techniques.

Table 7. Steady-state indices in case II.

Performance Index	Δf_1				
	Z–N	PSO	GWO	SSA	SSAGWO
ITAE	22.71×10^{-2}	17.88×10^{-2}	14.71×10^{-2}	12.45×10^{-2}	4.658×10^{-2}
IAE	5.845×10^{-2}	4.97×10^{-2}	4.356×10^{-2}	3.899×10^{-2}	2.069×10^{-2}
ITSE	10.1×10^{-4}	7×10^{-4}	5.3×10^{-4}	4.3×10^{-4}	1.1×10^{-4}
ISE	5.2×10^{-4}	4.2×10^{-4}	3.6×10^{-4}	3.3×10^{-4}	1.4×10^{-4}
Performance Index	Δf_2				
	Z–N	PSO	GWO	SSA	SSAGWO
ITAE	21.72×10^{-2}	17.07×10^{-2}	14.03×10^{-2}	11.91×10^{-2}	4.393×10^{-2}
IAE	5.146×10^{-2}	4.32×10^{-2}	3.747×10^{-2}	3.33×10^{-2}	1.679×10^{-2}
ITSE	8.5×10^{-4}	5.7×10^{-4}	4.1×10^{-4}	3.1×10^{-4}	0.801×10^{-4}
ISE	3.7×10^{-4}	2.8×10^{-4}	2.2×10^{-4}	1.9×10^{-4}	0.792×10^{-4}

Case III: In this case, WTPG resources are integrated into both areas 1 and area 2. Figure 10 shows the frequency deviation and tie-line power of the HPS model. The figures show that the ST of the frequency deviation when using SSAGWO is less than the other optimization-tuned PID controller parameters mentioned above. The detailed results of this case are shown in Table 7. With the same sequence of comparing the proposed SSAGWO with other optimization techniques illustrated in case I and case II, the ST was improved by 75.40%, 73.66%, 71.88%, and 70.03%, respectively. In addition, the RT of the frequency deviation was reduced by 88.10%, 84.01%, 79.56%, and 75.15%, respectively. There is also a reduction in the peak undershoot value of the frequency deviation of 8.37%, 7.89%, 7.89%, and 7.41, respectively, compared to the other techniques. Table 8 further demonstrates that using the proposed SSAGWO to tune PID controller parameters decreased the steady-state error of the frequency deviation of the power system by 88.41%, 83.97%, 79.52%, and 75.10%, respectively.

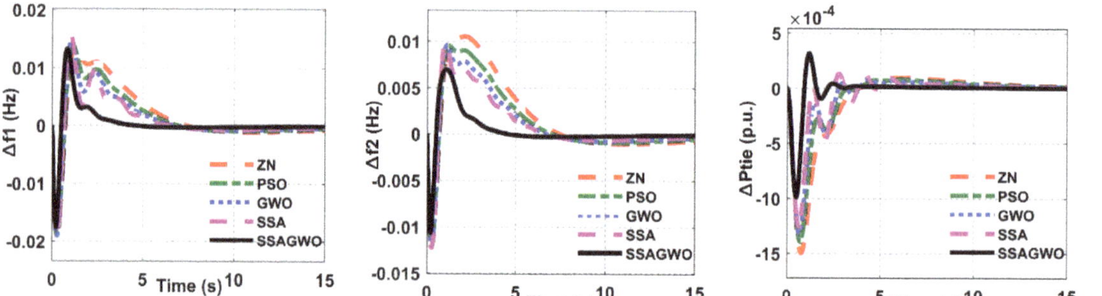

(**a**) (**b**) (**c**)

Figure 10. The frequency deviation responses of the optimization techniques and conventional Ziegler–Nichols method for tuning PID controller parameters for case III: (**a**) The frequency deviation of area 1; (**b**) The frequency deviation of area 2; (**c**) The power deviation of the tie line.

Table 8. Frequency and power deviation in the tie-line power of case III.

Optimization Technique	Controller Effort	ST (s)	RT (s)	Undershoot	Steady-State Error
			Δf_1		
Z-N	12.84×10^{-2}	15.5405	13.000×10^{-4}	-1.91×10^{-2}	-2.5010×10^{-4}
PSO	14.98×10^{-2}	14.5127	9.6740×10^{-4}	-1.90×10^{-2}	-1.8080×10^{-4}
GWO	17.32×10^{-2}	13.5950	7.5658×10^{-4}	-1.90×10^{-2}	-1.4150×10^{-4}
SSA	12.73×10^{-2}	12.7563	6.2227×10^{-4}	-1.89×10^{-2}	-1.1640×10^{-4}
SSAGWO	10.25×10^{-2}	3.8229	1.5464×10^{-4}	-1.75×10^{-2}	-0.2898×10^{-4}
			Δf_2		
Z-N	8.55×10^{-2}	16.7930	19×10^{-4}	-1.22×10^{-2}	-2.469×10^{-4}
PSO	7.93×10^{-2}	15.9452	14×10^{-4}	-1.22×10^{-2}	-1.783×10^{-4}
GWO	8.88×10^{-2}	15.1851	11×10^{-4}	-1.22×10^{-2}	-1.395×10^{-4}
SSA	9.99×10^{-2}	14.5028	8.7753×10^{-4}	-1.22×10^{-2}	-1.148×10^{-4}
SSAGWO	6.48×10^{-2}	8.8496	2.1741×10^{-4}	-1.06×10^{-2}	-0.285×10^{-4}
			ΔP_{tie}		
Z-N	7.2857×10^{-4}	12.0973	58.7×10^{-3}	-15×10^{-4}	6.501×10^{-6}
PSO	6.2357×10^{-4}	11.2433	34.6×10^{-3}	-14×10^{-4}	5.083×10^{-6}
GWO	8.4564×10^{-4}	10.5482	7.9×10^{-3}	-13×10^{-4}	4.173×10^{-6}
SSA	5.4×10^{-4}	9.9408	9.1×10^{-3}	-12×10^{-4}	3.536×10^{-6}
SSAGWO	1.7×10^{-4}	3.8952	3.2793×10^{-3}	-9×10^{-4}	1.022×10^{-6}

Compared to the best performance index of the optimization strategies presented in Table 9, the proposed SSAGWO improves the steady-state values of the ISE, ITSE, IAE, and ITAE by 50%, 74.42%, 49.88%, and 68.96%, respectively. In addition, SSAGWO reduced controller effort by 20.17%, 31.57%, 40.82%, and 19.48%, respectively, compared to other optimization methods.

Table 9. Steady-state indices in case III.

Performance Index			Δf_1		
	Z-N	PSO	GWO	SSA	SSAGWO
ITAE	23.42×10^{-2}	18.45×10^{-2}	15.17×10^{-2}	12.87×10^{-2}	3.995×10^{-2}
IAE	5.882×10^{-2}	5.01×10^{-2}	4.401×10^{-2}	3.947×10^{-2}	1.978×10^{-2}
ITSE	10.3×10^{-4}	7.1×10^{-4}	5.3×10^{-4}	4.3×10^{-4}	1.1×10^{-4}
ISE	5.1×10^{-4}	4.1×10^{-4}	3.5×10^{-4}	3.2×10^{-4}	1.6×10^{-4}
Performance Index			Δf_2		
	Z-N	PSO	GWO	SSA	SSAGWO
ITAE	22.43×10^{-2}	17.65×10^{-2}	14.51×10^{-2}	12.29×10^{-2}	3.749×10^{-2}
IAE	5.18×10^{-2}	4.357×10^{-2}	3.787×10^{-2}	3.366×10^{-2}	1.524×10^{-2}
ITSE	8.7×10^{-4}	5.9×10^{-4}	4.2×10^{-4}	3.2×10^{-4}	0.671×10^{-4}
ISE	3.6×10^{-4}	2.7×10^{-4}	2.2×10^{-4}	1.8×10^{-4}	0.731×10^{-4}

A comparison of the performance of the proposed hybrid SSAGWO optimization technique with the performance of the Grasshopper Optimization Algorithm (GOA) [17], Marine Predator Algorithm (MPA) [20], and Salp Swarm Optimization (SSO) [25] has been implemented in order to demonstrate the robustness of the suggested approach. Within the context of this discussion, case I has been utilized as a case study. Figure 11 depicts the dynamic response of the HPS model to the frequency deviation and the change in tie-line power. When compared with GOA, MPA, and SSO, the results demonstrate that the ST, overshoot, and oscillations of the frequency deviation, as well as the tie-line power, are improved when SSAGWO is utilized.

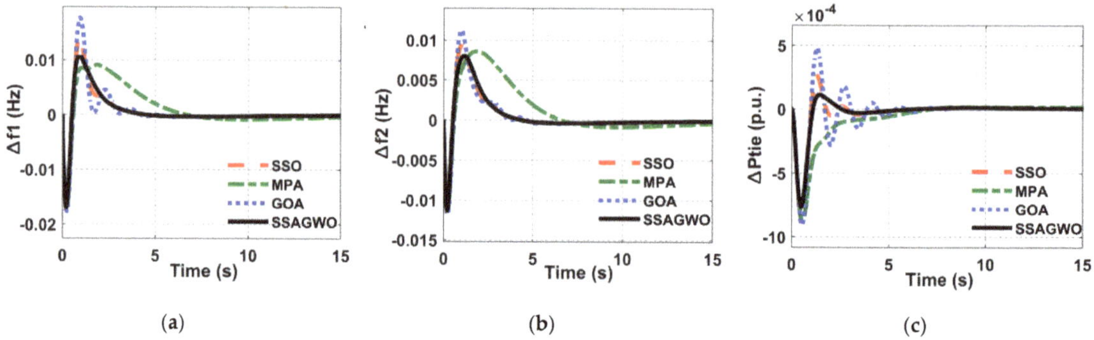

Figure 11. The frequency deviation responses and change in tie line power of SSAGWO compared with GOA, MPA, and SSO for tuning PID controller parameters for case I: (**a**) The frequency deviation of area 1; (**b**) The frequency deviation of area 2; (**c**) The power deviation of the tie line.

To study the effectiveness of the proposed algorithm, a random step change in load demand is applied in area 1, as shown in Figure 12. The frequency response of area 1, area 2, and tie-line power for this scenario are depicted in Figure 13. According to the findings, the proposed algorithm tunes the PID controller response more quickly for a sudden change in load than other optimization techniques do. Additionally, the power supplied by the tie-line changed in response to the changing load demands of the system while keeping a constant level of output despite these changes.

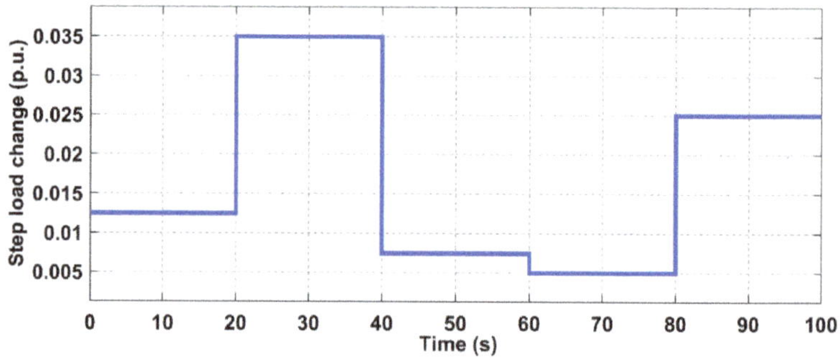

Figure 12. Random change load.

To further prove the robustness of the proposed technique, solar radiation data collected at Universiti Putra Malaysia, the output power from the SPV, is integrated with area 1 [3]. As shown in Figure 14, the solar irradiation values were measured from January through December 2014 for a total of 200 time-slots of 0.5 s. The multi-area power system in case I has been chosen to assess the dynamic response for real-time solar power fluctuations with all optimization techniques used in this study. The frequency deviation of area 1 and area 2 is shown in Figure 15 for this scenario. It can be deduced that the proffered SSAGWO algorithm-tuned PID controller shows a better and smoother response than other tuning methods, along with minimal undershoot, less ST, RT, steady-state error, and controller efforts.

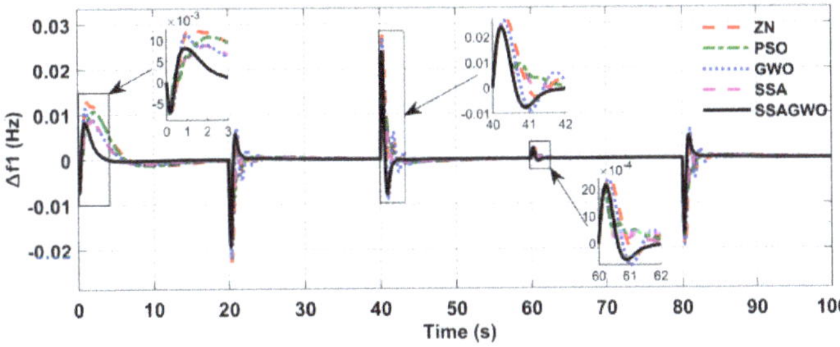

(a)

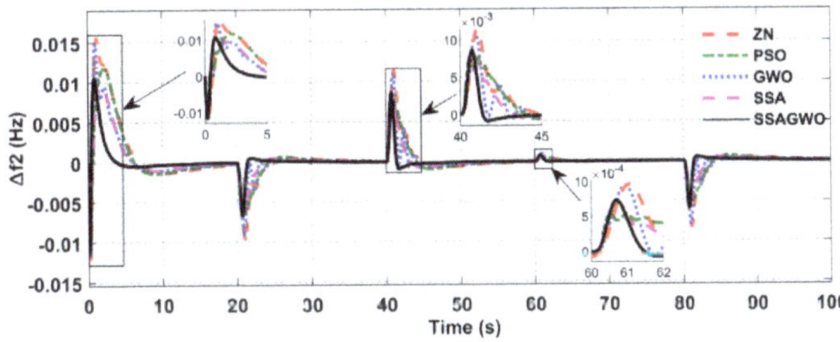

(b)

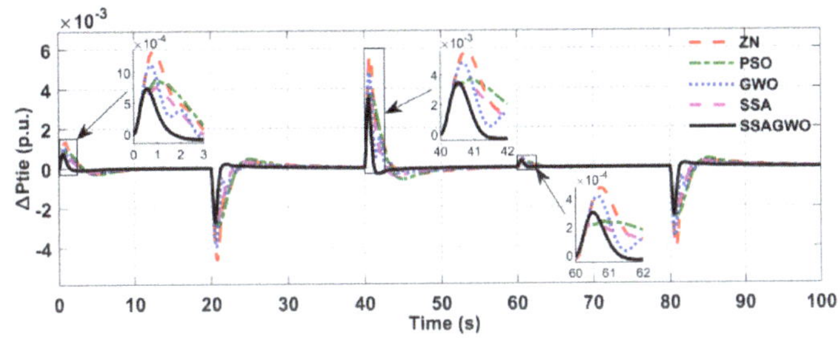

(c)

Figure 13. Dynamic response of the multi-area power system at a random change load: (**a**) The frequency deviation of area 1; (**b**) The frequency deviation of area 2; (**c**) The power deviation of the tie line.

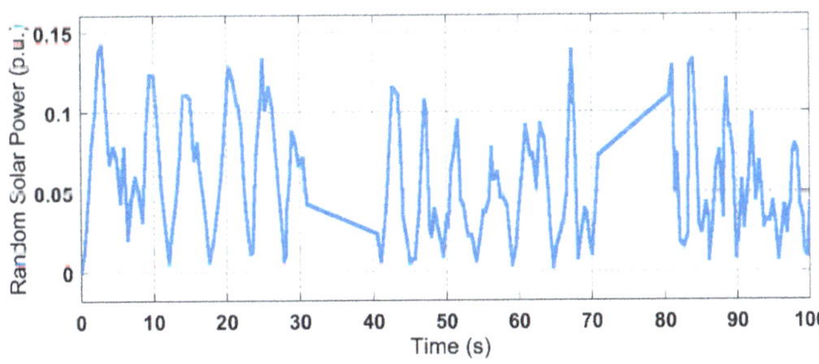

Figure 14. Realize solar input power.

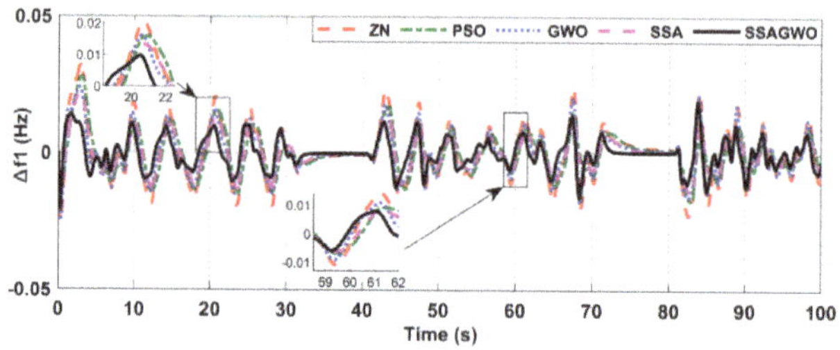

(a)

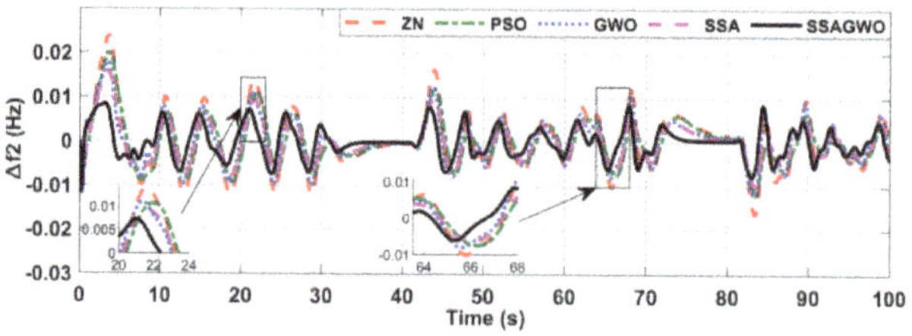

(b)

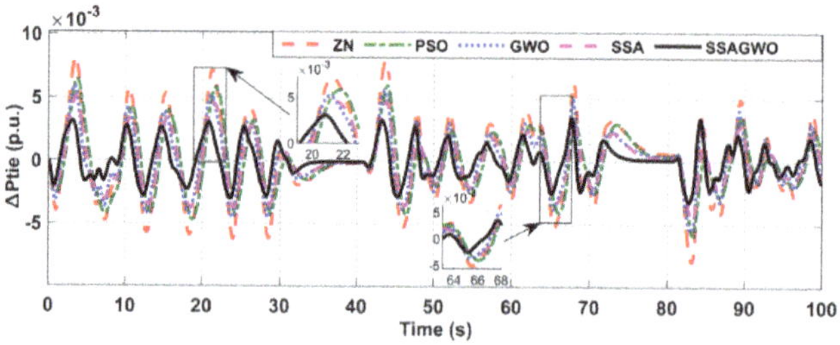

(c)

Figure 15. Dynamic response comparison of the multi-area power system using various optimization techniques: (**a**) The frequency deviation of area 1; (**b**) The frequency deviation of area 2; (**c**) The power deviation of the tie line.

The sensitivity analysis of the proposed controller technique has been performed to prove the robustness of the controller parameters obtained under a variation of the inertia constant and change in the nominal loading of the power system.

The value of the inertia constant of the power system changed from 75% to 125% of its nominal value (H = 5), and the loading condition changed by ±25% compared with the nominal system loading (50% loading). The controller's efficacy is demonstrated by using the value of the PID controller parameters under the nominal condition. Changing the inertia constant of the power system will affect the value of the time constant Tps. Likewise, the time constant of the power system block Tps and the gain constant Kps are affected by changes in load conditions. Figure 16 illustrates the dynamic response of the power system with variation in the inertia constant value. Figure 17 shows the dynamic response of the power system under change in the nominal loading. The result demonstrates a high tolerance for a wide range of changes in system parameters, as measured by the gain values obtained under nominal conditions. Because of this, one can draw the conclusion that the parameters do not need to be re-tuned even if there is a large amount of change in the system's conditions and parameters.

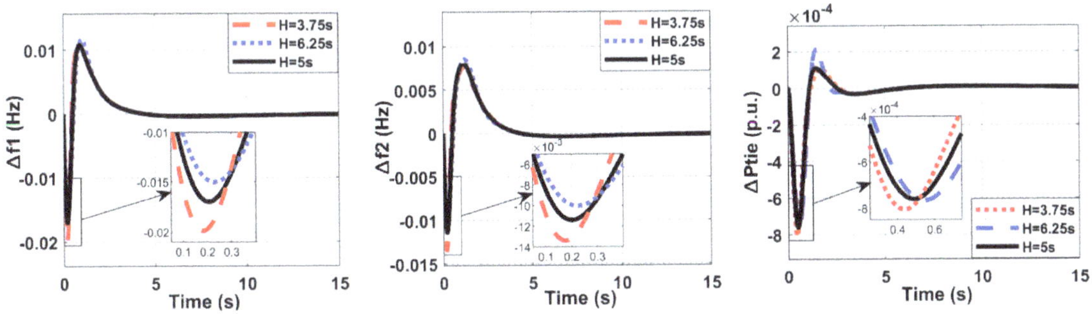

Figure 16. The dynamic response of the power system under different inertia constant (75% and 125% of the nominal value H = 5 s).

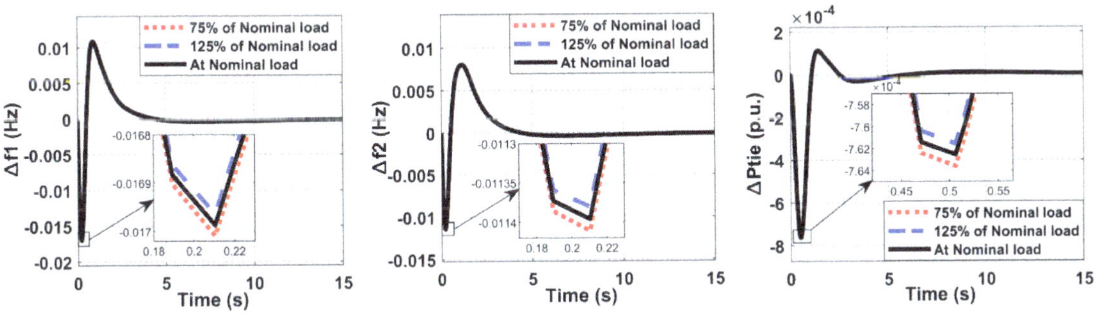

Figure 17. The dynamic response of the power system under different system loading (±25% of the nominal load).

This section presents the convergence curves obtained by the proposed SSAGWO algorithm and other optimization algorithms. The convergence curve showing the minimum ITAE of ACE for the model is presented in Figure 18. The hybrid SSAGWO approach is shown to converge faster than other optimization methods, proving that the proposed SSAGWO has a better balance between exploration and exploitation than the standard GWO and SSA algorithms.

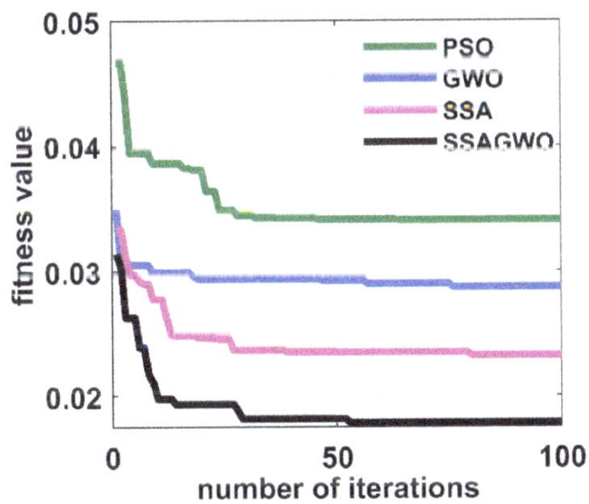

Figure 18. Convergence curve of the algorithms.

5.2. Stability Analysis

It is essential to take into consideration the effects of optimization techniques for PID tuning and validation of the stability analysis by calculating the state space of the system based on the Simulink model of the system. To assess the frequency stability of the power system and the HPS model optimized with SSAGWO-tuned PID controller parameters, two scenarios are considered. In scenario 1, Area I is considered as the input signal of the transfer function, while in scenario 2, Area II is considered as the input signal of the transfer function. The state-space equations with estimation of the Close Loop Transfer Function (CLTF) are illustrated as follows:

Let, $\Delta f_i = X_1$, $\Delta P_{Ti} = X2$, $\Delta T_T = X3$, $\Delta P_{tie,i} = X5$, $\Delta P_{WT} = X6$, $\Delta P_{PV} = X7$, and $\Delta P_{RFB} = X8$

The vector form of the state variables of the proposed model is present in Equation (12):

$$X = [X_1, X_2, X_3, X_4, X_5, X_6, X_7, X_8]^T \tag{12}$$

System parameters and their associated state variables are shown as below:

$$X = [\Delta f_i \Delta X_{1T} \Delta X_{1RE} \Delta P_{tie,i}]^T \tag{13}$$

where

$$\Delta X_{1T} = [\Delta P_T \Delta T_T \Delta G_T] \tag{14}$$

$$\Delta X_{1RE} = [\Delta P_{WT} \Delta P_{PV} \Delta P_{RFB}] \tag{15}$$

The state-space equation of the HPS is illustrated as follows:

$$X_1^{\bullet} = \frac{K_{pi}}{T_{pi}}[-X_1 + X_2 + X_3 + X_4 + X_5 + X_6 + X_7 - X_8 - \Delta P_{Di}] \tag{16}$$

The state-space modeling of the reheat turbine generator is also derived in Equations (17)–(19):

$$X_2^{\bullet} = -\frac{X_2}{T_{ti}} + \frac{X_3}{T_{ti}} \tag{17}$$

$$X_3^{\bullet} = -\frac{X_3}{T_{ri}} + \left[\frac{1}{T_{ri}} - \frac{K_r}{T_{Gi}}\right] X_4 + \frac{1}{T_{ri}} \Delta P_{ci} - \frac{X_1}{T_{ri} R_i} \tag{18}$$

$$X_4^{\bullet} = -\frac{X_4}{T_{Gi}} + \frac{\Delta P_{ci}}{T_{Gi}} - \frac{1}{R_i T_{Gi}} X_1 \tag{19}$$

In general, the state-space of the tie line power can be emulated as follows:

$$X_5^{\bullet} = \Delta P_{tie,i} = 2\pi \left[\sum_{i=1, i \neq j}^{n} T_{ij} X_i \right] \tag{20}$$

The wind turbine model in state space is given in Equation (21):

$$X_6^{\bullet} = -\frac{X_6}{T_{WT}} + K_{WT} P_W \tag{21}$$

The PV model in state space is obtained in Equation (22):

$$X_7^{\bullet} = -\frac{X_7}{T_{PV}} + K_{PV} P_V \tag{22}$$

The state-space model of the RFB is shown below:

$$X_8^{\bullet} = -\frac{X_8}{T_{RFB}} + K_{RFB} P_{RFB} \tag{23}$$

The general state-space representation is given below:

$$X^{\bullet} = AX + Bu \tag{24}$$

where A is the state matrix, B is the input matrix, and C is the output matrix. According to the previous mathematical steps, A, B, and C can be calculated.

The control input can be presented as:

$$u = [\Delta P_{ci}]^T \tag{25}$$

The definition of the output matrix of the proposed system is:

$$Y = CX \tag{26}$$

The closed loop control of the HPS shown in Figure 6, can be simplified as shows in Figure 19 in order to estimate the CLTF equation:

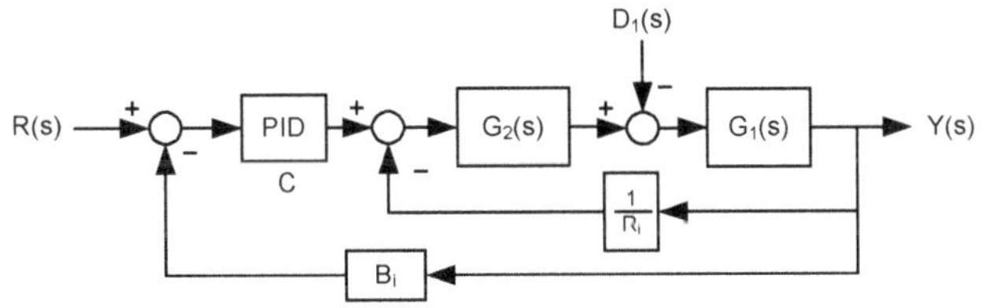

Figure 19. The simplified closed loop control system for stability analysis.

The derivation of the total response of the closed-loop transfer function of the interconnected power system was modeled with PID controller as the following:

$$R(s)G_2(s)G_1(s)C - B_iY(s)CG_2(s)G_1(s) - Y(s)\frac{1}{R_i}G_2(s)G_1(s) - D_1(s)G_1(s) = Y(s) \tag{27}$$

$$Y(s) = \frac{R(s)G_1(s)G_2(s)C}{Z} - \frac{D_1(s)G_1(s)}{Z} \tag{28}$$

$$\therefore Y(s) = Y_{11}(s) + Y_{12}(s) \tag{29}$$

where

$$Z = 1 + B_iCG_1(s)G_2(s) + \frac{1}{R_i}G_1(s)G_2(s) \tag{30}$$

The change in load demand (ΔP_D) is addressed for using the following expression for the CLTF of the system for the proposed controller:

$$\Delta f = \frac{-G_1}{1 + G_1G_2\left[CB_i + \frac{1}{R}\right]}\Delta P_D \tag{31}$$

The CLTF can be applied to variations in tie-line power, as indicated by the definition of power variation:

$$\Delta f = \frac{CG_1G_2}{1 + G_1G_2\left[CB_i + \frac{1}{R}\right]}\Delta P_{tie} \tag{32}$$

By using superposition theorem, the CLTF for the total variations of the frequency response can be defined as:

$$\Delta f = \frac{-G_1\Delta P_D + CG_1G_2\Delta P_{tie}}{1 + G_1G_2\left[CB_i + \frac{1}{R}\right]} \tag{33}$$

By using the state-space analysis, the CLTF for the two scenarios (i.e., area 1 and area 2) can be presented as:

$$CLTF = \frac{num}{den} \quad (34)$$

Appendix B shows the state-space matrix, definition of the num, and den of the proposed power system model.

The proposed SSAGWO-tuned PID controller is of a higher order (11th), making stability analysis challenging. Therefore, the higher-order CLTF is reduced to a second-order transfer function using the Hankel matrix (HM) norm approximation method [48]. Detailed steps for reducing the higher-order transfer function are given below.

From the state-space in Appendix B, HM can be obtained. The general form of the HM can be presented as:

$$H_{nn}^{0} = \sum_{i=1}^{n}\sum_{j=1}^{n}\left(CA^{-n+1}B\right)_{i,j} \quad (35)$$

The value of n = 13, therefore the HM is shown in Equation (35), which can be expressed as below:

$$H_{1313}^{0} = \sum_{i=1}^{13}\sum_{j=1}^{13} e_{i,j} \quad (36)$$

More detail about the HM technique to reduce the order of the CLTF of the proposed power system model has been given in [48]. To stabilize power generators, grid frequency should be controlled based on the droop characteristic that relates to the generator output [49]. Figure 20 depicts the frequency response of the Bode plot for the HPS model considering the two scenarios proposed in stability analysis, with a gain margin of 1.57 and 8.71 dB and a gain cross-over frequency of 3.65 and 0.81 (rad/s) for Area I and Area II, respectively. According to the Bode analysis response, the closed-loop system for the proposed SSAGWO optimized PID controller of the HPS network is stable.

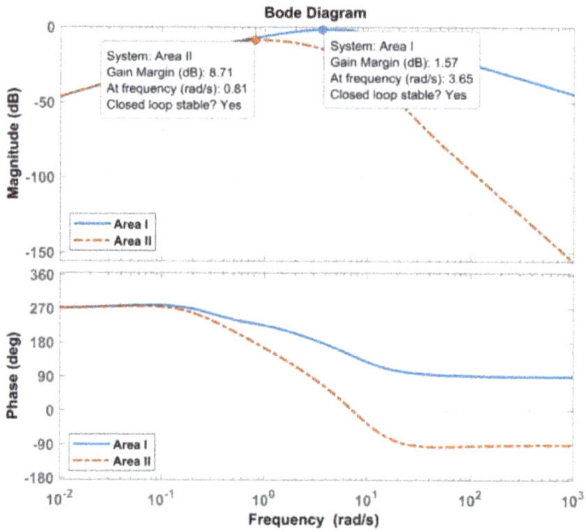

Figure 20. Bode analysis of frequency stability.

6. Conclusions

In this manuscript, a hybrid SSAGWO algorithm is developed for optimal tuning of a PID controller for the load frequency control of multi-area interconnected power system. The proposed power system includes a high nonlinearity reheat steam station integrated with various RERs. The proposed SSAGWO optimization algorithm was utilized to enhance the parameters of the PID controller to minimize the frequency deviation in the presence of a PV system, wind turbine system generation, and load disturbance and to improve the robustness of the PID controller with SSAGWO contrary to the penetration of RERs and the effect of nonlinearities of the conventional plants of the HPS. Since the performance of the SSAGWO algorithm depends on minimizing the complexity of the fitness functions, the constraints and the boundaries of the proposed algorithm was modified by using the SSA algorithm and the GWO algorithm. This modified approach improves the dynamic performance of the LFC, while considering reduced-order of the entire transfer function for the sensitivity and stability analyses. In order to validate the dynamic performance and rigidity of the proposed optimization algorithm, a state-space model of the high-order close loop transfer function was computed in this work. Furthermore, the dynamic performance of the optimization algorithm was compared with PSO, GWO, SSA, GOA, MPA, and SSO. It is concluded that the SSAGWO achieved the best performance (i.e., settling time is 3.8834 s) in comparison with PSO, GWO and SSA; where the settling time values are 15.1049 s, 13.6758 s, and 13.4111 s, respectively.

Author Contributions: Conceptualization, B.A.F.: proposed the hybrid optimization algorithm, methodology, formal analysis, and authored the paper; N.I.A.W.: supervision and investigation; A.J.M.: validation and proofreading; M.P. and A.X.R.I.: writing, review and proofreading; M.A.B.M.R.: review and editing; A.B.C.S.: review and editing; V.V.: review and investigation. All authors have read and agreed to the published version of the manuscript.

Funding: This research project is supported by Putra Grant (GP-GPB/2021/9706100) and The APC was funded by Universiti Putra Malaysia.

Acknowledgments: The authors gratefully acknowledge Department of Electrical and Electronic Engineering, Universiti Putra Malaysia, for providing space to conduct this research.

Conflicts of Interest: The authors declare no conflict of interest.

Abbreviations

ACE	Area Control Error
ACO	Ant Colony Optimization
ANN	Artificial Neural Network
ALFC	Automatic Load Frequency Control
BD	Boiler Dynamics
BES	Battery Energy System
CLTF	Closed-Loop Transfer Function
CTO	Class Topper Optimization
DE	Differential Evolutionary
FLC	Fuzzy Logic Controller
GDB	Governor Dead Band
GOA	Grasshopper Optimization Algorithm
GRC	Generator Rate Constraints
GWO	Grey Wolf Optimizer
HIO	Hybrid Intelligent Optimization
HPS	Hybrid Power System
IAE	Integral Absolute Error
ISE	Integral Square Error
ITAE	Integral Time Absolute Error
ITSE	Integral Time Square Error
MPA	Marine Predator Algorithm
MPC	Model Predictive Control
MRFO	Manta-Ray Foraging Optimizer
PFMPID	Predictive Functional Modified Proportional Integral Derivative
PID	Proportional Integral Derivative
PIDA	Proportional-Integral-Derivative-Acceleration
PSO-GSA	Particle Swarm Optimized-Gravitational Search Algorithm
PV	Photovoltaic Cell

RER	Renewable Energy Resources
RFB	Redox Flow Battery
RT	Rise Time
SMC	Sliding Mode Control
SMES	Superconducting Magnetic Energy Storage
SPV	Solar PV
SSAGWO	Sparrow Search Algorithm-Grey Wolf Optimizer
SSO	Salp Swarm Optimization
ST	Settling Time
STPP	Solar Thermal Power Plant
WOA	Whale Optimization Algorithm
WTPG	Wind Turbine Power Generator
Z–N	Ziegler-Nichols

Appendix A

Table A1. The parameters of the proposed HPS and RERs.

System Parameters	Value
Base Rated Power of area 1 and area 2 $P_{R1} = P_{R2}$	2000 MW
The gains of power system $K_{P1} = K_{P2}$	120 Hz/p.u.MW
The time constant of the power system $T_{P1} = T_{P2}$	0.08 s
The turbine time constant $T_{T1} = T_{T2}$	0.3 s
The time constant of the Reheat $T_{r1} = T_{r2}$	10 s
The gains of the Reheat $K_{r1} = K_{r2}$	0.5
The governor adjustment deviation coefficients $R_1 = R_2$	2.4 Hz/p.u.MW
The frequency response coefficients $B_1 = B_2$	0.425 p.u.MW/Hz
The system damping coefficient $D_1 = D_2$	0.00833 p.u.MW/Hz
The time constants of tie-line flow $T_{12} = T_{21}$	0.08674 p.u.MW/rad
System inertia $H_1 = H_2$	5 s
Solar PV time constant T_{PV}	1.3
Wind turbine time constant T_{WT}	1.5

Table A2. The parameters of the steam boiler for the power plant.

Boiler Parameters	Value
K_1	0.85
K_2	0.095
K_3	0.92
C_B	200
T_D	0
K_{IB}	0.03
T_{IB}	26
T_{RB}	69
T_F	10

Appendix B

$$A = \begin{bmatrix} -4.07 & 0 & -0.44 & 4.81 & 0.04 & 0 & 0 & 0 & 0 & 0 & 0 & 0.77 & 0 \\ 0 & -4.07 & 0.44 & 0 & 0 & 4.81 & 0.04 & 0 & 0 & 0 & 0 & 0 & 0.67 \\ 6 & -6 & 0 & 0 & 0 & 0 & 0 & 0 & 0 & 0 & 0 & 0 & 0 \\ -10.2 & 0 & -1.43 & -12.5 & 0 & 0 & 0 & -0.22 & 0.8 & 0 & 0 & 0 & 0 \\ 0 & 0 & 0 & 12.5 & -0.1 & 0 & 0 & 0 & 0 & 0 & 0 & 0 & 0 \\ 0 & -10.2 & 1.43 & 0 & 0 & -12.5 & 0 & 0 & 0 & -0.22 & 0.8 & 0 & 0 \\ 0 & 0 & 0 & 0 & 0 & 12.5 & -0.1 & 0 & 0 & 0 & 0 & 0 & 0 \\ -0.05 & 0 & -0.01 & 0 & 0 & 0 & 0 & -0.28 & 0 & 0 & 0 & 0 & 0 \\ -10.2 & 0 & -1.78 & 0 & 0 & 0 & 0 & 0 & 0 & 0 & 0 & 0 & 0 \\ 0 & -0.05 & 0.01 & 0 & 0 & 0 & 0 & 0 & 0 & -0.28 & 0 & 0 & 0 \\ 0 & -10.2 & 1.78 & 0 & 0 & 0 & 0 & 0 & 0 & 0 & 0 & 0 & 0 \\ 0 & 0 & 0 & 0 & 0 & 0 & 0 & 0 & 0 & 0 & 0 & -0.77 & 0 \\ 0 & 0 & 0 & 0 & 0 & 0 & 0 & 0 & 0 & 0 & 0 & 0 & -0.67 \end{bmatrix} \tag{A1}$$

$$B = \begin{bmatrix} -1 & 0 & 0 & 0 & 0 & 0 & 0 & 0 & 0 & 0 & 0 & 0 & 0 \end{bmatrix}^T \tag{A2}$$

$$C = \begin{bmatrix} 6 & 0 & 0 & 0 & 0 & 0 & 0 & 0 & 0 & 0 & 0 & 0 & 0 \end{bmatrix} \tag{A3}$$

$$\text{num} = -6s^{10} - 179s^9 - 1991s^8 - 9844s^7 - 1.591 \times 10^4 s^6 - 1.231 \times 10^4 s^5 - 5013s^4 - 1088s^3 - 117s^2 + 4.747s \tag{A4}$$

$$\begin{aligned} \text{den} = \; & s^{11} + 33.9s^{10} + 505s^9 + 4006s^8 + 1.739 \times 10^4 s^7 + 3.727 \times 10^4 s^6 + 4.061 \times 10^4 s^5 + 2.468 \times 10^4 s^4 + \\ & 8800s^3 + 1837s^2 + 208.3s + 9.932 \end{aligned} \tag{A5}$$

References

1. Irudayaraj, A.X.R.; Wahab, N.I.A.; Premkumar, M.; Radzi, M.A.M.; Bin Sulaiman, N.; Veerasamy, V.; Farade, R.A.; Islam, M.Z. Renewable sources-based automatic load frequency control of interconnected systems using chaotic atom search optimization. *Appl. Soft Comput.* **2022**, *119*, 108574. [CrossRef]
2. Shouran, M.; Anayi, F.; Packianather, M. The bees algorithm tuned sliding mode control for load frequency control in two-area power system. *Energies* **2021**, *14*, 5701. [CrossRef]
3. Veerasamy, V.; Wahab, N.I.A.; Ramachandran, R.; Othman, M.L.; Hizam, H.; Kumar, J.S.; Irudayaraj, A.X.R. Design of single- and multi-loop self-adaptive PID controller using heuristic-based recurrent neural network for ALFC of hybrid power system Veerapandiyan. *Expert Syst. Appl.* **2022**, *192*, 116402. [CrossRef]
4. Mokhtar, M.; Marei, M.I.; Sameh, M.A.; Attia, M.A. An Adaptive Load Frequency Control for Power Systems with Renewable Energy Sources. *Energies* **2022**, *15*, 573. [CrossRef]
5. Arya, Y. AGC performance enrichment of multi-source hydrothermal gas power systems using new optimized FOFPID controller and redox flow batteries. *Energy* **2017**, *127*, 704–715. [CrossRef]
6. Huddar, A.P.; Kulkarni, P.S. Load frequency control of a multi-area power system using linear quadratic regulator. *J. Inst. Eng. Electr. Eng. Div.* **2009**, *90*, 69–76.
7. Wang, Z.; Liu, Y.; Yang, Z.; Yang, W. Load frequency control of multi-region interconnected power systems with wind power and electric vehicles based on sliding mode control. *Energies* **2021**, *14*, 2288. [CrossRef]
8. Mohamed, A.; Diab, A.A.Z.; Rezk, H.; Jin, T. A novel adaptive model predictive controller for load frequency control of power systems integrated with DFIG wind turbines. *Neural Comput. Appl.* **2020**, *32*, 7171–7181. [CrossRef]
9. Mohseni, N.A.; Bayati, N. Robust Multi-Objective H2/H∞ Load Frequency Control of Multi-Area Interconnected Power Systems Using TS Fuzzy Modeling by Considering Delay and Uncertainty. *Energies* **2022**, *15*, 5525. [CrossRef]
10. Soliman, M.S.; Belkhier, Y.; Ullah, N.; Achour, A.; Alharbi, Y.M.; Al Alahmadi, A.A.; Abeida, H.; Khraisat, Y.S.H. Supervisory energy management of a hybrid battery/PV/tidal/wind sources integrated in DC-microgrid energy storage system. *Energy Rep.* **2021**, *7*, 7728–7740. [CrossRef]
11. Al Alahmadi, A.A.; Belkhier, Y.; Ullah, N.; Abeida, H.; Soliman, M.S.; Khraisat, Y.S.H.; Alharbi, Y.M. Hybrid wind/PV/battery energy management-based intelligent non-integer control for smart DC-microgrid of smart university. *IEEE Access* **2021**, *9*, 98948–98961. [CrossRef]
12. Ramachandran, R.; Madasamy, B.; Veerasamy, V.; Saravanan, L. Load frequency control of a dynamic interconnected power system using generalised Hopfield neural network based self-adaptive PID controller. *IET Gener. Transm. Distrib.* **2018**, *12*, 5713–5722. [CrossRef]
13. Safari, A.; Babaei, F.; Farrokhifar, M. A load frequency control using a PSO-based ANN for micro-grids in the presence of electric vehicles. *Int. J. Ambient. Energy* **2021**, *42*, 688–700. [CrossRef]

14. Yesil, E. Interval type-2 fuzzy PID load frequency controller using Big Bang-Big Crunch optimization. *Appl. Soft Comput. J.* **2014**, *15*, 100–112. [CrossRef]
15. Veerasamy, V.; Wahab, N.I.A.; Ramachandran, R.; Vinayagam, A.; Othman, M.L.; Hizam, H.; Satheeshkumar, J. Automatic load frequency control of a multi-area dynamic interconnected power system using a hybrid PSO-GSA-Tuned PID controller. *Sustainability* **2019**, *11*, 6908. [CrossRef]
16. Guha, D.; Roy, P.K.; Banerjee, S. Load frequency control of interconnected power system using grey Wolf optimization. *Swarm Evol. Comput.* **2016**, *27*, 97–115. [CrossRef]
17. SNosratabadi, M.; Bornapour, M.; Gharaei, M.A. Grasshopper optimization algorithm for optimal load frequency control considering Predictive Functional Modified PID controller in restructured multi-resource multi-area power system with Redox Flow Battery units. *Control Eng. Pract.* **2018**, *89*, 204–227. [CrossRef]
18. Rai, A.; Das, D.K. Optimal PID Controller Design by Enhanced Class Topper Optimization Algorithm for Load Frequency Control of Interconnected Power Systems. *Smart Sci.* **2020**, *8*, 125–151. [CrossRef]
19. Paliwal, N.; Srivastava, L.; Pandit, M. Application of grey wolf optimization algorithm for load frequency control in multi-source single area power system. *Evol. Intell.* **2022**, *15*, 563–584. [CrossRef]
20. Sobhy, M.A.; Abdelaziz, A.Y.; Hasanien, H.M.; Ezzat, M. Marine predators algorithm for load frequency control of modern interconnected power systems including renewable energy sources and energy storage units. *Ain Shams Eng. J.* **2021**, *12*, 3843–3857. [CrossRef]
21. Gupta, D.K.; Jha, A.V.; Appasani, B.; Srinivasulu, A.; Bizon, N.; Thounthong, P. Load frequency control using hybrid intelligent optimization technique for multi-source power systems. *Energies* **2021**, *14*, 1581. [CrossRef]
22. Wang, X.; Wang, Y.; Liu, Y. Dynamic load frequency control for high-penetration wind power considering wind turbine fatigue load. *Int. J. Electr. Power Energy Syst.* **2020**, *117*, 105696. [CrossRef]
23. Yakout, A.H.; Attia, M.A.; Kotb, H. Marine Predator Algorithm based Cascaded PIDA Load Frequency Controller for Electric Power Systems with Wave Energy Conversion Systems. *Alexandria Eng. J.* **2021**, *60*, 4213–4222. [CrossRef]
24. Guha, D.; Roy, P.K.; Banerjee, S. Whale optimization algorithm applied to load frequency control of a mixed power system considering nonlinearities and PLL dynamics. *Energy Syst.* **2020**, *11*, 699–728. [CrossRef]
25. Hasanien, H.M.; El-Fergany, A.A. Salp swarm algorithm-based optimal load frequency control of hybrid renewable power systems with communication delay and excitation cross-coupling effect. *Electr. Power Syst. Res.* **2019**, *176*, 105938. [CrossRef]
26. Lu, P.; Ye, L.; Zhao, Y.; Dai, B.; Pei, M.; Tang, Y. Review of meta-heuristic algorithms for wind power prediction: Methodologies, applications and challenges. *Appl. Energy* **2021**, *301*, 117446. [CrossRef]
27. Almotairi, K.H.; Abualigah, L. Hybrid Reptile Search Algorithm and Remora Optimization Algorithm for Optimization Tasks and Data Clustering. *Symmetry* **2022**, *14*, 458. [CrossRef]
28. Sundararaju, N.; Vinayagam, A.; Veerasamy, V.; Subramaniam, G. A Chaotic Search-Based Hybrid Optimization Technique for Automatic Load Frequency Control of a Renewable Energy Integrated Power System. *Sustainability* **2022**, *14*, 5668. [CrossRef]
29. Pozna, C.; Precup, R.E.; Horvath, E.; Petriu, E.M. Hybrid Particle Filter-Particle Swarm Optimization Algorithm and Application to Fuzzy Controlled Servo Systems. *IEEE Trans. Fuzzy Syst.* **2022**, *30*, 4286–4297. [CrossRef]
30. Sureshkumar, V.; Balasubramaniam, S.; Ravi, V.; Arunachalam, A. A hybrid optimization algorithm-based feature selection for thyroid disease classifier with rough type-2 fuzzy support vector machine. *Expert Syst.* **2021**, *39*, 137–145. [CrossRef]
31. Gharehpasha, S.; Masdari, M.; Jafarian, A. Power efficient virtual machine placement in cloud data centers with a discrete and chaotic hybrid optimization algorithm. *Cluster Comput.* **2021**, *24*, 1293–1315. [CrossRef]
32. Dorigo, M.; Socha, K. Ant colony optimization. In *Handbook of Approximation Algorithms and Metaheuristics*; Chapman and Hall/CRC: New York, NY, USA, 2007; pp. 28–39. [CrossRef]
33. Zhao, W.; Zhang, Z.; Wang, L. Manta ray foraging optimization: An effective bio-inspired optimizer for engineering applications. *Eng. Appl. Artif. Intell.* **2018**, *87*, 103300. [CrossRef]
34. Castelli, M.; Manzoni, L.; Mariot, L.; Nobile, M.S.; Tangherloni, A. Salp Swarm Optimization: A critical review. *Expert Syst. Appl.* **2021**, *189*, 116029. [CrossRef]
35. Xue, J.; Shen, B. A novel swarm intelligence optimization approach: Sparrow search algorithm. *Syst. Sci. Control Eng.* **2020**, *8*, 22–34. [CrossRef]
36. Okwu, M.O.; Tartibu, L.K. Grey Wolf Optimizer. *Stud. Comput. Intell.* **2014**, *927*, 43–52. [CrossRef]
37. Yuan, J.; Zhao, Z.; Liu, Y.; He, B.; Wang, L.; Xie, B.; Gao, Y. DMPPT Control of Photovoltaic Microgrid Based on Improved Sparrow Search Algorithm. *IEEE Access* **2021**, *9*, 16623–16629. [CrossRef]
38. Irudayaraj, A.X.R.; Wahab, N.I.A.; Umamaheswari, M.G.; Radzi, M.A.M.; Bin Sulaiman, N.; Veerasamy, V.; Prasanna, S.C.; Ramachandran, R. A Matignon's theorem based stability analysis of hybrid power system for automatic load frequency control using atom search optimized FOPID controller. *IEEE Access* **2020**, *8*, 168751–168772. [CrossRef]
39. Sahu, R.K.; Gorripotu, T.S.; Panda, S. A hybrid DE-PS algorithm for load frequency control under deregulated power system with UPFC and RFB. *Ain Shams Eng. J.* **2015**, *6*, 893–911. [CrossRef]
40. El-Ela, A.A.A.; El-Sehiemy, R.A.; Shaheen, A.M.; Diab, A.E.G. Enhanced coyote optimizer-based cascaded load frequency controllers in multi-area power systems with renewable. *Neural Comput. Appl.* **2021**, *33*, 8459–8477. [CrossRef]

41. Oshnoei, S.; Oshnoei, A.; Mosallanejad, A.; Haghjoo, F. Novel load frequency control scheme for an interconnected two-area power system including wind turbine generation and redox flow battery. *Int. J. Electr. Power Energy Syst.* **2021**, *130*, 107033. [CrossRef]
42. Dhandapani, L.; Abdulkareem, P.; Muthu, R. Two-area load frequency control with redox ow battery using intelligent algorithms in a restructured scenario. *Turkish J. Electr. Eng. Comput. Sci.* **2018**, *26*, 330–346. [CrossRef]
43. Sun, X.; Hu, C.; Lei, G.; Guo, Y.; Zhu, J. State feedback control for a PM hub motor based on gray Wolf optimization algorithm. *IEEE Trans. Power Electron.* **2020**, *35*, 1136–1146. [CrossRef]
44. Singh, N.; Singh, S.B. A Modified Mean Gray Wolf Optimization Approach for Benchmark and Biomedical Problems. *Evol Bioinform.* **2017**, *13*, 1176934317729413. [CrossRef] [PubMed]
45. Xie, Q.; Guo, Z.; Liu, D.; Chen, Z.; Shen, Z.; Wang, X. Optimization of heliostat field distribution based on improved Gray Wolf optimization algorithm. *Renew. Energy* **2021**, *176*, 447–458. [CrossRef]
46. Liu, G.; Shu, C.; Liang, Z.; Peng, B.; Cheng, L. A modified sparrow search algorithm with application in 3D route planning for UAV. *Sensors* **2021**, *21*, 1224. [CrossRef]
47. Zhu, Y.; Yousefi, N. Optimal parameter identification of PEMFC stacks using Adaptive Sparrow Search Algorithm. *Int. J. Hydrogen Energy* **2021**, *46*, 9541–9552. [CrossRef]
48. Veerasamy, V.; Wahab, N.I.A.; Ramachandran, R.; Othman, M.L.; Hizam, H.; Irudayaraj, A.X.R.; Guerrero, J.M.; Kumar, J.S. A Hankel Matrix Based Reduced Order Model for Stability Analysis of Hybrid Power System Using PSO-GSA Optimized Cascade PI-PD Controller for Automatic Load Frequency Control. *IEEE Access* **2020**, *8*, 71422–71446. [CrossRef]
49. Idan, R.F.; Mahdi, A.J.; Wahhab, T.M.A. Optimized proportional-integral controller for a photovoltaic-virtual synchronous generator system. *Int. J. Power Electron. Drive Syst.* **2022**, *13*, 509–519. [CrossRef]

energies MDPI

Review

Fault Detection, Isolation and Service Restoration in Modern Power Distribution Systems: A Review

Ishan Srivastava [1], Sunil Bhat [1], B. V. Surya Vardhan [1] and Neeraj Dhanraj Bokde [2,3,*]

1 Department of Electrical Engineering, Visvesvaraya National Institute of Technology, Nagpur 440010, India
2 Center for Quantitative Genetics and Genomics, Aarhus University, 8000 Aarhus, Denmark
3 iCLIMATE Aarhus University Interdisciplinary Centre for Climate Change, Foulum, 8830 Tjele, Denmark
* Correspondence: neerajdhanraj@qgg.au.dk

Abstract: This study examines the conceptual features of Fault Detection, Isolation, and Restoration (FDIR) following an outage in an electric distribution system.This paper starts with a discussion of the premise for distribution automation, including its features and the different challenges associated with its implementation in a smart grid paradigm. Then, this article explores various concepts, control schemes, and approaches related to FDIR. Service restoration is one of the main strategies for such distribution automation, through which the healthy section of the power distribution network is re-energized by changing the topology of the network. In a smart grid paradigm, the presence of intelligent electronic devices can facilitate the automatic implementation of the service restoration scheme. The concepts of service restoration and various approaches are thoroughly presented in this article. A comparison is made among various significant approaches reported for distribution automation. The outcome of our literature survey and scope for future research concludes this review.

Keywords: distribution automation; reliability; service restoration; smart grid; multi-agent system

check for updates

Citation: Srivastava I.; Bhat S.; Vardhan, B.V.S.; Bokde, N.D. Fault Detection, Isolation and Service Restoration in Modern Power Distribution Systems: A Review. *Energies* **2022**, *15*, 7264. https://doi.org/10.3390/en15197264

Academic Editor: Shailendra Singh

Received: 29 August 2022
Accepted: 29 September 2022
Published: 3 October 2022

1. Introduction

The year 2020 has seen the largest decline in global energy demand since World War II and the highest absolute decline ever. The effects of the COVID-19 pandemic on global energy use are shown by the recent statistics on energy usage in 2021 [1]. In 2021, electricity demand is anticipated to rise by 4.5%. With a forecast GDP growth of 9% in China and 12% in India in 2021, energy demand in both nations is likely to increase by about 8% over 2020 [1].

From the installation of the first electrical grid on 4 September 1882, in Manhattan, supplying power to 85 customers and 400 lamps, to the present grid today, many aspects of electrical power have changed. These include the deregulation of the energy market, changes in metering, changes in power generation, distributed energy decentralization, and the rise in micro-generation and isolated micro-grids, including renewable sources of energy and electric vehicles in the grid. The complexity of the current grid is growing, along with the number of consumers and loads. To effectively manage this extremely complex power network, the idea of the Smart Grid (SG) has been introduced. The SG is defined by NIST, USA [2] as "a modernized grid that enables bidirectional flows of energy and uses two-way communication and control capabilities that will lead to an array of new functionalities and applications". Power electronics, communication engineering, and material science advancements have all contributed to the development of SG. The following requirements must be met by the modern grid's design, expansion, and operation.

1. Energy distribution and transmission must be cost-effective.
2. The cost of electricity for consumers must be lower.
3. The security of the system must be ensured.
4. In the event of a fault event, service restoration must be rapid.

5. By connecting more Renewable Energy Sources (RES) to the grid, the environmental implications must be properly monitored.
6. RES management and integration must be improved.

The conventional grid differs from an SG in numerous ways, as noted in [3]. In comparison to an SG infrastructure, the conventional grid has fewer sensors. These sensors are responsible for providing the grid's real-time data for monitoring and control of the system. The SG's communication network needs to be set up for two-way communication between customers and control centres. Additionally, the introduction of SG Advanced Metering Infrastructure (AMI) depends on this two-way connection. A standard power grid's electro-mechanical metering method precludes the use of real-time data transfer in this setup. Consequently, a digital metering system is a crucial necessity for SG in order to adopt automation solutions such as Automatic Meter Reading (AMR) or AMI. Energy meter data collection and secure transmission to the control station for billing, troubleshooting, and analysis are the tasks of AMR or AMI. SG consists of various domains or operational units, as shown in Figure 1. Each unit's functionality is directly or indirectly dependent on every other unit in the grid. The operation under this scenario is known as interoperability; to have this feature in SG, a proper two-way communication infrastructure is required.

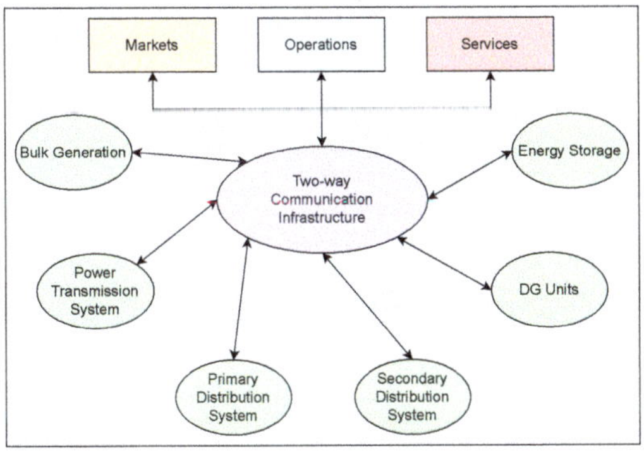

Figure 1. Smart grid building blocks.

Along with fault detection and isolation strategies, service restoration forms one the key feature of distribution automation. Most previous review articles focus on individual strategies rather than presenting the problems and solutions of FDIR as a single strategy.

In [4], a review is presented for different techniques used to solve service restoration problems. The methodologies for fault isolation and detection are not covered in this article. Similarly, Ref. [5] covers various conventional and artificial intelligence-based methods for fault location and detection techniques in a power distribution system. In review [6], the authors have presented various centralized and distributed strategies for fault localization and service restoration. Multi-agent strategies have been highlighted by the authors as well However, the features of distribution automation and novel Fault Detection, Isolation, and Restoration (FDIR) approaches such as IoT-based FDIR were not discussed. In [7], various features of smart distribution systems are mentioned, including smart metering systems, security and outage management aspects, and of service restoration solutions. However, the discussion is restricted to the performance of smart distribution system under different scenarios, and the details of FDIR are not presented by the authors.

The principles of distribution automation are discussed in this paper, along with the associated advantages and challenges. The communication infrastructure required to implement various distribution automation strategies is discussed as well. The explanation of FDIR concepts and their background research work is presented in the following section of this paper. Various aspects of FDIR problem, as well as centralized and decentralized approaches to solving them, are discussed. The evolution of various service restoration algorithms is addressed, along with their benefits and drawbacks.

2. Distribution Automation

Any automatic process is defined as a process that performs a task in an automated manner to achieve a faster rate of operation. The concept of distribution automation was proposed for the first time in the 1970s [8], when the goal was to enhance the operating performance of the power distribution system. However, due to high implementation costs and immature communications technology, the concept was not fully adopted by the electric utilities. In recent years, with the advancement of communications technology, data acquisition systems, data analysis tools and power electronics, distribution automation is again becoming the focus of research and improvement. The Institute of Electrical and Electronics Engineers (IEEE) has defined Distribution Automation Systems (DAS) as systems that enable an electric utility to monitor, coordinate, and operate distribution components in real time from remote locations [9]. Distribution automation enables more flexible network control operations. As a result, the power supply is more reliable, efficient, and of higher quality. The speed, cost, and accuracy of multiple key distribution processes can be improved with distribution automation. These include failure detection, feeder switching, outage management, voltage monitoring and control, reactive power management, feeder line equipment maintenance, and grid integration of distributed energy resources. In the SG paradigm, the management of power flow can be carried out by automatically monitoring the grid and controlling switching operations through intelligent devices.

2.1. Benefits of Distribution Automation

There are several benefits of implementing distribution automation strategies.

1. Operational and maintenance benefits
    - Increased reliability by lowering outage time through the use of an automated restoration strategy
    - Reduced man-hours and manpower costs
    - Improved fault detection and diagnostics
    - Improved system and component loading management
2. Financial benefits
    - Quick restoration results in increased revenue
    - Improved usage of system capacity
    - Customer retention to boost supplier quality
    - Improved service reliability
    - For industrial and commercial users, lower interruption costs

2.2. Features of Distribution Automation

(a) Automation Devices: The automation device must be able to perform multiple tasks. For example, the device deployed in the field to control various equipment must gather the information and perform local decision-making when required. In the modern grid, this multi-task is carried out by Intelligent Electronic Devices (IEDs).

(b) Data Analysis Tools: Automation requires processing the large amount of data collected by different sensors in the network. These data are required in order to (i) estimate the current state of the grid, (ii) predict its load patterns, and (iii) identify vulnerable buses in the network. Therefore, data processing tools must be present in substations and central control stations. Data availability, integrity, and security must be maintained in order to carry out proper data analysis.

(c) Real Time Implementation: High-speed data transfer communication infrastructure and computational resources are required for the real time implementation of a distribution automation strategy. The current state of the network is determined by the different parameters of the components, topology of the network, status of various switches/breakers, and measured data from nodes in the network. Subsequently, based on the type of contingency, the optimal control plan is deployed through distributed or centralized control schemes.

(d) Sensors: In the contemporary era, the advancement of sensor technology is reducing the cost of sensors. In a complex grid, more sensors are required to collect all of the information relevant to the current state of the grid. Therefore, it is economically possible to place the required number of sensors at critical locations throughout the grid.

(e) Asset Management: Distribution automation can play a vital role in asset management for the utility. Advanced sensors can regularly monitor the condition of various equipment present in the network, which is useful for the utility in preparing a proper maintenance schedule. Real time analysis of the grid can provide the loading details of different types of equipment. Based on this information, proper utilization of these assets is possible.

(f) Communication Infrastructure: Recent advances in communications technology are key to the evolution of smart grids. A two-way communication channel ensures the proper exchange of information among various components of the grid. All the domains of the SG are interconnected through the internet via various Local Area Network (LAN), Wide Area Network (WAN), Home Area Network (HAN), Building Area Network (BAN), Industry Area Network (IAN), etc. The efficiency of DAS is mostly dependent on the performance of the implicit communication network. With advanced communications technology, protocols such as IEC61850, which were proposed originally for substation automation can be implemented for feeder automation and service restoration. In the SG, various sensors and IEDs are deployed to gather information, which then needs to be transferred as quickly as possible through a low-latency communication network. This information is essential for decision-making processes in DAS. A typical architecture of a Wireless Mesh Network (WMN) used in SG communication is shown in Figure 2.

Improvements in grid management and monitoring are possible with the deployment of Information and Communication Technology (ICT) in the electricity distribution system. This infrastructure is highly interconnected, and is responsible for the exchange of information among substations, control centers, power markets, and various measurement devices [10]. The merits of an ICT-enabled network are shown in Figure 3.

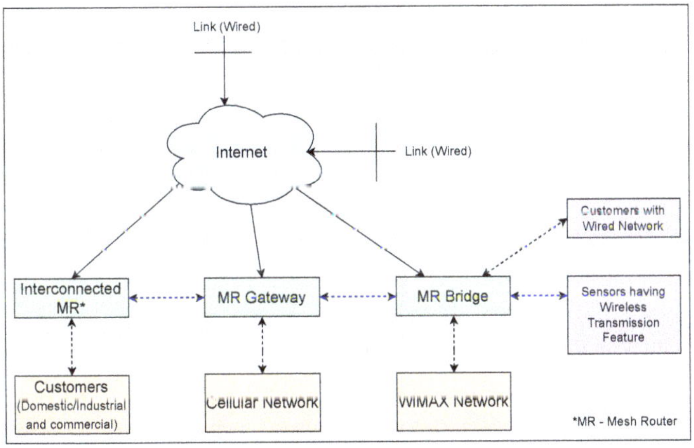

Figure 2. Architecture of wireless mesh network.

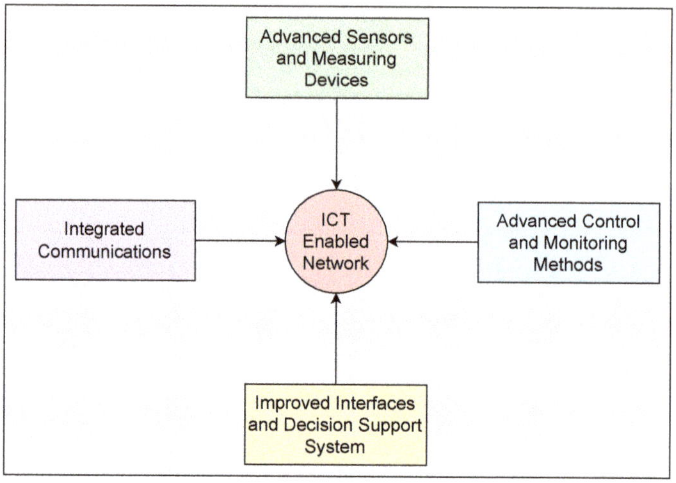

Figure 3. Merits of ICT-enabled network.

2.3. Technical Challenges for Distribution Automation Systems

There are several challenges for a distribution automation system, including the following:

- High penetration of DER
- Common protocols for the communication system
- Data loss, data security, and data integrity issues
- Estimation of real-time pricing of electricity in a market-driven framework
- Requirement of high-speed computational resources

One of the features of the modern power distribution system is the presence of Distributed Generation (DG) units. As the complexity of the network increases, the placement and control of DG units becomes challenging with respect to the deployment of distribution automation strategies. Additionally, the dynamic load variations in the case of modern distribution systems need to be considered. In most of the reported works, the daily load variation in case distribution automation has not been considered.

To examine the dynamic behaviour of automated distribution systems, the authors in [11] included daily load variations in the analysis. In this work, a methodology based on the Manta Ray Foraging Optimization (MRFO) algorithm has been proposed to control and allocate distributed generators and capacitor banks in the system and to deploy the network reconfiguration strategy.

In another recently reported work [12], a novel Improved Heap-Based Optimizer (IHBO) was proposed to solve the network reconfiguration problem and ensure optimal sizing and placement of DG units. Minimization of power loss and maximization of voltage stability was the objectives considered for network reconfiguration. This problem of integrating network reconfiguration with DG sizing and placement was solved by another variant of heap-based optimizer, the Heap-Based Optimizer with Deeper Exploitative Improvement (HODEI), in [13]. In this case, the authors considered different loading conditions for analysis.

3. Fault Detection, Isolation, and Service Restoration (FDIR)

In order to retain consumers in a deregulated market, power providers must enhance their service quality. As a result, several electric utilities have introduced DAS as a smart method to improve distribution system reliability and efficiency. FDIR is regarded as the most crucial of all distribution automation functions. The goal of FDIR is to minimize service restoration time for a permanent fault event of distribution feeders from an average of 58 min to less than 5 min [14]. When a fault occurs in the distribution side, different intelligent components such as IEDs that are linked via a communication network transmit fault information across the affected network segment. In addition to protection, IEDs perform functions such as control, measurement, disturbance recording, and condition monitoring. The data transfer rate through IEDs is much faster than with conventional methods [15]. When these IEDs are networked together, they can provide almost all of the necessary information to the system operators. In case of fault, IEDs associated with circuit breakers send control messages to the other network IEDs. Based on this information exchange, necessary steps are taken to clear the fault and restore the power as quickly as possible.

The communication network plays a critical role in identifying the fault and restoring power. Substation protocols such as IEC61850 can be used in modern power systems to exchange information. The advent of the IEC61850 standard enables a range of protection and control capabilities and benefits, including peer to peer communication through Generic Object Oriented Substation Event (GOOSE) messaging [15]. GOOSE wireless communication can be used to improve current Medium Voltage (MV) substation automation applications. GOOSE enables more extensive monitoring of the system. All GOOSE connections are checked continually and if any abnormalities are found, they are reported to the operator. Fast and reliable data transfer is crucial for the implementation of the FDIR strategy.

In FDIR strategies, the following steps are implemented sequentially:

A. Fault Detection: Different types of faults can affect distribution networks. Fault detection is the process of determining whether or not a fault has occurred. High currents and/or low voltages reported by the sensors or any other measurement device can automatically trigger alarms, which can traced to identify the fault's location. Feeder Circuit Breakers (FCBs) and Fault Passage Indicators (FPIs) can operate in case of a fault event. The primary role of a fault passage indication system is to identify faults in the section downstream of the site of their occurrence. When a current exceeds a preset value, FCBs are tripped, triggering a temporal reaction that is dependent on the measured current value [16]. The DAS is promptly notified by these devices.
B. Fault Isolation: After the DAS has detected the faulty section, the next step is to isolate the faulted portion by operating the relevant upstream and downstream sectionalizing switches. In this step, the DAS tries to isolate the smallest section of the network.
C. Fault Classification: The DAS attempts to identify the type of fault that occurred in the distribution network during this step. The fault classification details can be useful in assessing the required fault clearing time, which can help the operator to choose a suitable restoration plan.
D. Estimation of Capacity: The ampacity of a conductor is determined by the physical and electrical properties of the conductor's material and construction, its insulation and ambient temperature, and the environmental conditions proximal to the conductor. Before restoring power, the capacity of adjacent feeders are calculated. This can be achieved by performing load flow analysis. The restoration solutions which exceed the feeder capacity are discarded.
E. Service Restoration: Following fault isolation, the distribution network operator's goal is to restore as many out-of-service loads as possible with the minimum of switching operations. Service restoration is the process of finding suitable backup feeders and laterals (branches of the network) to re-energize the out-of-service area by changing the topology of the network [17]. After estimating the feeder capacity, the distribution network operator decides whether the adjacent healthy feeders are capable of supplying total or partial out-of-service loads.

Control Schemes for FDIR

The desirable characteristic of a modern power system is the self-healing capability of the network. By following self-healing methods and protocols, the network is able to perform corrective action on its own when an abnormality occurs in the system. To alter the distribution system state to a better one, control actions are key parts of self-healing. The available control strategies can be categorized as centralized control or distributed (de-centralized) control.

1. Centralized Control Scheme:
   In this scheme, all data are routed to a centralized control station and the control strategy is determined depending on the requirements. In the modern grid, a huge amount of data are collected by various sensors and measurement devices. All these data are transferred to the centralized control station through a two-way communication channel. As a centralized controller frequently requires a low-latency system in order for large amounts of data to be transferred between various power system components and control centers. This kind of control scheme is very efficient in the case of a small system. However, in the case of a network with a large number of nodes there are various disadvantages associated with this kind of control. The deployment of a centralized control strategy can be time-consuming for to various reasons, including high computational burden, lack of automation throughout the system, human error, etc. Furthermore, the inclusion of distributed energy resources and electric vehicles in the power grid results in more complexity, thus requiring more complex network management. The risk of single points of failure is a major issue associated with centralized control schemes [18].
2. Distributed Control Scheme:
   Decentralized methods rely on direct peer-to-peer communication between monitoring and control devices present in the network [19]. Data are collected by various sensors and IEDs, and data processing is performed by IEDs locally. Decentralized approaches primarily focus on simultaneously resolving a problem. Control/protection measures that govern the overall system response are made in each distribution or primary substation depending on the contingency. This scheme requires distributed agents able to perform different tasks and an efficient communication system through which information exchange is carried out among these agents [20]. Agents can conduct fault detection, isolation, classification, and service restoration independently, and all are connected via a single platform. Distributed strategies can substantially increase the reliability of power monitoring and control and improve the flexibility and efficiency of the system [21]. In a centralized control scheme, data must be transferred from all nodes to the central control station regardless of how far apart they are in the network. These long communication links require significant investment, and the failure of one of these communication links can lead to a considerable delay in data collection at control stations, increasing the deployment time of the control strategy. Therefore, the presence of a well-structured two-way communication network in the distributed control scheme makes it more economically efficient than a centralized control scheme.

From the above discussion, it can be concluded that one of the main aspects of DAS is to enhance the reliability of the power supply by introducing self-healing features to the grid. FDIR is the basis of this self-healing characteristic. Fault detection and isolation is based on efficient information exchange between different power system components. The optimal service restoration strategy, on the other hand, is a combination of several factors, which are addressed in the next portion of this chapter.

4. Literature Review

This section provides a comprehensive review of the literature on various FDIR and service restoration optimization approaches. This section is divided into multiple

subsections based on the type of approach and the computational techniques used to address the restoration problem.

4.1. FDIR Methods for Distribution System

A fully automated distribution automation system that includes all customers, substations, and generating stations was proposed by the US Department of Energy in its "Grid 2030" vision [20]. The key feature of this smart distribution network is the two-way flow of information and electricity. The ability to respond to disturbances, reduce their impact, and restore power quickly is required to achieve the goal of this smart distribution network. Its establishment is expected to result in better system stability during disturbances and be the indicator for the self-healing capability of the grid [20,22]. Reducing outage times and raising system reliability as a whole are the main goals of the FDIR algorithm. The system reliability is measured by indicators such as the System Average Interruption Duration Index (SAIDI), the System Average Interruption Frequency Index (SAIFI), and the Customer Average Interruption Duration Index (CAIDI). These indicators are utilised to assess the distribution automation performance during outages that are sustained and extend for five minutes or more [23]. ENMAX Power Corporation in Calgary, Alberta, Canada has deployed an FDIR strategy to part of its network; the benefits obtained are reported in [24]. The authors reported that from 2004–2010, the total estimated customer savings were approximately $ 7,500,000. The FDIR strategy was applied to 73 circuits out of 160 in the network. It has been shown that through FDIR, the annual customer benefits ($ 10,067,000) are more than the annual cost of automation ($ 3,100,000).

As mentioned earlier, there are various benefits to using a distributed control strategy over centralized control in the deployment of an FDIR scheme. The main features of this distributed control scheme are mentioned in [25] and are given below.

1. All processing units are linked together through a communications network.
2. Processing units such as IEDs can make the control decisions based on the local information available to them.
3. Single points of failure or malfunctioning of one control unit is not a major issue, as control can be managed by an adjacent unit.
4. Information transfer among different units occurs rapidly, making proper communication links of the utmost importance.
5. Data security measures are required, as the data network is widely dispersed and vulnerable to cyberattacks.

Different communication protocols have been enacted to communicate information among devices at different levels, including the central controller level, I/O level, bay level, and station level. IEC61850 is a contemporary communications protocol utilized for substation automation as well as distribution automation. In contrast to traditional protocols, it is able to exchange information among various devices through peer-to-peer communication via GOOSE messages. GOOSE signals are received by IEDs and take action accordingly without the help of a central controller [26].

Three FDIR approaches are outlined in [27]: loop control, peer-to-peer schemes, and decentralized scheme. These schemes are implemented through IEDs using IEC61850 communication protocol. The measured data are transmitted as GOOSE signal and used to develop the logic for fault isolation and service restoration. Another peer-to-peer GOOSE messaging scheme for FDIR using IEC61850 and DNP3 protocols is proposed in [28]. The authors devised a way to update information in Remote Terminal Units (RTUs) utilising programmable automation controllers, on which they based the logic for fault detection and isolation.

In [29], the authors presented a "close-before-opening" method, in which the idea is to close the tie-switches before opening the corresponding sectionalizing switches, thereby causing the elimination of power blinks or momentary power outages during switching operations. According to the findings of this study, the proposed strategy improves the overall reliability of the distribution power supply. In the fiber optic-based communication technology used by the authors, each recloser can communicate with an adjacent one through a fiber optic link [29]. The SCADA system is integrated in the same way.

4.1.1. Multi-Agent Systems for FDIR

A multi-agent system is one of the most prominent ways of applying decentralized power systems management; it is applicable to multiple areas, including fault diagnosis, voltage stability, electricity pricing, coordination of protection schemes, and restructuring and restoration of the power system. Distributed control schemes are increasingly popular at present thanks to their more efficient performance for large and complex systems. As the system size increases, the computational burden of solving the problem of distribution automation is substantial. Researchers have proposed various agent-based schemes or multi-agent systems to implement distributed (decentralized) control for the FDIR problem. By performing many tasks simultaneously via defined agents, this multi-agent system can alleviate the computational burden. The multi-agent system is subject to change, and acts as a decision-making support system that includes all decision-making components within the system.

The authors of [30] proposed a multi-agent approach for service restoration. They first defined the load agents that collect local information; the feeder agents, which are the decision-making entities, are connected to the load agents via communications link. The local information guides the multi-agent system to build a logic for service restoration and implement it through the feeder agents. The proposed framework works well for medium and large system sizes. However, the authors did not include the logic for fault detection and isolation in the scheme or discuss the protocols required for communication. To address this issue, subsequent research works have incorporated fault detection and isolation logic and proposed the necessary communication connections.

In 2009, Lin et al. [14] presented a multi-agent system to deploy a distribution automation strategy for the 43 feeders of the Taipower distribution system. This multi-agent architecture uses four agents for FDIR, namely, a Main Transformer Agent (MTR), Remote Terminal Unit Agent (RTU), Feeder Circuit Breaker Agent (FCB), and Feeder Terminal Unit Agent (FTU). FTUs and FCBs are similar to the load agents and feeder agents used in [30], and information exchange among these agents happens locally. The MTR and RTU agents are used in decision-making. All of these agents are integrated through a Java-based JADE platform. A multi-agent approach using feeder agents and bus agents based on the JADE platform is reported in [31].

4.1.2. FDIR with DG Penetration

In subsequent analyses of the FDIR problem utilizing a multi-agent paradigm, researchers have considered the penetration of Distributed Generation (DG) in the distribution network. It has been reported in [32] that the total processing time required by the centralized controller is more than that of the distributed controller with or without DG in the network. In [33], Khamphanchai et al. incorporated a DG agent to efficiently control the use of DGs in the network in both regular and emergency situations. During restoration, the FDIR approach is able to re-energize critical loads while performing load-shedding for non-critical loads. The authors used JADE to design the multi-agent environment, and the simulation model was developed in MATLAB/Simulink, with communication between these platforms carried out via MACSimJX [33].

In [34], the authors suggested a hybrid control architecture in which both centralized and decentralized control techniques are incorporated. The agents in this work were split into two levels. Load agents and feeder agents at the lower level acquire information and send it to agents at the higher level. Decisions regarding service restoration are then made by higher-level agents, which communicate with one another during the decision-making process. The DG security and voltage quality issues are addressed in this work, and parallel simulation MATLAB/PSAT-JADE platforms are integrated together. The work reported by Ghorbani et al. [35] presents a new learning-based approach for higher level agents.

The service restoration scheme is obtained using the information provided by the lower level agents and the knowledge acquired by learning (termed as Q-learning). Due to this learning capability, fewer messages are required by the agents for decision-making, which reduces the overall computational burden, cost of computation, and the processing time required to obtain the restoration scheme. The authors in [36] proposed a decentralized multi-agent control scheme that uses a controlled DG islanding methodology as additional input for the service restoration strategy. Integration of electric vehicles is considered and the corresponding vehicle-to-grid facility is studied from the perspective of service restoration.

In [37], a service restoration scheme for an active distribution system was suggested. This scheme introduces a "region agent" that is integrated with the feeder agent and is responsible for performing restoration using local information. The load priority is considered as one of the constraints for service restoration. The restoration scheme was tested for an IEEE-14 bus test system modeled in DIGSILENT.

The authors in [38] described a microgrid-based FDIR scheme that makes use of a multi-agent framework. The fault is identified by comparing current phasors recorded by two agents stationed at the extremities of the network section under study. To deploy a fault detection and isolation strategy, they reported that the data transfer between local agents is sufficient. Furthermore, the service restoration agent and generator agent are decision-making entities for the purpose of restoring power to the network's healthy section.

In recent research work [39], a multi-agent based service restoration scheme was proposed which integrates static and mobile (movable) energy storage systems for service restoration planning. To manage the mobile energy storage device, an additional transportation layer is included along with the existing two-layer multi-agent schemes. The proposed method was tested for single fault conditions as well as multiple fault conditions on an IEEE-33 bus test system. A comparison of different agent-based methods is shown in Table 1.

Table 1. A comparison of several agent-based distribution automation techniques.

Articles	Tools/Software Incorporated	Merits of Proposed Method	Limitations of Proposed Method
[30]	JAVA	(1) Multi-agent method for restoration is presented. (2) Local and global search operation are performed by agents.	(1) Intelligence and decision making feature at local level is missing. (2) Failure of facilitator agent can cause single point failure.
[32]	NS-2	(1) Comparison is made between centraized and distributed MAS. (2) DG penetration taken into account.	(1) Decision making agent is present at a fixed loacation. (2) Voltage limit is not considered.
[34]	MATLAB, JADE, PSAT	(1) Hybrid centralized-decentralized control scheme is presented. (2) Load priority, Voltage quality and load priority is considered.	(1) Membership function and fuzzy logic rules are difficult to formulate. (2) Time requirement is more. (3) Difficult to implement in real time application.
[40]	dNetSim, JADE	(1) Distributed energy storage is considered for restoration of power. (2) FDIR scheme has the feature of islanding coordination.	(1) Communication solutions not disclosed.
[35]	MATLAB, CYME	(1) Easy communication among agents. (2) Local agent can do decision making.	(1) Problem formulation is difficult.

Table 1. *Cont.*

Articles	Tools/Software Incorporated	Merits of Proposed Method	Limitations of Proposed Method
[41]	MATLAB, Opal-RT	(1) IoT based FDIR approach presented. (2) Approach compatible with various protocols.	(1) Data loss aspect is not discussed. (2) FDIR agent working is not disclosed.
[14]	JADE	(1) MAS scheme applied to Taipower network.	(1) Not suitable for complex system.
[33]	JADE, MATLAB, MACSimJX	(1) Load shedding technique is considered with service restoration scheme. (2) Less time requirement.	(1) Communication delay not mentioned. (2) Small test system size.

4.1.3. IoT-Based FDIR Approach

The development of the Energy Internet (EI), Internet of Things (IoT), and data-driven analysis has aided in the development of smart distribution network communications infrastructure [42]. The most crucial and challenging duty in a smart distribution system is energy management [43–45]. A suitable communication infrastructure is necessary for EI applications [46]. There are additional challenges with SGs or cyber-enabled networks, such as the lack of an acceptable technique to calculate the rate of return on SG capital investments [47]. Additionally, the grid's data flow is susceptible to a variety of dangers. Data from smart metres (SM) is essential for effective power network monitoring [48], which can be altered by a potential hacker for various reasons. Li et al. [49] covered several grid-related cyberthreats. The correct operation of the grid can be seriously hampered by data loss through the communication link. In a wireless communications network, factors including hardware breakdown, radio interference, and data congestion can cause data loss [50]. Weather issues, incorrect antenna alignment, and long transmission distances are other causes of data loss via wireless networks [51]. In addition, unlike a wired network, a wireless network's link capacity, which defines the continuation of the communications link, may not remain constant due to interference [52].

Outage management schemes are becoming more reliable and flexible in the SG paradigm thanks to interoperability among the various components of the grid. IoT has wide applications in SG systems. Interoperability refers to communication ability among the various devices present in all the domains and sub-domains of the SG. The three layers of information collection, transmission, and processing in an IoT-based SG are shown in Figure 4 [53]. For FDIR, Estebsari et al. [41] have proposed an IoT-based framework. For modelling of physical components of the grid, they used Opal-RT integrated with MATLAB to develop a service restoration logic. A Message Queuing Telemetry Transport (MQTT)-based communication link emulates real-time communications delays based on type of the IEDs used, bandwidth, amount of traffic in the network, type of protocol, etc. This IoT-based technology can be implemented along with the application of grid computing. FDIR requires distributed computational resources; therefore, grid computing which facilitates distributed parallel computing can play a major role in reducing overall FDIR time requirements [54]. The self-healing capability, observability, and controllability of the network have all been reported to be enhanced in an IoT-based SG [55]. A study using a multi-agent framework for power restoration using an IoT-based system has been reported in [55].

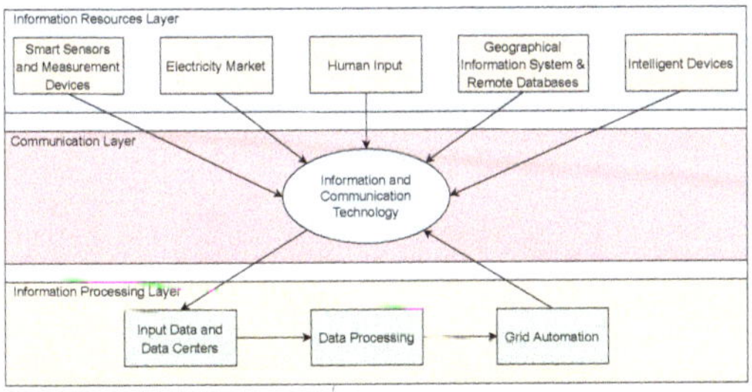

Figure 4. Three layers of information exchange and processing for grid automation.

4.2. Methods to Solve the Service Restoration Problem

Service restoration is a combinatorial optimization problem with multiple objectives and constraints. Each discrete member in the set that makes up the possible solutions corresponds to a switching sequence. Different methods have been suggested in the literature to identify an optimum switching sequence that must be implemented in the network for power restoration. The restoration process starts after a fault is detected in the system; see Figure 5. In the case of a permanent fault condition, part of the network remains in outage condition, and is recovered only when the fault is cleared. In the next scenario, where the fault clearing time is relatively low, the power utility tries to implement a restoration strategy to supply power to the healthy section of the network. On failure of implementation of the service restoration strategy, the number of disconnected loads is higher compared to the condition in which service restoration is properly implemented [56]. Two network segments can be joined or disconnected for service restoration using sectionalizing switches. These switches are normally closed. The distribution network includes tie-switches, which are typically left open, to regulate the use of tie lines for power transfer between two nodes. During a fault, a tie line can provide electricity to loads in nearby areas based on its capacity. The problem of service restoration can be expressed as a multi-objective constrained optimization problem. The total out-of-service area is provided as follows:

$$f_1 = \min \frac{\beta_{\text{int}}(n1)}{\beta_{total}(n2)} \tag{1}$$

where $\beta_{int}(n2)$ denotes the total number of interrupted customers in the healthy section of the network or the out-of-service area and $\beta_{total}(n2)$ represents the total load which is connected in the network obtained after restoration.

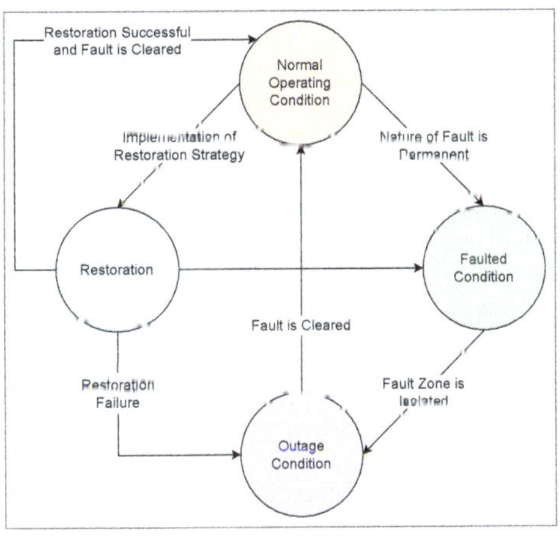

Figure 5. System operating conditions during restoration process.

Tie switches and sectionalizing switches can be operated manually or remotely. The time-consuming procedure of operating manually-operated switches can delay the restoration process. Remotely Controlled Switching (RCS) operation, on the other hand, is rapid, although it requires sufficient communication infrastructure in the network and is significantly more expensive than manually controlled switches.

Minimizing the number of switching operations is the other main objective considered by researchers. The problem can be written as

$$f_2 = \min \frac{\alpha_{mcs}(n1, n2)}{\alpha_{mt}} \tag{2}$$

The network topology is changed from $(n1)$ to $(n2)$ during the restoration process; the corresponding total manual switching operations is represented by α_{mcs}. For normalization, α_{mt} (representing the total number of manually controlled switches in the network) is taken into consideration as follows:

$$f_3 = \min \frac{\alpha_{rcs}(n1, n2)}{\alpha_{rt}} \tag{3}$$

Here, α_{rcs} is the total number of remotely controlled switching operations required for service restoration and α_{rt} denotes the total number of RCS present in the network.

4.2.1. Expert System Approach

Expert systems are developed based on the knowledge of a particular system acquired by the operator over a period of time. This knowledge is utilized to create the rules that guide the decision-making process. Various researchers have used an expert system method to identify the best switching sequence for service restoration.

The rules of the expert system may change for different types of systems. In [57], the suggested expert system comprised generic rules for service restoration which are independent of the system type.

Zhang et al. [58] developed an expert system on the PROLOG platform for the general task of network re-organization; for validation, an optimal switching sequence is obtained in the electric substation. The work in [57] was further expanded in [59], where the knowledge base for service restoration was strengthened by the addition of certain new relevant rules. The authors of [60] proposed a methodology for maintaining an online rule-based expert system. The proposed work was tested for a real-time expert system.

In [61], the formation of two knowledge bases was suggested by the authors for an expert system to solve the service restoration problem. These knowledge bases were (1) a network state knowledge base and (2) a restoration knowledge base. These two knowledge bases were integrated in order to provide a solution for total and partial service restoration. A best search technique-based expert system strategy employing heuristic rules for power restoration was presented by the authors of [62].

In [63], Tsai et al. presented an expert system which takes load variations into account; they suggested a multiple service restoration solution for the same fault instance, allowing the operator to choose the most suitable plan for real-time implementation.

The development of an expert system's knowledge base is a tedious task because the information gathering process takes a long time and necessarily requires the expertise of a system operator. Furthermore, expert system rules may not be generalized for the majority of scenarios, and it is difficult to formulate them for complex systems such as power distribution systems. Therefore, expert systems can be substituted by alternative ways to more quickly address service restoration issues.

4.2.2. Fuzzy Logic Approach

In this approach, various parameters of the distribution network are formulated as fuzzy variables. These variables include the number of disconnected loads in the network, bus voltages, line currents, the total number of switching operations, etc.

Before solving the restoration problem, earlier studies required load estimation techniques. This step is required when the operator does not know the exact loading at various nodes in the network due to the lack of sensors and other measuring devices in the network. The authors of [64] proposed a load estimation method as the initial step and afterwards presented a heuristic-based service restoration method to obtain the optimal switching scheme.

Huang et al. in [65] used an analytical hierarchy process to obtain the weights for each service restoration objective. Consequently, the multi-objective problem is formulated as a single objective problem using the weighted sum approach. In the same work, the authors used a fuzzy cause–effect network to obtain the optimal switching sequence for service restoration. The proposed method was tested on a real distribution network of the Taiwan Power Company. In [66], the multi-objective restoration problem was solved using fuzzy evaluation of feasible candidates through the application of different restoration schemes. The concept of a local network was introduced by the authors of [67] as a way to reduce the dimensionality of the problem, and the fuzzy logic method was used to solve the service restoration problem.

4.2.3. Artificial Neural Network Approach

Artificial neural networks (ANNs) are best suited for solving complicated, ambiguous, highly nonlinear, multi-faceted, and stochastic issues. The goal of an ANN is to simulate the behaviour of biological neural networks. Over the years, numerous ANN models and network-related learning methods have been created. In neural networks, the complicated nonlinear link between the load levels of the zone and system topologies can be mapped, which is an important task that needs to be done before deploying the service restoration scheme for the network.

Load level estimation using ANN was presented in [68]. Network reconfiguration to reduce power losses was performed with varying load patterns. The ANN was trained using different inputs from each zone and system topology.

The researchers in [69] proposed an ANN and pattern recognition-based approach for service restoration. The developed approach was tested for a typical Taiwan-based distribution system. In further studies, Bretas and Phadke [70] developed a back-propagation learning-based ANN method for service restoration. The authors tested the proposed method on a 162-bus system and obtained feasible solutions for various fault scenarios in a short time.

Although the ANN approach is generalized and offers high processing speed, there are several drawbacks, and training the algorithm for the various loading conditions present in a complex modern power distribution network becomes tedious and difficult.

4.2.4. Heuristic Methods

Heuristic methods are relatively fast compared to conventional optimization methods. A heuristic approach is governed by the rules derived for a particular system based on the system characteristics. In combination with conventional computational methods, these rules can solve a system-specific problem. The heuristic rules are framed from the knowledge of the distribution system operator, similar to the case of the expert system knowledge base formation method. Consequently, these approaches are likewise subject to the same constraints as the expert system approach. For large networks, these approaches are not very effective. In the case of service restoration, the problem is a multi-objective combinatorial one. The goal is to find a switching sequence for the changing network topology such that all the objectives are met and all the constraints are satisfied. The conventional way of solving this problem (uninformed search algorithms) is to check every feasible switching sequence. Practically, the deployment of a service restoration strategy is time-sensitive, and the conventional approach fails in the case of real-time scenarios. Therefore, heuristic methods (informed search algorithms) are used to guide the search process. Heuristic algorithms are more efficient, as they employ data feedback to guide the search direction. With the help of a good heuristic algorithm, an informed search can significantly outperform any uninformed search. Although heuristic algorithms do not always guarantee the best feasible solution, such strategies aid in identifying a solution in an acceptable amount of time and computation.

Heuristic search approaches determine the operators based on domain knowledge. In the search process, the algorithm moves through the search space using these operators. The transition from one point in search space to another depends on the fitness value of objectives at these two points. When compared to a point far from the optimal situation, the one closer to it has better fitness.

In the case of the service restoration problem, the starting point of all algorithms is to assume an initial condition, such as a feasible arbitrary switching sequence. The goal is to find the switching sequence with the best fitness value corresponding to the objectives. Based on the heuristic algorithm, the search space is explored; instead of checking all feasible solutions in the search space, the search process is guided from diversification to intensification through the heuristic algorithm to find the global optimal solution. During the search process, the fitness value corresponding to each visited point is calculated. For example, if the objective is to minimize the out-of-service area, then the total number of disconnected loads in the healthy section of the network corresponding to each switching sequence (each point in the search space) is determined. In the case of multi-objective optimization formulation, the concepts of dominance and Pareto optimality of the solution are used. In [71], a heuristic method using a depth-first search method along with practical rules was proposed for service restoration. This study conducted research to determine the relationship between the rules of the proposed approach and the optimality of the solution. The rules governing the heuristic method were acquired by researchers in [72] from an experienced operator at the Taiwan Power Company, and service restoration was conducted after the faulty zone was isolated.

Nahman and Sťrbac [73] performed minimization of total interruption costs during service restoration. The number of switching operations was minimized using MV busbars to facilitate load transfer among transformers.

A search technique for service restoration based on heuristic rules was reported in [68]. The objective of this study was to minimize the number of switching operations required to re-energize the interrupted loads. This study is primarily relevant for developing countries, as most of the switching was manual and components such as transformers and feeders were operated near the rated values.

In [74], a heuristic method to minimize the number of switching operations during service restoration was developed by Shirmohammadi. In this approach, the isolated section is first connected to the adjacent feeders by closing the corresponding tie-switch, thus making the system weakly meshed. Afterwards, load flow analysis is performed to determine the current through each branch. Finally, the switching status is changed from ON to OFF for the branch with minimum current. The same procedure is repeated for the complete network until the configuration of the network becomes radial.

The authors of [75] developed a ranking-based search algorithm to perform service restoration of a large-scale distribution network. Serving a maximum number of priority customers was the additional objective for the restoration problem in this work. Other heuristic based restoration methods have been suggested by researchers in [76,77].

4.2.5. Meta-Heuristic Optimization Methods

What exactly are meta-heuristics? In essence, the phrase heuristic refers to a resource that assists in discovering 'something.' The word 'meta' is often used to highlight the presence of a higher-level strategy that guides the search process [78]. The meta-heuristics may differ depending on the problem [79]. Many meta-heuristics are centered on converting natural events or physical processes into computing tools [80]. Large-scale complex nonlinear optimization problems with a large number of variables and multiple local optima can be solved using meta-heuristics.

Service restoration is a complex optimization problem and requires fast computation to detect the switching scheme after the fault event. The obtained scheme must be implemented by the operator to quickly re-energize the interrupted loads. Reliability indicators such as SAIDI and CAIDI are directly related to outage duration. Therefore, quick implementation of the service restoration scheme is needed in order to reduce the total outage duration. Although heuristic methods are faster than conventional methods, they are governed by system-dependent rules. Therefore, the generalized framework of meta-heuristic methods have been implemented by researchers to solve the service restoration problem.

Tabu-Search Method

In this approach, local search is carried out to obtain the optimal solution. Local search techniques are usually entrapped in sub-optimal areas, in other words, the algorithm reports the local optimal solution. However, the Tabu search technique solves this problem by forbidding local search methods from returning to the same solution more than once. As a result, there is a high likelihood of obtaining the global optimum solution.

The authors of [81] devised a reactive Tabu search technique for the service restoration problem, which was framed as a single objective problem using the weighted sum method. The suggested technique outperforms the standard Tabu search algorithm, Genetic Algorithm (GA), and Parallel Simulated Annealing methods.

In [82], the researchers presented a parallel Tabu search method combined with ordinal optimization for service restoration. Parallel Tabu search has a lower computational time compared to the conventional Tabu search method. The time was further decreased by the use of the ordinal optimization approach in the selection of solution candidates.

Genetic Algorithm

GA can simulate genetic evolution by expressing individual traits through genotypes. The major driving operators of a GA are selection, which represents the "survival of the fittest", and recombination via the use of a crossover operator, which models reproduction. Both GA and its updated form, Non-Dominated Sorting Genetic Algorithm (NSGA), are highly effective for solving combinatorial optimization problem. The parent and offspring population is expressed in the form of strings. Therefore, formulating the service restoration problem through GA and its variant is relatively simple. The status of all the switches (either 0 or 1) can form one single string for a particular network topology. These strings can be changed through crossover and mutations. The conventional GA has the disadvantage of slow response. As a result, various fast variants of GA have been developed by researchers.

Generally, a GA-based approach has input strings coded in binary form. The authors of [83] adopted an integer permutation encoding method to reduce the computational burden. In this encoding method, each chromosome is a list of indices of switches present in the distribution network.

A parallel GA method which is capable of providing a trade-off solution between computational time and hardware cost for service restoration problem was reported in [84].

A hybrid approach combining GA and tabu search techniques to solve the service restoration problem was proposed by the authors of [85]. Cost minimization was taken as the primary objective for the restoration problem. A comparative analysis between the reactive tabu search method, parallel simulated annealing, and GA for service restoration optimization was performed by authors of [86]. To address the multi-objective nature of the service restoration challenge, the traditional GA has been modified to provide a faster and more efficient solution. The concepts of dominance, front formation, and selection have been included in newer GA versions such as NSGA, NSGA-II, and NSGA-III.

In [87,88], Kumar et al. have reported an NSGA-II optimization method for service restoration. The problem is solved as a multi-objective optimization problem using the concepts of dominance and Pareto optimality. This approach alleviates the problem of determining weights for each objective of the restoration problem. Both articles present an approximate trade-off solution for service restoration. The downside of employing

NSGA-II is that the method is stochastic, which means that it does not always provide the global optimal solution.

The work reported in [89] included DG penetration in the distribution network and presented a solution to the service restoration problem using the NSGA-II algorithm. An improved multi-objective evolutionary algorithm based on NSGA-II was proposed in [90] for service restoration. Through Remotely Controlled Switches (RCS), the switching operations are prioritized in order to obtain a faster and more economical solution.

Particle Swarm Optimization

Particle Swarm Intelligence is a swarm intelligence-based method in which an initial random population of particles proceeds through guided search in the search space. Similar to GA, in Particle Swarm Optimization (PSO) a random starting population is created for service restoration after encoding the particle's position in binary or integer forms. A PSO algorithm can provide faster and more efficient solutions to both continuous and discrete optimization problems.

The authors of [91] developed a PSO-based decision support tool to help system operators decide on a suitable restoration plan after a fault event. The developed tool provides the optimal switching scheme which requires minimum number of switching operations and can re-energize the maximum number of loads.

In their research work [92], Wu and Tsai have used binary coded PSO to solve the feeder reconfiguration problem. This work has been extended by the same authors, who reported an Enhanced Integer Coded Particle Swarm Optimization (EICPSO) method to deal with service restoration problem in [93]. In this approach, the *pbest* variable in the PSO velocity update equation is substituted with a local optimum list. This list is generated by picking the optimal particle position in successive iterations. This improves the probability of finding the global optimal solution. In [94], a modified PSO approach was reported for obtaining the optimal switching sequence after a fault event. The authors of [95] used binary coded PSO to solve the service restoration problem with the objectives of maximizing the supplied load and voltage regulation while minimizing power loss and switching operations.

PSO–ANN-based optimization is well suited for optimization problems with large system size. Therefore, the authors of [96] proposed a restoration plan for a large-scale distribution network using a PSO–ANN algorithm.

In 2020, a Boolean PSO method developed by integrating Boolean algebra and a PSO algorithm in binary space was reported in [97] for solving the service restoration problem for an islanded section of a network. This work investigated the effect of restoration schemes and the use of distributed energy resources and capacitor banks on the stability and reliability of power distribution system.

Ant Colony Optimization

This approach is based on the natural of behavior ants, which locate the shortest path from their colony to their food source. While travelling, ants leave a chemical substance called a pheromone along the route. Other ants are guided in their path choices by the intensity of this pheromone. This kind of search methodology is utilized by researchers to find the optimal solution for combinatorial problems such as the service restoration problem.

The research proposed in [98] discusses two Ant Colony Optimization (ACO) methods for solving the service restoration problem, namely, Absolute Switch Position-ACO (ASP-ACO) and Relative Switch Position-ACO (RSP-ACO). These algorithms were applied for restoration in a scenario involving cold load pickup conditions, which occur due to long-duration interruptions.

A hyper-cube framework-based ACO for service restoration was proposed by the authors of [99]. The "energy not supplied" value was minimized in order to ensure a lower number of interrupted loads in the restored network. The same objective was considered by the authors in [100] along with the additional goals of providing electricity to priority consumers, loss minimization, and distinct formulation for both manually controlled switching operations and remotely controlled switching operations.

The research presented by the authors of [101] discusses an ACO method combined with stochastic a spanning tree algorithm to improve search efficiency in restoration optimization. The authors used three IEEE test systems to test their proposed method.

4.2.6. Hybrid Methods

These methods are developed to improve efficiency and reduce the required computational time of an algorithm. A hybrid algorithm is obtained when two or more algorithms are combined together in order to utilize the merits of all the integrated algorithms. Several hybrid techniques for service restoration optimization have been suggested by researchers in recent years.

A hybrid method combining GA and an expert system was reported in [102]. When the total power source capacity is insufficient to restore all out-of-service regions, the expert system calculates switch operations to increase the supply margin of power sources. For each power source, the GA calculates a portion of the out-of-service area.

The authors of [103] proposed a fuzzy GA approach for service restoration with multiple objectives, viz., maximization of load supplied, minimization of number of switching operations, voltage deviation, loading of the transformer, and feeder current. These objectives were formulated as fuzzy sets. A membership function was associated with each objective, the value of which determines the satisfaction level of that objective in the restoration plan obtained by the GA.

A technique involving chaotic optimization and an immune algorithm for service restoration was discussed in [104]. The suggested method uses a global search method known as a thick search to boost convergence speed by first initialising the immune algorithm's antibodies via chaotic optimization. In order to increase population variation and avoid the algorithm becoming stuck in local minima, local search then uses the artificial immune method to finish a thin search for chaotic optimization. A non-dominated sorting fuzzy evolution strategy as a new multi-objective method was presented by the authors of [105] to solve the service restoration problem.

The combination of Evolutionary Programming (EP) with ACO was suggested in [106]. The goal was to reduce system actual power loss with the least number of switching operations by considering the changes in load currents. EP was used to identify a collection of acceptable solutions using a specially constructed mutation operator. The authors suggested employing the grey correlation grade instead of constant values for the desirability of state transitions during the ACO search process.

4.2.7. Mathematical Programming Methods

Mathematical programming refers to mathematical models that are used to solve problems such as decision problems. In contrast to computer programming, which uses algorithms designed for specific problems to tackle complex optimization problems, mathematical programming employs declarative techniques. This indicates that a distinction is made between the representation of a problem via a mathematical model and its solution.

Khushalani et al. [107] formulated the multi-objective service restoration problem as a mixed-integer nonlinear programming problem. They proposed an unbalanced distribution network and tested it for the IEEE-13 bus and IEEE-37 bus test systems. The main characteristics of this formulation are that a separate computation of the unbalanced power flow in three phases is not required, the equations are not linearized, and mutually coupled cables are considered.

In [108], a two-stage technique to restore power in electrical distribution network was proposed. In the first stage, the original problem is expressed as a mixed integer linear programming problem using linearization. This mixed integer linear programming model results in an optimal solution with the use of a conventional optimization tool. In the second stage, a nonlinear programming model is obtained by fixing the decision binary variables obtained in the first stage, resulting in a steady-state operating point of the solution.

The authors of [109] used a mixed-integer nonlinear programming model to find the best switching scheme for restoring service in an unbalanced distribution network. The suggested mixed-integer nonlinear programming model determines the state of RCS and dispatchable DG units, which are utilized to de-energize the faulted section of the network and try to supply the maximum number of interrupted customers in the healthy section of the network.

In [110], Romero et al. proposed a detailed mathematical model for solving the restoration problem in balanced radial distribution systems. The restoration problem is presented as a mixed integer second-order cone programming problem that can be solved effectively utilizing a variety of commercial solvers.

To increase the self-healing capabilities of advanced distribution networks, the authors of [111] proposed a mixed-integer second order cone programming approach for obtaining the optimal coordination of multiple soft open points.

4.2.8. Miscellaneous Methods

A service restoration plan after the occurrence of substation outages due to extreme events was developed by the authors of [112]. In this technique, the DERs are regulated such that the maximum load pickup is realized while the constraints are satisfied. The network is described with a linearized optimal power flow and the restoration problem is expressed as model predictive control. In another work [113], island partitioning as a restoration technique after extreme events was proposed. For this objective, a graph theory-based island partitioning system and optimal load shedding were studied. To rejoin the nodes on the island with main grid after fault clearance, an optimal multi-step reconnection technique is used.

A Machine Learning (ML)-based approach was proposed by Kalysh et al. in [114] to obtain an optimal restoration plan for a smart distribution system.

The authors of [115] tested a fault location, isolation, and service restoration algorithm using an advanced distribution management system tool. In [116], an adaptive restoration plan employing a modified sequential opening branches was reported. This approach searches the system's loop configuration for the optimal tie-switch or sectionalize switch for operation. In [117], the authors proposed a hybrid ML framework integrating four meta-heuristic algorithms for service restoration. The reported accuracy of this ML framework is promising. A comparison of various methods proposed for solving the service restoration problem is shown in Table 2. A block diagram representing various methods for solving the service restoration problem is shown in Figure 6.

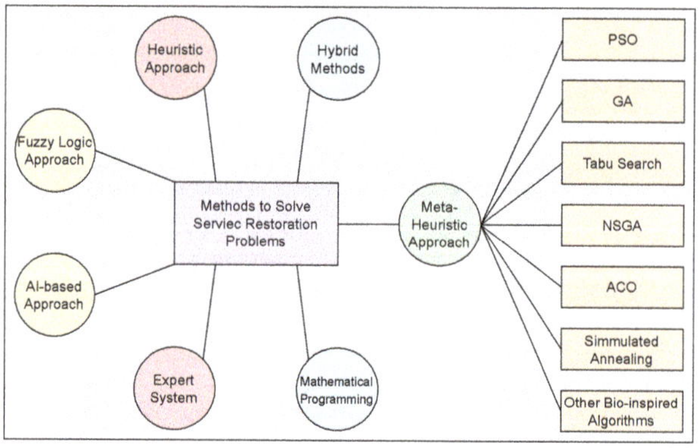

Figure 6. Various methods of solving the service restoration problem.

Table 2. Merits and demerits of various methods proposed for solving the service restoration problem.

Method	Advantages	Drawbacks	Articles
Expert system	1. System attributes are taken into account while finding the solution	1. Knowledge base formation is difficult. 2. Not generalized.	[57–63]
Fuzzy logic approach	1. Can provide solution to complex problem like service restoration. 2. Logic system construction is simple	1. Depends on human expertise. 2. Fails in case of inaccurate data input.	[64–67]
ANN	1. Method efficiency is independent of system size. 2. High reliability.	1. Weight assignment for layers is complex.	[68–70]
Heuristic approach	1. Less computational burden. 2. Fast execution.	1. Optimal solution not guaranteed. 2. System specific	[71–77]
Tabu search	1. Can be used for multi-objective optimization. 2. The final solution obtained is either optimal or close to optimal solution.	1. As the system size increases, the time for obtaining the solution also increases.	[81,82]
Genetic algorithm	1. Very effective for combinatorial problems.	1. Weight factor calculation is difficult for multi-objective optimization problem. 2. Computational time depends upon system size.	[83–86]
NSGA-II	1. Suitable for multi-objective optimization. 2. No convergence issue.	1. Global optimal solution not guaranteed.	[87–90]
PSO	1. Suitable for multi-objective optimization. 2. Easy formulation.	1. Needs modifications for various search strategies.	[91–97]
ACO	1. Global search efficiency is good. 2. Robust	1. Convergence time depends on system size.	[98–101]
Mathematical Optimization	1. Captures all the key features like objectives, constraints and decision variables. 2. No algorithm update required. 3.Highly accurate	1. High computational burden.	[107–111]

5. Conclusions

The management of a smart distribution grid is a challenging task and requires a great deal of computational resources. Distribution automation uses smart sensors and two-way communication technologies to automate the feeder switching process, monitor various grid parameters at various locations, monitor the health of installed equipment, and manage the reactive power at various buses. In this paper, different strategies related to the outage management aspect of distribution automation have been discussed.

The following are the main observations derived from previous research work on solveing the FDIR and service restoration optimization problems:

- For proper deployment of an FDIR scheme in modern power distribution systems, a decentralized control architecture is required.
- Enabling the grid to handle two-way flow of power and information is of the utmost importance in a modern grid.
- The issues of data availability, data integrity, data loss, and data security may arise during information exchange in modern distribution systems. These issues must be addressed by utilities to ensure smooth operation of any smart grid.
- Protecting the grid from cyberattacks is one of the key future challenges associated with power grids.
- High penetration of renewable energy sources and inclusion of micro grids results in a more complex control strategy for the grid.
- A better solution for service restoration can be obtained through a multi-objective approach using the concept of Pareto optimality.
- The use of machine learning and artificial intelligence methods for planning and management of contemporary power grids represents a viable option in future research work.
- Deterministic methods are very accurate compared to stochastic methods. Although there is no guarantee that a stochastic or probabilistic technique will always result in a global optimal solution, these methods are nonetheless popular because they yield approximate solutions most of the time without requiring many processing resources.

Based on the study carried out in this review, the following areas can be identified as promising future aspects of research:

- Load variations in DAS must be considered during problem formulation, as this can make an impact in the real-time implementation of proposed methodologies.
- The systems taken into consideration in most extant research works are of a radial type, whereas in the present SG paradigm the distribution grids can be either radial or weakly meshed in nature. Therefore, previously proposed methods can be extended to address weakly meshed system configurations.
- There are various challenges present related to communication aspects in smart grids, including issues pertaining to data availability, data integrity, and data security. These problems can be addressed in the multi-agent framework by including proper agents.
- Proposed multi-agent frameworks by the researchers can be tested using the IEC61850 protocol to ensure better suitability in practical smart grid systems.

Author Contributions: Conceptualization, I.S. and S.B.; methodology, I.S., B.V.S.V. and S.B.; software, I.S.; validation, I.S. and B.V.S.V.; formal analysis, I.S., B.V.S.V. and S.B.; investigation, I.S. and B.V.S.V.; resources, S.B. and N.D.B.; data curation, I.S. and B.V.S.V.; writing—original draft preparation, I.S., B.V.S.V., S.B. and N.D.B.; writing—review and editing, I.S., B.V.S.V. and N.D.B.; visualization, N.D.B.; supervision, S.B. and N.D.B.; project administration, N.D.B.; funding acquisition, S.B. and N.D.B. All authors have read and agreed to the published version of the manuscript.

Funding: This research received no external funding.

Institutional Review Board Statement: Not applicable.

Informed Consent Statement: Not applicable.

Data Availability Statement: Not applicable.

Conflicts of Interest: The authors declare no conflict of interest.

Nomenclature

SG	smart grid
RES	renewable energy sources
AMI	advanced metering infrastructure
FDIR	fault detection, isolation, and service restoration
DAS	distribution automation system
IED	intelligent electronic device
DG	distributed generation
FCB	feeder circuit breaker
f_n	n_{th} objective of service restoration
α_{rcs}	number of remotely controlled switching operations
α_{mcs}	number of manually controlled switching operations
α_{rt}	number of remotely controlled switches
α_{mt}	number of manually controlled switches
β	out-of-service area

References

1. Global Energy Review 2021. p. 36. 2021. Available online: https://www.iea.org/reports/global-energy-review-2021 (accessed on 17 April 2022).
2. Arnold, G. W.; Wollman, D. A.; FitzPatrick, G.; Prochaska, D.; Holmberg, D.; Su, D. H.; Hefner Jr, A. R.; Golmie, N. T.; Brewer, T. L.; Bello, M.; et al. *NIST Framework and Roadmap for Smart Grid Interoperability Standards, Release 1.0*; NIST Special Publication; National Institute of Standards and Technology: Gaithersburg, MD, USA, 2010; p. 145. [CrossRef]
3. Nafi, N.S.; Ahmed, K.; Gregory, M.A.; Datta, M. A Survey of Smart Grid Architectures, Applications, Benefits and Standardization. *J. Netw. Comput. Appl.* **2016**, *76*, 23–36. [CrossRef]
4. Latare, N. A.; Bhat, S. S.; Srivastava, I. Literature Review of Service Restoration in Distribution System. In Proceedings of the 2017 Second International Conference on Electrical, Computer and Communication Technologies (ICECCT), Coimbatore, India, 22–24 February 2017; pp. 1–6. [CrossRef]

5. Gururajapathy, S.S.; Mokhlis, H.; Illias, H.A. Fault Location and Detection Techniques in Power Distribution Systems with Distributed Generation: A Review. *Renew. Sustain. Energy Rev.* **2017**, *74*, 949–958. [CrossRef]
6. Mwifunyi, R.J.; Kissaka, M.M.; Mvungi, N.H. Distributed Approach in Fault Localisation and Service Restoration: State-of-the-Art and Future Direction. *Cogent Eng.* **2019**, *6*, 1628424. [CrossRef]
7. Jiang, Y.; Liu, C.-C.; Xu, Y. Smart Distribution Systems. *Energies* **2016**, *9*, 297. [CrossRef]
8. Pahwa, A. Planning and Analysis Tools to Evaluate Distribution Automation Implementation and Benefits. *IEEE Power Eng. Soc. Gen. Meet.* **2005**, *3*, 2622–2623. [CrossRef]
9. Gruenemeyer, D. Distribution Automation: How Should It Be Evaluated? In Proceedings of the 1991 Rural Electric Power Conference, the 35th Annual Conference, Dearborn, MI, USA, 28–30 April 1991; p. C3/1–C310. [CrossRef]
10. Sanchez, J.; Caire, R.; Hadjsaid, N. ICT and Power Distribution Modeling Using Complex Networks. In Proceedings of the 2013 IEEE Grenoble Conference, Grenoble, France, 16–20 June 2013; pp. 1–6. [CrossRef]
11. Elattar, E.E.; Shaheen, A.M.; El-Sayed, A.M.; El-Sehiemy, R.A.; Ginidi, A.R. Optimal Operation of Automated Distribution Networks Based-MRFO Algorithm. *IEEE Access* **2021**, *9*, 19586–19601. [CrossRef]
12. Shaheen, A.; Elsayed, A.; Ginidi, A.; El-Sehiemy, R.; Elattar, E. Improved Heap-Based Optimizer for DG Allocation in Reconfigured Radial Feeder Distribution Systems. *IEEE Syst. J.* **2022**, 1–10. [CrossRef]
13. Shaheen, A.M.; Elsayed, A.M.; Ginidi, A.R.; El-Sehiemy, R.A.; Elattar, E. A Heap-Based Algorithm with Deeper Exploitative Feature for Optimal Allocations of Distributed Generations with Feeder Reconfiguration in Power Distribution Networks. *Knowl.-Based Syst.* **2022**, *241*, 108269. [CrossRef]
14. Chia-Hung Lin; Hui-Jen Chuang; Chao-Shun Chen; Chung-Sheng Li; Chin-Ying Ho. Fault Detection, Isolation and Restoration Using a Multiagent-Based Distribution Automation System. In Proceedings of the 2009 4th IEEE Conference on Industrial Electronics and Applications, Xi'an, China, 25–27 May 2009; pp. 2528–2533. [CrossRef]
15. Apostolov, A.; Vandiver, B. IEC 61850 GOOSE Applications to Distribution Protection Schemes. In Proceedings of the 2011 64th Annual Conference for Protective Relay Engineers, College Station, TX, USA, 4–7 April 2016; pp. 178–184. [CrossRef]
16. Milioudis, A.N.; Andreou, G.T.; Labridis, D.P. Enhanced Protection Scheme for Smart Grids Using Power Line Communications Techniques—Part I: Detection of High Impedance Fault Occurrence. *IEEE Trans. Smart Grid* **2012**, *3*, 1621–1630. [CrossRef]
17. Li, J.; Ma, X.-Y.; Liu, C.-C.; Schneider, K.P. Distribution System Restoration With Microgrids Using Spanning Tree Search. *IEEE Trans. Power Syst.* **2014**, *29*, 3021–3029. [CrossRef]
18. Zidan, A.; El-Saadany, E.F. A Cooperative Multiagent Framework for Self-Healing Mechanisms in Distribution Systems. *IEEE Trans. Smart Grid* **2012**, *3*, 1525–1539. [CrossRef]
19. Shahin, M. A. Smart Grid Self-Healing Implementation for Underground Distribution Networks. In Proceedings of the 2013 IEEE Innovative Smart Grid Technologies-Asia (ISGT Asia), Bangalore, India, 10–13 November 2013; pp. 1–5. [CrossRef]
20. Zidan, A.; Khairalla, M.; Abdrabou, A.M.; Khalifa, T.; Shaban, K.; Abdrabou, A.; El Shatshat, R.; Gaouda, A.M. Fault Detection, Isolation, and Service Restoration in Distribution Systems: State-of-the-Art and Future Trends. *IEEE Trans. Smart Grid* **2017**, *8*, 2170–2185. [CrossRef]
21. Shahidehpour, M.; Wang, Y. *Communication and Control in Electric Power Systems: Applications of Parallel and Distributed Processing*; IEEE Press Series on Power Engineering; IEEE Press: Piscataway, NJ, USA; Wiley-Interscience: Hoboken, NJ, USA, 2003.
22. Arefifar, S.A.; Mohamed, Y.A.-R.I.; EL-Fouly, T.H.M. Comprehensive Operational Planning Framework for Self-Healing Control Actions in Smart Distribution Grids. *IEEE Trans. Power Syst.* **2013**, *28*, 4192–4200. [CrossRef]
23. IEEE Guide for Electric Power Distribution Reliability Indices-Redline. IEEE Std 1366-2012 (Revision of IEEE Std 1366-2003)-Redline 2012, pp. 1–92. Available online: https://ieeexplore.ieee.org/document/6329910 (accessed on 17 April 2022).
24. Craig, D.; Spare, J. Benefits of Automatic Fault Detection, Isolation and Service Restoration on Power Distribution Circuits. In Proceedings of the PES T&D 2012, Orlando, FL, USA, 7–10 May 2012; pp. 1–4. [CrossRef]
25. Tanenbaum, A.S.; Van Steen, M. *Distributed Systems: Principles and Paradigms*; Prentice-Hall: Upper Saddle River, NJ, USA, 2007.
26. Mohagheghi, S.; Stoupis, J.; Wang, Z. Communication Protocols and Networks for Power Systems-Current Status and Future Trends. In Proceedings of the 2009 IEEE/PES Power Systems Conference and Exposition, Seattle, WA, USA, 15–18 March 2009; pp. 1–9. [CrossRef]
27. Mekic, F.; Alloway, K.; Angelo, C.; Goodin, R. Fault Detection Isolation and Restoration on the Feeder (FDIR): Pick Your Technology. In Proceedings of the CIRED, 21ST International Conference on Electricity Distribution, Frankfurt, Germany, 6–9 June 2011; pp. 6–9.
28. Jenkins, R.; Dolezilek, D. Case Study: Using IEC 61850 Methods for RTU Replacement and Distributed Automation. In Proceedings of the 10th Annual Western Power Delivery Automation Conference, Spokane, WA, USA, April 2008.
29. Wernsing, R.; Hubertus, J.; Duffy, M.; Hataway, G.; Conner, D.; Nelson, E. Implementation of an Innovative Advanced Loop Scheme to Provide Distribution Reliability Improvement. In Proceedings of the 2010 IEEE Rural Electric Power Conference (REPC), Orlando, FL, USA, 16–19 May 2010, pp. B2–B2 8. [CrossRef]
30. Nagata, T.; Tao, Y.; Sasaki, H.; Fujita, H. A Multiagent Approach to Distribution System Restoration. In Proceedings of the 2003 IEEE Power Engineering Society General Meeting (IEEE Cat. No.03CH37491), Toronto, ON, Canada, 13–17 July 2003; pp. 655–660. [CrossRef]

31. Chen, X.; Kong, B.; Liu, F.; Gong, X.; Shen, X. System Service Restoration of Distribution Network Based on Multi-Agent Technology. In *Proceedings of the 2013 Fourth International Conference on Digital Manufacturing & Automation*; IEEE: Qingdao, China, 2013; pp. 1371–1374. [CrossRef]
32. Lo, Y. L.; Wang, C. H.; Lu, C. N. A Multi-Agent Based Service Restoration in Distribution Network with Distributed Generations. In Proceedings of the 2009 15th International Conference on Intelligent System Applications to Power Systems, Curitiba, Brazil, 8–12 November 2009; pp. 1–5. [CrossRef]
33. Khamphanchai, W.; Pipattanasomporn, M.; Rahman, S. A Multi-Agent System for Restoration of an Electric Power Distribution Network with Local Generation. In Proceedings of the 2012 IEEE Power and Energy Society General Meeting, San Diego, CA, USA, 22–26 July 2012; pp. 1–8. [CrossRef]
34. Elmitwally, A.; Elsaid, M.; Elgamal, M.; Chen, Z. A Fuzzy-Multiagent Service Restoration Scheme for Distribution System With Distributed Generation. *IEEE Trans. Sustain. Energy* **2015**, *6*, 810–821. [CrossRef]
35. Ghorbani, M.J.; Choudhry, M.A.; Feliachi, A. A Multiagent Design for Power Distribution Systems Automation. *IEEE Trans. Smart Grid* **2016**, *7*, 329–339. [CrossRef]
36. Sharma, A.; Srinivasan, D.; Trivedi, A. A Decentralized Multiagent System Approach for Service Restoration Using DG Islanding. *IEEE Trans. Smart Grid* **2015**, *6*, 2784–2793. [CrossRef]
37. Dong, Z.; Lin, L.; Guan, L.; Chen, H.; Liang, Q. A Service Restoration Strategy for Active Distribution Network Based on Multiagent System. In Proceedings of the 2018 IEEE Innovative Smart Grid Technologies—Asia (ISGT Asia), Singapore, 22–25 May 2018; pp. 384–389. [CrossRef]
38. Habib, H.F.; Youssef, T.; Cintuglu, M.H.; Mohammed, O.A. Multi-Agent-Based Technique for Fault Location, Isolation, and Service Restoration. *IEEE Trans. Ind. Applicat.* **2017**, *53*, 1841–1851. [CrossRef]
39. Prabawa, P.; Choi, D.-H. Multi-Agent Framework for Service Restoration in Distribution Systems With Distributed Generators and Static/Mobile Energy Storage Systems. *IEEE Access* **2020**, *8*, 51736–51752. [CrossRef]
40. Nguyen, C.P.; Flueck, A.J. Agent Based Restoration With Distributed Energy Storage Support in Smart Grids. *IEEE Trans. Smart Grid* **2012**, *3*, 1029–1038. [CrossRef]
41. Estebsari, A.; Patti, E.; Barbierato, L. Fault Detection, Isolation and Restoration Test Platform Based on Smart Grid Architecture Model Using Intenet-of-Things Approaches. In Proceedings of the 2018 IEEE International Conference on Environment and Electrical Engineering and 2018 IEEE Industrial and Commercial Power Systems Europe (EEEIC / I&CPS Europe), Palermo, Italy, 12–15 June 2018; pp. 1–5. [CrossRef]
42. Hussain, S.M.S.; Nadeem, F.; Aftab, M.A.; Ali, I.; Ustun, T.S. The Emerging Energy Internet: Architecture, Benefits, Challenges, and Future Prospects. *Electronics* **2019**, *8*, 1037. [CrossRef]
43. Gundu, V.; Simon, S.P. A Novel Energy Routing Technique with Hybrid Energy Storage for Residential Electricity Cost Minimization in a Smart Distribution Network. *Energy Sources Part A Recovery Util. Environ. Eff.* **2020**, 1–18. [CrossRef]
44. Shewale, A.; Mokhade, A.; Funde, N.; Bokde, N.D. An Overview of Demand Response in Smart Grid and Optimization Techniques for Efficient Residential Appliance Scheduling Problem. *Energies* **2020**, *13*, 4266. [CrossRef]
45. Shewale, A.; Mokhade, A.; Funde, N.; Bokde, N.D. A Survey of Efficient Demand-Side Management Techniques for the Residential Appliance Scheduling Problem in Smart Homes. *Energies* **2022**, *15*, 2863. [CrossRef]
46. Kabalci, Y.; Kabalci, E.; Padmanaban, S.; Holm-Nielsen, J.B.; Blaabjerg, F. Internet of Things Applications as Energy Internet in Smart Grids and Smart Environments. *Electronics* **2019**, *8*, 972. [CrossRef]
47. Mets, K.; Ojea, J. A.; Develder, C. Combining Power and Communication Network Simulation for Cost-Effective Smart Grid Analysis. *IEEE Commun. Surv. Tutor.* **2014**, *16*, 1771–1796. [CrossRef]
48. Avancini, D.B.; Rodrigues, J.J.P.C.; Rabêlo, R.A.L.; Das, A.K.; Kozlov, S.; Solic, P. A New IoT-based Smart Energy Meter for Smart Grids. *Int. J. Energy Res.* **2021**, *45*, 189–202. [CrossRef]
49. Li, X.; Liang, X.; Lu, R.; Shen, X.; Lin, X.; Zhu, H. Securing Smart Grid: Cyber Attacks, Countermeasures, and Challenges. *IEEE Commun. Mag.* **2012**, *50*, 38–45. [CrossRef]
50. Ni, K.; Ramanathan, N.; Chehade, M.N.H.; Balzano, L.; Nair, S.; Zahedi, S.; Kohler, E.; Pottie, G.; Hansen, M.; Srivastava, M. Sensor Network Data Fault Types. *ACM Trans. Sen. Netw.* **2009**, *5*, 1–29. [CrossRef]
51. Zou, Z.; Bao, Y.; Li, H.; Spencer, B.F.; Ou, J. Embedding Compressive Sensing-Based Data Loss Recovery Algorithm Into Wireless Smart Sensors for Structural Health Monitoring. *IEEE Sens. J.* **2015**, *15*, 797–808. [CrossRef]
52. Showail, A.; Jamshaid, K.; Shihada, B. Buffer Sizing in Wireless Networks: Challenges, Solutions, and Opportunities. *IEEE Commun. Mag.* **2016**, *54*, 130–137. [CrossRef]
53. Davoody-Beni, Z.; Sheini-Shahvand, N.; Shahinzadeh, H.; Moazzami, M.; Shaneh, M.; Gharehpetian, G. B. Application of IoT in Smart Grid: Challenges and Solutions. In Proceedings of the 2019 5th Iranian Conference on Signal Processing and Intelligent Systems (ICSPIS), Shahrood, Iran, 18–19 December 2019; pp. 1–8. [CrossRef]
54. Qi Huang; Kaiyu Qin; Wenyong Wang. Development of a Grid Computing Platform for Electric Power System Applications. In Proceedings of the 2006 IEEE Power Engineering Society General Meeting, Montreal, QC, Canada, 18–22 June 2006; p. 7. [CrossRef]
55. Rea, P.; Ottaviano, E.; Machado, J.; Antosz, K. (Eds.) *Design, Applications, and Maintenance of Cyber-Physical Systems*; Advances in Systems Analysis, Software Engineering, and High Performance Computing; IGI Global: Hershey, PA, USA, 2021. [CrossRef]

56. Kousar, S.; Zafar, N.A.; Ali, T.; Alkhammash, E.H.; Hadjouni, M. Formal Modeling of IoT-Based Distribution Management System for Smart Grids. *Sustainability* **2022**, *14*, 4499. [CrossRef]
57. Liu, C.-C.; Lee, S.J.; Venkata, S.S. An Expert System Operational Aid for Restoration and Loss Reduction of Distribution Systems. *IEEE Trans. Power Syst.* **1988**, *3*, 619–626. [CrossRef]
58. Zhang, Z.Z.; Hope, G.S.; Malik, O.P. A Knowledge-Based Approach to Optimize Switching in Substations. *IEEE Trans. Power Deliv.* **1990**, *5*, 103–109. [CrossRef]
59. Ahmed, S.S.; Rabbi, S.; Mostafa, R.; Bhuiya, A.R. Development of an Expert System for Restoring the Service Interrupted by Sustained Line Faults in a Distribution System. *Electr. Power Syst. Res.* **1993**, *26*, 101–108. [CrossRef]
60. Ma, T.-K.; Liu, C.-C.; Tsai, M.-S.; Rogers, R.; Muchlinski, S. L.; Dodge, J. Operational Experience and Maintenance of an On-Line Expert System for Customer Restoration and Fault Testing. In *1991 Power Industry Computer Application Conference*; IEEE: Baltimore, MD, USA, 1991; pp. 350–357. [CrossRef]
61. Teo, C.Y.; Gooi, H.B. Restoration of Electrical Power Supply through an Algorithm and Knowledge Based System. *Electr. Power Syst. Res.* **1994**, *29*, 171–180. [CrossRef]
62. Kim, H.; Ko, Y.; Jung, K.-H. Algorithm of Transferring the Load of the Faulted Substation Transformer Using the Best-First Search Method. *IEEE Trans. Power Deliv.* **1992**, *7*, 1434–1442. [CrossRef]
63. Tsai, M.-S. Development of an Object-Oriented Service Restoration Expert System With Load Variations. *IEEE Trans. Power Syst.* **2008**, *23*, 219–225. [CrossRef]
64. Kuo, H.-C.; Hsu, Y.-Y. Distribution System Load Estimation and Service Restoration Using a Fuzzy Set Approach. *IEEE Trans. Power Deliv.* **1993**, *8*, 1950–1957. [CrossRef]
65. Huang, C.M. Multiobjective Service Restoration of Distribution Systems Using Fuzzy Cause-Effect Networks. *IEEE Trans. Power Syst.* **2003**, *18*, 867–874. [CrossRef]
66. Lee, S.-J.; Lim, S.-I.; Ahn, B.-S. Service Restoration of Primary Distribution Systems Based on Fuzzy Evaluation of Multi-Criteria. *IEEE Trans. Power Syst.* **1998**, *13*, 1156–1163. [CrossRef]
67. Popovic, D.S.; Popovic, Z.N. A Risk Management Procedure for Supply Restoration in Distribution Networks. *IEEE Trans. Power Syst.* **2004**, *19*, 221–228. [CrossRef]
68. Kim, H.; Ko, Y.; Jung, K.-H. Artificial Neural-Network Based Feeder Reconfiguration for Loss Reduction in Distribution Systems. *IEEE Trans. Power Deliv.* **1993**, *8*, 1356–1366. [CrossRef]
69. Hsu, Y.-Y. Distribution System Service Restoration Using the Artificial Neural Network Approach and Pattern Recognition Method. *IEE Proc., Gener. Transm. Distrib.* **1995**, *142*, 251–256. [CrossRef]
70. Bretas, A.S.; Phadke, A.G. Artificial Neural Networks in Power System Restoration. *IEEE Trans. Power Deliv.* **2003**, *18*, 1181–1186. [CrossRef]
71. Morelato, A.L.; Monticelli, A.J. Heuristic Search Approach to Distribution System Restoration. *IEEE Trans. Power Deliv.* **1989**, *4*, 2235–2241. [CrossRef]
72. Hsu, Y.-Y.; Huang, H.-M.; Kuo, H.-C.; Peng, S.K.; Chang, C.W.; Chang, K.J.; Yu, H. S.; Chow, C.E.; Kuo, R.T. Distribution System Service Restoration Using a Heuristic Search Approach. *IEEE Trans. Power Deliv.* **1992**, *7*, 734–740. [CrossRef]
73. Nahman, J.; St'rbac, G. A New Algorithm for Service Restoration in Large-Scale Urban Distribution Systems. *Electr. Power Syst. Res.* **1994**, *29*, 181–192. [CrossRef]
74. Shirmohammadi, D. Service Restoration in Distribution Networks via Network Reconfiguration. *IEEE Trans. Power Deliv.* **1992**, *7*, 952–958. [CrossRef]
75. Miu, K.N.; Chiang, H.-D.; McNulty, R.J. Multi-Tier Service Restoration through Network Reconfiguration and Capacitor Control for Large-Scale Radial Distribution Networks. In Proceedings of the 21st International Conference on Power Industry Computer Applications. Connecting Utilities. PICA 99. to the Millennium and Beyond (Cat. No.99CH36351), Santa Clara, CA, USA, 21–21 May 1999; pp. 153–159. [CrossRef]
76. Miu, K.N.; Chiang, H.-D.; Yuan, B.; Darling, G. Fast Service Restoration for Large-Scale Distribution Systems with Priority Customers and Constraints. *IEEE Trans. Power Syst.* **1998**, *13*, 789–795. [CrossRef]
77. Singh, S.P.; Raju, G.S.; Rao, G.K.; Afsari, M. A Heuristic Method for Feeder Reconfiguration and Service Restoration in Distribution Networks. *Int. J. Electr. Power Energy Syst.* **2009**, *31*, 309–314. [CrossRef]
78. Chicco, G.; Mazza, A. Metaheuristic Optimization of Power and Energy Systems: Underlying Principles and Main Issues of the 'Rush to Heuristics'. *Energies* **2020**, *13*, 5097. [CrossRef]
79. Sörensen, K.; Sevaux, M.; Glover, F. A History of Metaheuristics. In *Handbook of Heuristics*; Martí, R., Panos, P., Resende, M.G.C., Eds.; Springer International Publishing: Cham, Switzerland, 2018; pp. 791–808. [CrossRef]
80. Salcedo-Sanz, S. Modern Meta-Heuristics Based on Nonlinear Physics Processes: A Review of Models and Design Procedures. *Phys. Rep.* **2016**, *655*, 1–70. [CrossRef]
81. Toune, S.; Fudo, H.; Genji, T.; Fukuyama, Y.; Nakanishi, Y. A Reactive Tabu Search for Service Restoration in Electric Power Distribution Systems. In Proceedings of the 1998 IEEE World Congress on Computational Intelligence (Cat. No.98TH8360), Anchorage, AK, USA, 4–9 May 1998; pp. 763–768. [CrossRef]
82. Mori, H.; Tani, H. A Hybrid Method of PTS and Ordinal Optimization for Distribution System Service Restoration. In Proceedings of the 2003 IEEE International Conference on Systems, Man and Cybernetics. Conference Theme-System Security and Assurance (Cat. No.03CH37483), Washington, DC, USA, 17 November 2003; Volume 4, pp. 3476–3483. [CrossRef]

83. Luan, W.P.; Irving, M.R.; Daniel, J.S. Genetic Algorithm for Supply Restoration and Optimal Load Shedding in Power System Distribution Networks. *IEEE Proc., Gener. Transm. Distrib.* **2002**, *149*, 145–151. [CrossRef]
84. Fukuyama, Y.; Chiang, H.-D.; Miu, K.N. Parallel Genetic Algorithm for Service Restoration in Electric Power Distribution Systems. *Int. J. Electr. Power Energy Syst.* **1996**, *18*, 111–119. [CrossRef]
85. Shin, D.-J.; Kim, J.-O.; Kim, T.-K.; Choo, J.-B.; Singh, C. Optimal Service Restoration and Reconfiguration of Network Using Genetic-Tabu Algorithm. *Electr. Power Syst. Res.* **2004**, *71*, 145–152. [CrossRef]
86. Toune, S.; Fudo, H.; Genji, T.; Fukuyama, Y.; Nakanishi, Y. Comparative Study of Modern Heuristic Algorithms to Service Restoration in Distribution Systems. *IEEE Trans. Power Deliv.* **2002**, *17*, 173–181. [CrossRef]
87. Kumar, Y.; Das, B.; Sharma, J. Service Restoration in Distribution System Using Non-Dominated Sorting Genetic Algorithm. *Electr. Power Syst. Res.* **2006**, *76*, 768–777. [CrossRef]
88. Kumar, Y.; Das, B.; Sharma, J. Multiobjective, Multiconstraint Service Restoration of Electric Power Distribution System With Priority Customers. *IEEE Trans. Power Deliv.* **2008**, *23*, 261–270. [CrossRef]
89. Huang Xianchao; Zhang Lizi; Taylor, G. A. Service Restoration of Distribution System with Distributed Generation. In *2010 International Conference on Power System Technology*; IEEE: Hangzhou, China, 2010; pp. 1–5. [CrossRef]
90. Marques, L.T.; Fernandes, J.P.R.; London, J.B. A. Improved Multi-Objective Evolutionary Algorithm in Subpopulation Tables with Features from NSGA-II for the Service Restoration Problem. In *2019 IEEE Milan PowerTech*; IEEE: Milan, Italy, 2019; pp. 1–5. [CrossRef]
91. Lambert-Torres, G.; Martins, H.G.; Coutinho, M.P.; Salomon, C.P.; Vieira, F.C. Particle Swarm Optimization Applied to System Restoration. In Proceedings of the 2009 IEEE Bucharest PowerTech, Bucharest, Romania, 28 June–2 July 2009; pp. 1–6. [CrossRef]
92. Wu, W.-C.; Tsai, M.-S. Feeder Reconfiguration Using Binary Coding Particle Swarm Optimization. *Int. J. Control. Autom. Syst.* **2008**, *6*, 488–494.
93. Wu, W.-C.; Tsai, M.-S. Application of Enhanced Integer Coded Particle Swarm Optimization for Distribution System Feeder Reconfiguration. *IEEE Trans. Power Syst.* **2011**, *26*, 1591–1599. [CrossRef]
94. Abdelaziz, A.Y.; Mekhamer, S.F.; Badr, M.A.L.; Mohamed, F. M.; El-Saadany, E. F. A Modified Particle Swarm Algorithm for Distribution Systems Reconfiguration. In Proceedings of the 2009 IEEE Power & Energy Society General Meeting, Calgary, AB, Canad, 26–30 July 2009; pp. 1–8. [CrossRef]
95. Oliveira, L.W.; Oliveira, E.J.; Silva, I.C.; Gomes, F.V.; Borges, T.T.; Marcato, A.L. M.; Oliveira, A.R. Optimal Restoration of Power Distribution System through Particle Swarm Optimization. In Proceedings of the 2015 IEEE Eindhoven PowerTech, Eindhoven, The Netherlands, 29 June–2 July 2015; pp. 1–5. [CrossRef]
96. Soomro, D.M.; Al-Izzi, M.Y.; Soodi, H.A.; Elasager, N.M.; Chong, S.C. Optimal Restoration for Distribution System Using PSO and ANN. *JOurnal Telecommun. Electron. Comput. Eng. (JTEC)* **2017**, *9*, 1–6.
97. ElDesouky, A.A.; Reyad, E.M.; Mahmoud, G.A. Implementation of Boolean PSO for Service Restoration Using Distribution Network Reconfiguration Simultaneously with Distributed Energy Resources and Capacitor Banks. *Int. J. Renew. Energy Res.* **2020**, *10*, 354–365.
98. Mohanty, I.; Kalita, J.; Das, S.; Pahwa, A.; Buehler, E. Ant Algorithms for the Optimal Restoration of Distribution Feeders during Cold Load Pickup. In Proceedings of the 2003 IEEE Swarm Intelligence Symposium. SIS'03 (Cat. No.03EX706), Indianapolis, IN, USA, 26 April 2003; pp. 132–137. [CrossRef]
99. Watanabe, I. An ACO Algorithm for Service Restoration in Power Distribution Systems. In Proceedings of the 2005 IEEE Congress on Evolutionary Computation, Scotland, UK, 2–5 September, 2005; Volume 3, pp. 2864–2871. [CrossRef]
100. Karn, R.; Kumar, Y.; Agnihotri, G. Development of ACO Algorithm for Service Restoration in Distribution System. *Int. J. Emerg. Technol.* **2011**, *2*, 71–77.
101. Lu, Z.; Wen, Y.; Yang, L. An Improved ACO Algorithm for Service Restoration in Power Distribution Systems. In Proceedings of the 2009 Asia-Pacific Power and Energy Engineering Conference, Wuhan, China, 28–30 March 2009; pp. 1–4. [CrossRef]
102. Fukuyama, Y.; Endo, H.; Nakanishi, Y. A Hybrid System for Service Restoration Using Expert System and Genetic Algorithm. In Proceedings of the 1996 International Conference on Intelligent System Application to Power Systems, Orlando, FL, USA, 28 January–2 February 1996; pp. 394–398. [CrossRef]
103. Hsiao, Y.-T.; Chien, C.-Y. Enhancement of Restoration Service in Distribution Systems Using a Combination Fuzzy-GA Method. *IEEE Trans. Power Syst.* **2000**, *15*, 1394–1400. [CrossRef]
104. Lei, S.; Li, S.; Yang, J.; Jiang, D. Distribution Service Restoration Using Chaotic Optimization and Immune Algorithm. In Proceedings of the 2011 International Conference on Information Science and Technology, Nanjing, China, 23–24 July 2011; pp. 1129–1133. [CrossRef]
105. Augugliaro, A.; Dusonchet, L.; Sanseverino, E.R. Evolving Non-Dominated Solutions in Multiobjective Service Restoration for Automated Distribution Networks. *Electr. Power Syst. Res.* **2001**, *59*, 185–195. [CrossRef]
106. Tsai, M.S.; Chu, C.C. Applications of Hybrid EP-ACO for Power Distribution System Loss Minimization under Load Variations. In Proceedings of the 2011 16th International Conference on Intelligent System Applications to Power Systems, Hersonissos, Greece, 25–28 September 2011; pp. 1–7. [CrossRef]
107. Khushalani, S.; Solanki, J.M.; Schulz, N.N. Optimized Restoration of Unbalanced Distribution Systems. *IEEE Trans. Power Syst.* **2007**, *22*, 624–630. [CrossRef]

108. Cavalcante, P.L.; Lopez, J.C.; Franco, J.F.; Rider, M.J.; Garcia, A.V.; Malveira, M.R.R.; Martins, L.L.; Direito, L.C.M. Centralized Self-Healing Scheme for Electrical Distribution Systems. *IEEE Trans. Smart Grid* **2016**, *7*, 145–155. 54436. [CrossRef]
109. Lopez, J.C.; Franco, J.F.; Rider, M.J.; Romero, R. Optimal Restoration/Maintenance Switching Sequence of Unbalanced Three-Phase Distribution Systems. *IEEE Trans. Smart Grid* **2018**, *9*, 6058–6068. [CrossRef]
110. Romero, R.; Franco, J.F.; Leao, F.B.; Rider, M.J.; de Souza, E.S. A New Mathematical Model for the Restoration Problem in Balanced Radial Distribution Systems. *IEEE Trans. Power Syst.* **2016**, *31*, 1259–1268. [CrossRef]
111. Picioroaga, I.I.; Tudose, A.M.; Sidea, D.O.; Bulac, C.; Toma, L. Application of Soft Open Points for Increasing the Supply Restoration in Active Distribution Networks. In Proceedings of the 2021 12th International Symposium on Advanced Topics in Electrical Engineering (ATEE), Bucharest, Romania, 25–27 March 2021; pp. 1–6. [CrossRef]
112. Eseye, A.T.; Knueven, B.; Zhang, X.; Reynolds, M.; Jones, W. Resilient Operation of Power Distribution Systems Using MPC-Based Critical Service Restoration. In Proceedings of the 2021 IEEE Green Technologies Conference (GreenTech), Denver, CO, USA, 7–9 April 2021; pp. 292–297. [CrossRef]
113. Jacob, R.A.; Zhang, J. Outage Management in Active Distribution Network with Distributed Energy Resources. In Proceedings of the 2020 52nd North American Power Symposium (NAPS), Tempe, AZ, USA, 11–13 April 2021; pp. 1–6. [CrossRef]
114. Kalysh, I.; Kenzhina, M.; Kaiyrbekov, N.; Kumar Nunna, H.S.V.S.; Dadlani, A.; Doolla, S. Machine Learning-Based Service Restoration Scheme for Smart Distribution Systems with DGs and High Priority Loads. In Proceedings of the 2019 International Conference on Smart Energy Systems and Technologies (SEST), Porto, Portugal, 11 September 2019; pp. 1–6. [CrossRef]
115. Velaga, Y.N.; Prabakar, K.; Baggu, M.; Schneider, K.P. Evaluation of Centralized Model Based FLISR in a Lab Setup. In Proceedings of the 2021 IEEE Rural Electric Power Conference (REPC), Savannah, GA, USA, 27–29 April 2021; pp. 8–14. [CrossRef]
116. Lorm, R.; Eng, S.; Suk, S.; Chhlonh, C.; Bun, L.; Vai, V. Service Restoration in Distribution System Using Modified Sequential Opening Branches. In Proceedings of the 2021 9th International Electrical Engineering Congress (iEECON), Pattaya, Thailand, 10–12 March 2021; pp. 69–72. [CrossRef]
117. Srivastava, I.; Bhat, S.; Thadikemalla, V.S.G.; Singh, A.R. A Hybrid Machine Learning and Meta-heuristic Algorithm Based Service Restoration Scheme for Radial Power Distribution System. *Int. Trans. Electr. Energy Syst.* **2021**, 31, e12894. [CrossRef]

energies
MDPI

Article

Neural Network Controlled Solar PV Battery Powered Unified Power Quality Conditioner for Grid Connected Operation

Okech Emmanuel Okwako [1], Zhang-Hui Lin [1,*], Mali Xin [1], Kamaraj Premkumar [2] and Alukaka James Rodgers [1]

1 Department of Electrical Engineering, School of Mechanical Engineering, University of Shanghai for Science and Technology, Shanghai 200093, China
2 Department of Electrical and Electronics Engineering, Rajalakshmi Engineering College, Chennai 602105, India
* Correspondence: zhanghuilin@usst.edu.cn

Abstract: The Unified Power Quality Conditioner (UPQC) is a technology that has successfully addressed power quality issues. In this paper, a photovoltaic system with battery storage powered Unified Power Quality Conditioner is presented. Total harmonic distortion of the grid current during extreme voltage sag and swell conditions is more than 5% when UPQC is controlled with synchronous reference frame theory (SRF) and instantaneous reactive power theory (PQ) control. The shunt active filter of the UPQC is controlled by the artificial neural network to overcome the above problem. The proposed artificial neural network controller helps to simplify the control complexity and mitigate power quality issues effectively. This study aims to use a neural network to control a shunt active filter of the UPQC to maximise the supply of active power loads and grid and also used to mitigate the harmonic problem due to non-linear loads in the grid. The performance of the model is tested under various case scenarios, including non-linear load conditions, unbalanced load conditions, and voltage sag and voltage swell conditions. The simulations were performed in MATLAB/Simulink software. The results showed excellent performance of the proposed approach and were compared with PQ and SRF control. The percent total harmonic distortion (%THD) of the grid current was measured and discussed for all cases. The results show that the %THD is within the acceptable limits of IEEE-519 (less than 5%) in all test case scenarios by the proposed controller.

Keywords: shunt converter; unified power quality conditioner; total harmonic distortion; artificial intelligence; renewable energy system

check for updates

Citation: Okwako, O.E.; Lin, Z.-H.; Xin, M.; Premkumar, K.; Rodgers, A.J. Neural Network Controlled Solar PV Battery Powered Unified Power Quality Conditioner for Grid Connected Operation. *Energies* **2022**, *15*, 6825. https://doi.org/10.3390/en15186825

Academic Editors: Veerapandiyan Veerasamy, Shailendra Singh and Sunil Kumar Singh

Received: 21 August 2022
Accepted: 14 September 2022
Published: 18 September 2022

1. Introduction

Life depends on energy, which is its main source of sustenance. Electricity is one of the most important forms of energy available today. Consequently, the market for this type of energy is growing rapidly. In recent years, researchers, especially in the field of electrical engineering, have paid great attention to the quality of electricity. The decarbonisation of the power grid is promoted by fixing power quality problems, which enables the smooth generation of electrical energy. Harmonics generated by non-linear loads such as electronic device high frequency switching components greatly distort the sinusoidal nature of the grid voltage and current; this poses a major concern to the utility. In this sense, voltage interruptions are of major concern to customers using the power from the main grid. Power quality issues can interact with nearby communication lines and cause interference and interruptions to various end users; this subsequently leads to power losses in the grid, undesirable or abnormal equipment operation as well as shortening the lifespan of the electronic gadgets and may also lead to safety issues. IEEE Standard-1159 [1] describes the behaviour of a standard waveform and categorises different types of interference. A minor periodic distortion, such as sags or surges, can cause a complicated power quality scenario resulting from the combination of two or more different power disturbances. To protect equipment from the hazardous effect of these power disturbances, industrial customers have built a number of active power filters (APFs) [2–7]. Shunt APFs have been designed to mitigate the harmonic generated into the grid due to the increasing use of power electronic components in modern plants; they reduce the harmonic effect as well as protect sensitive load in the system [8–10].

The increased connection of renewable energy to the grid has greatly contributed to the power quality issues in the grid. For instance, the output power of a photovoltaic system is highly susceptible to changes in environmental weather; this implies that the amount of power added to the grid fluctuates with the changes in weather conditions, hence the power quality issues. A lot of research is underway to reduce the intensity of these fluctuations. In [11], fuel cell technology (FC) was presented as effectively addressing power quality issues by providing a backup for power during voltage interruptions. However, in this work, there was little emphasis on integrating renewables with UPQC or decarbonizing grid-connected systems. A UPQC-integrated PV system was presented in [12] to generate clean energy and alleviate power quality issues. In their experiment, a Superconducting Magnetic Energy Storage System (SMES) and a Dynamic Voltage Restorer (DVR) were used to keep the load running during long-term outages. However, interruptions such as severe voltage sags as well as harmonic distortion were not considered. A Battery Energy Storage System (BESS) are very critical for renewable energy systems that use UPQC in islanding mode. Battery storage in conjunction with the PV-UPQC delivers a steady power supply to the load, especially when supplying power to important loads such as semiconductor fabs or hospitals, where reliable, high-quality power supply must be guaranteed. In this project, we use a BESS to support the UPQC. According to our simulation design, the PV is connected to the DC-link via a DC–DC boost converter, [13] whereas the BESS is connected directly to the DC-link. In case of the occurrence of a prolonged voltage interruption, the BESS kicks in and provides active energy to stabilise the distribution grid. In [11,14–16], researchers have tried to develop the DC-link voltage regulation algorithm to obtain a stable and continuous DC-link capacitor voltage. However, the UPQC controller has become increasingly complicated and computationally intensive. Since the DC-link capacitor can be externally supported by PV-BESS, this is a better option than the internal DC-link capacitor.

Synchronisation is a critical factor when controlling a UPQC system. In order to provide a reference current and voltage, both the APF shunt compensator must be precisely phase synchronised [17]. If the UPQC injects voltage and current in step with the grid, it should be able to successfully perform the synchronisation process. The UPQC controllers developed in [3,11,17–21] use the conventional phase-locked loop (PLL) synchronisation method with a synchronous reference frame, which is complex to implement; it was found that the PQ theory works well when a sinusoidal source is used to generate tuned reference signals for compensation. When a source contains harmonics, the performance of any system is degraded. A control scheme based on SRF can provide better performance regardless of whether the source contains harmonics or not. Harmonic extraction techniques developed later in the theory of PQ have the potential to improve performance in the same way as SRF when the source contains harmonics. However, both techniques required extremely complicated mathematical calculations. Even under extreme harmonic conditions, performance was only moderate. The target of the new generation of control systems for machines and devices is artificial intelligence (AI). AI algorithms have the ability to respond quickly and dynamically while maintaining system stability over a wide operating range, according to statistics, artificial neural networks (ANN) have found wider application in power electronics compared to other AI techniques.

In recent years, the use of neural network (NN) control techniques has gained popularity. Recent advances in NNs have directed to a decrease in the mathematical effort and complexity of the algorithms. Owing to the precision offered by these approaches and the combined neural assembly used in approximation, NN-based control algorithms are used in a variety of applications, including aircraft landing control and electrical machines [17]. Furthermore, grid-integrated systems are increasingly relying on NN-based controls to improve control and decision-making capabilities. As an illustration, SRF-based NN control was described in [22,23] showed how to control distribution networks with a NN structure based on least mean squares (LMS); it also described a control strategy based on the backpropagation mechanism (BP). In contrast, the BP approach learns all weights at the same rate. Consequently, the rate of change is constant for all weights. Although the Adam optimisation-based approach is becoming increasingly popular, it has limited applicability [24]. Therefore, neural networks (NNs) can be integrated into traditional control systems in various ways. However, the fundamental flaw of these methods is that they do not evaluate performance under abnormal grid conditions, although this is a critical phenomenon that often occurs in distributed generation systems.

A neural network-based controller ensures the stability of an inverter system over a wide operating range while delivering a fast dynamic response; it is considered the most important tool for improving power quality through control loop design. In recent years, much work has been done on the development of UPQC control loops to provide reliable control algorithms and responsive techniques for providing switching control signals [17]. PV systems using ANN-based UPQC have been designed to maintain smooth operation during low voltage ride through (LVRT) and recover the quality of power [25]. Methods for identifying, detecting, and controlling power quality problems in a grid-connected micro-distribution system using an ANN with SWT have been reported [26]. A grid-connected microgrid with UPQC devices and an ANFIS-based adaptive control approach was described to prevent power quality distortions. ANFIS-based PV-integrated UPQC has been presented for handling balanced and unbalanced sources as well as balanced and unbalanced loads [27,28]. There are two types of control based on PQ and SRF; PQ uses a to b to zero transformations and SRF uses a to c to zero and d to q to zero. If you want to build complex controls with a lot of memory, you should use DSPs and FPGAs; these controllers can be used to perform mathematical calculations quickly. However, because of their complexity, they increase the cost and complexity of the overall system.

To address the above challenges, this study presents a solar photovoltaic system and a battery storage system powered by a UPQC controlled by an Artificial Neural Network to answer multifaceted quality of power problems, especially during prolonged voltage interruptions. Through various case studies, the dynamic performance of the proposed artificial neural network-based shunt converter Unified Power Quality Conditioner (ANN-SC-UPQC) is evaluated in conjunction with a photovoltaic solar system and a battery storage system.

In addition, active power control and harmonic control using neural networks are implemented for the shunt voltage source converter. To mitigate the sag and swell problems of the system, a series voltage source converter with voltage control using a dq0 reference frame is presented. Lastly, the performance of the recommended UPQC system is investigated using MATLAB/Simulink software under active conditions; this investigation is a contribution to the literature by reducing the complexity and mathematical computations of PV-UPQC systems and mitigating the harmonics in extreme voltage swell and sag conditions. The testing of the idea proposed in this paper has confirmed the ability of the proposed control system to mitigate power quality problems in grid-connected photovoltaic battery systems. The main contributions of the work are listed below:

1. The design and Implementation of the solar PV system and Energy storage system are provided for the UPQC.
2. Artificial Neural Network controller implemented for Solar PV Battery sourced UPQC.
3. The developed control algorithm is tested in MATLAB/Simulink software for various operating conditions.
4. The results of the Artificial Neural Network controller are compared with PQ and synchronous reference frame control methods.
5. Power quality is improved further by using an Artificial Neural Network controller compared to PQ and Synchronous reference frame control methods.

The paper is organised as follows: Section 2 presents the methodology used for the proposed ANN-based control of a solar PV battery-based powered UPQC. Section 3 discusses the design of the neural network-based shunt converter control and series converter voltage control. The Simulink model detail of the proposed work is presented in Section 4, and Section 5 shows the comprehensive discussions simulation results for PQ, SRF and proposed neural network control.

2. Methodology for System Integration with UPQC Configuration

The solar photovoltaic and battery storage system powered UPQC shown in Figure 1 is based on a two-stage UPQC model, a three-phase system designed for grid-connected solar photovoltaic and battery storage systems. A DC-link capacitor is utilised to link the active series and shunt filters. The photovoltaic (PV) panel is connected to the inverter via a step-up DC–DC converter, while the battery is linked directly to the DC link. The series voltage converter regulates the load voltage and mitigates the effects of voltage dips, voltage surges, voltage interruptions, and voltage harmonics. A shunt voltage source converter is used to mitigate line current harmonics caused by load non-linearity. The shunt voltage source is linked to the main power system through coupling reactors. A series-switched transformer connects the series voltage source converter to the main grid. The control logic for a shunt voltage source converter is designed to maximise the output power of the photovoltaic system while minimizing current fluctuations. Under normal operating circumstances, the PV system output transports real power to the loads and the grid and charges BESS. Therefore, the battery storage system can meet the entire demand even when the PV system is not producing power.

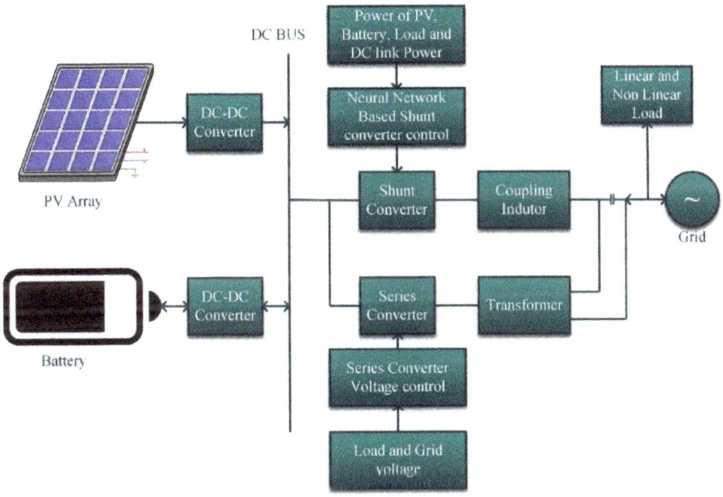

Figure 1. Block Diagram of the Proposed work.

2.1. PV Array Configuration

A solar cell uses the photovoltaic effect to convert sunlight into electricity. Figure 2a shows the basic single diode PV equivalent circuit. The model is constructed using the following mathematical expressions,

$$I = I_p - I_d - I_{sh} \tag{1}$$

$$I_p = (I_{sc} + K_i(T_k - T)) \times \frac{G}{1000} \tag{2}$$

Here, I_p and I_d represent the photovoltaic current and the diode current, respectively. I_{sh} and I represent the shunt current and the output current, respectively. In Equation (2), I_{sc}, K_i, T and G represent short-circuit current, Boltzmann constant, PV cell temperature and irradiance, respectively.

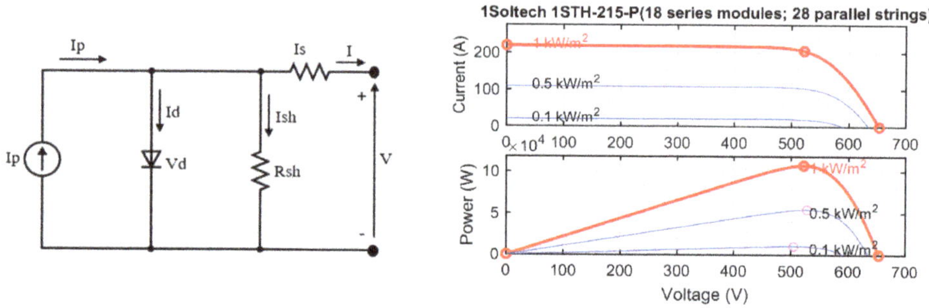

(**a**) (**b**)

Figure 2. (**a**) Fundamental single diode equivalent circuit; (**b**) Current-Voltage and Power-Voltage characteristics of the considered PV array for different irradiance.

In this work, the MATLAB Simulink PV model library is used. In the PV model, strings of PV modules are linked in parallel. Moreover, each string includes a number of PV modules to achieve the corresponding current, voltage and power values. The PV power of the single model is 213.15 W, 29 V and 7.35 A. In this work, 18 PV modules connected in series and 29 strings connected in parallel are considered. The total power of the PV generator is 107.4 kW, the voltage at the power peak is 522 V and the current at the power peak is 205.8 A. The current, voltage and power characteristics of the considered PV generator are shown in Figure 2b for different irradiation conditions.

The maximum power point tracking algorithm controls the peak power of the photovoltaic system. In this paper, the Perturb and Observe (P & O) procedure is used to track the peak power of the photovoltaic system. P & O is a simple strategy that allows more accurate tracking of the peak power point. The PV voltage (Vpv) and current (Ipv) are the input to the peak power point tracking algorithm.

For this approach to work, PV module current and voltage must be measured to estimate module power (Ppv). Once the operating point reaches the peak power point, the observation and perturbation methods are repeated until the operating point reaches the maximum power point. The algorithm compares the current instantaneous power and voltage to the previous instantaneous power and voltage and predicts the time required to reach the peak power point.

The MPPT voltage of the photovoltaic module is decreased or increased at regular intervals using the P & O algorithm. When a solar module experiences a positive power shift, even a very small voltage fluctuation can affect its performance, and this voltage fluctuation pattern persists. While it is possible to reach the peak power point by reducing disturbances, when a negative shift occurs, the peak power point is a peak power tracking algorithm that disturbs and observes the PWM generator that controls the DC–DC boost converter to extract peak power from the photovoltaic system. The pseudo-code of the MPPT algorithm used is presented in Appendix A.

2.2. PV Array Configuration

In this article, we use the Simulink library battery model. The nominal of the battery is 700 V its nominal capacity is 48 Ah, and the battery type is lithium-ion. The discharge characteristics of the battery under consideration are shown in Figure 3.

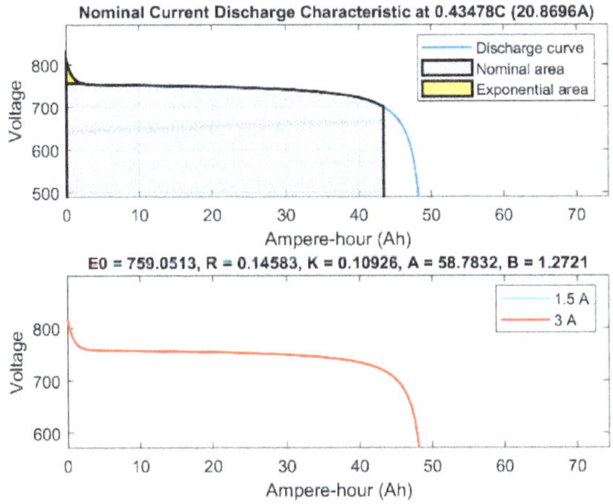

Figure 3. Discharging characteristics of the considered lithium-ion.

2.3. Design of DC–DC Boost Converter for PV Array Integration

The DC–DC boost converter acted as a peak power tracker for the PV system based on control inputs from the peak power point tracking algorithm (perturb and observe). The capacitor and inductor of the DC–DC boost converter were designed based on the following parameters: The input power of the DC–DC boost converter is 107.4 kW (PV rated power), the input voltage of the converter is 522 V (PV voltage), the output voltage of the converter is 700 V (DC link voltage requirement of series and shunt voltage source converter and battery rated voltage), the switching frequency of the converter is 10 kHz, the percentage of inductor current ripple is 1% and percentage of capacitor voltage ripple is 1%. The value of capacitor and inductor of DC–DC boost convert is 6668 µF and 3 mH, respectively.

2.3.1. Capacitor Rating of DC Link

Fluctuations of the modulation value and the system voltage per phase affect the DC link voltage DC (U_{dc_link}); it is recommended that the peak voltage per phase of the three-phase system should be more than twice the DC–DC link voltage as shown in Equation (3)

$$U_{dc\_link} = \frac{\left(V_{line-line} \times 2\sqrt{2}\right)}{\left(\gamma \times \sqrt{3}\right)} \tag{3}$$

$V_{line-line}$ is the line voltage and the modulation value γ is assumed to be 1. For a line voltage of 400 V, the minimum bus voltage requirement is DC 653.1 V. To achieve the operating conditions for the peak power of the PV generator DC–DC under standard test conditions, 700 V is set as the DC bus voltage (U_{dc_link}).

The power required and the voltage level of the DC bus determine the size of the DC-link capacitor. Equation (4) is the energy balance equation for the DC-bus capacitor,

$$C_{dc\_link} = \frac{\left(3 \times I_p \times V_p \times \gamma \times t \times O_{factor}\right)}{\left(\frac{1}{2}\left(U^2_{dclinkr} - U^2_{dc\_link}\right)\right)} \tag{4}$$

where O_{factor} is the overload factor, Vp is the phase voltage, t is the minimum time required to stabilise the voltage after a disturbance of the DC bus voltage, Ip is the phase voltage transformer current, and γ is the factor, that accounts for energy fluctuations during dynamics. The minimum DC link voltage requirement U_{dc_link} is 653.1 V, the required DC link voltage U_{dc_linkr} is 700 V, the phase current of the shunt voltage source converter is 132 A, the stabilisation time is 30 ms, the overload factor is 1.2, the change in dynamic energy is 10%, and the capacitor value of the DC link is 6668 μF.

2.3.2. Coupling Inductor of the Shunt Voltage Source Converter

The values of DC-intermediate circuit voltage, switching frequency and ripple current effect the size of the coupling inductor of a shunt voltage source converter. We calculate the value of the coupling inductor using Equation (5),

$$L_{cl} = \frac{\left(U_{dclinkr} \times \gamma \times \sqrt{3}\right)}{\left(I_{l\_ripple} \times F_{ssw} \times O_{factor} \times 12\right)} \tag{5}$$

Here, the switching frequency of the shunt voltage source converter F_{ssw} is 10 kHz, I_{l_ripple} is the ripple current in the coupling inductance and is 20% of the Ip. $\gamma = 1$, $O_{factor} = 1.2$, $U_{dclinkr} = 700$ V and I_{l_ripple} is 6.23 A, and the value of coupling inductance is 1.39 mH.

2.3.3. Injection Transformer for the Series Voltage Source Converter

The photovoltaic solar system and battery storage system are designed for sag or swell threshold of 0.2 pu or 46 V/phase, when the DC link voltage is 700 V, the series voltage source transformer only needs to feed 46 V/phase, resulting in a low modulation index for the series compensator. One must keep the modulation index of the series voltage source converter as close to unity as possible to minimise harmonic distortion. As a result, a series transformer is used with a turn ratio expressed as follows,

$$T_{ratio} = \frac{V_{pp}}{V_{svsc}} \tag{6}$$

Here T_{ratio} is the transformation ratio of the transformer, V_{pp} is the phase voltage, and V_{svsc} is the voltage of the series voltage converter. In this work, $V_{pp} = 230$ V, $V_{svsc} = 46$ gives a transformer ratio of ≈ 5. The feed transformer rating for the series voltage source converter is given by Equation (7),

$$R_{IT} = 3 \times V_{svsc} \times I_{svsc\_sag} \tag{7}$$

Here, R_{IT} is the power of the injection transformer rating, I_{svsc_sag} is the sag current of the series voltage transformer. In this work, $V_{SVSC} = 46$ V, $I_{svsc\_sag} = 30$ A which gives a feeder transformer power of 4140 VA.

3. Proposed Control Logic

Series and shunt voltage source converters are both essential components of the PV battery storage system sourced UPQC. By using the shunt voltage source converter, the system is intelligent to correct the quality of power problems of the load such as harmonics and reactive power. UPQC based on PV solar systems and battery storage system uses shunt voltage source converter for power supply in addition to the PV solar system. The neural network concept is used in the shunt voltage source converter to generate a reference current that provides active power and mitigates the harmonic problem owing to the non-linearity of the load and extreme swell and sag. The serial voltage source converter keeps the load from grid-side power quality problems such as voltage fluctuations and sag by adding a suitable voltage in phase with the grid voltage.

3.1. Control Logic of Shunt Voltage Source Converter

As shown in Figure 4, the inputs of the neural network are the current generated by the photovoltaic system, the root mean square current of the load, the leakage current from the DC-link voltage control, and the battery current. The output of the neural network is a reference for the current. The amplitude of the reference current is multiplied by three-phase sine waves in an a-b-c synchronisation frame and then compared to the shunt voltage source converter current. The hysteresis controller processes the compensation current and generates the pulse for the voltage source shunt converter; it regulates the current pumped into the grid and mitigates the harmonic disturbances caused by the non-linear load on the grid side.

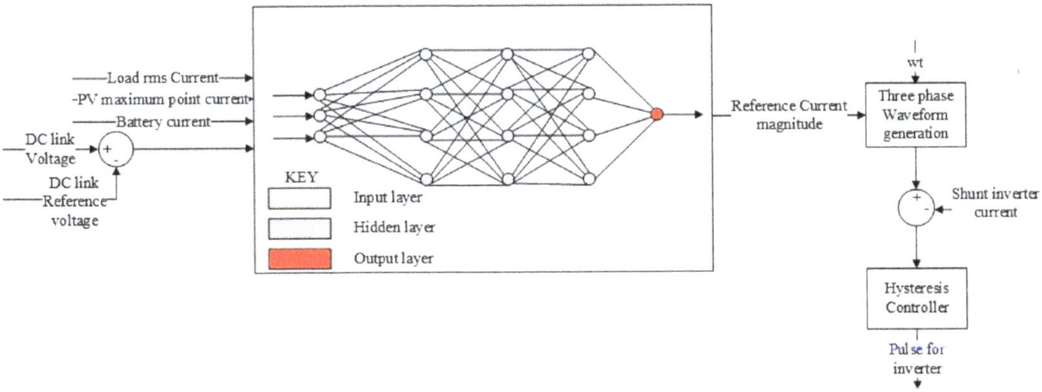

Figure 4. Shunt voltage source converter control logic model.

Training and Validation of Artificial Neural Network

To effectively generate the reference signal, this work uses an artificial neural network trained with the Levenberg-Marquardt backpropagation. The training is performed using Matlab software. First, a total of 4.2×10^6 data are collected from the grid-connected photovoltaic two-level UPQC model designed in Matlab/Simulink software. The Matlab neural network training software tool is used to train the ANN, using the collected data. By default, the neural network fitting tool considers 70% of samples for training. Here, the network is fitted according to its error, 15% of the samples are used for validation to measure the simplification of the net and stop training when the simplification stops improving, and 15% of the samples are used for testing. A total of 1000 iterations were considered during the training process. Figure 5 shows the validation performance, training state, error histogram, and regression plots. Figure 5a shows the best validation performance at the 1000th iteration is 0.21433. Variation of gradient during training is shown in Figure 5b. From the regression plots (Figure 5c), it can be seen that the correlation between the outputs and the value of the target for the training, validation and testing is close to 1. The error histogram of the target and outputs are depicted in Figure 5d. The final trained neural network controller is shown in Figure 5c.

3.2. Control Logic of Series Voltage Source Converter

The control technique of the serial voltage source converter includes in-phase compensation, pre-compensation, and energy optimisation. In this study, a minimum injection voltage is achieved by using a serial voltage source converter to add a voltage in phase with the mains voltage. The series voltage source converter has a control structure as shown in Figure 6, It helps maintain the rated load voltage when the line voltage is unstable. In this case, the series voltage source compensator produces an output voltage that is unstable. In this case, the series voltage source compensator produces an output voltage that is out of phase with the disturbance and corrects the problem.

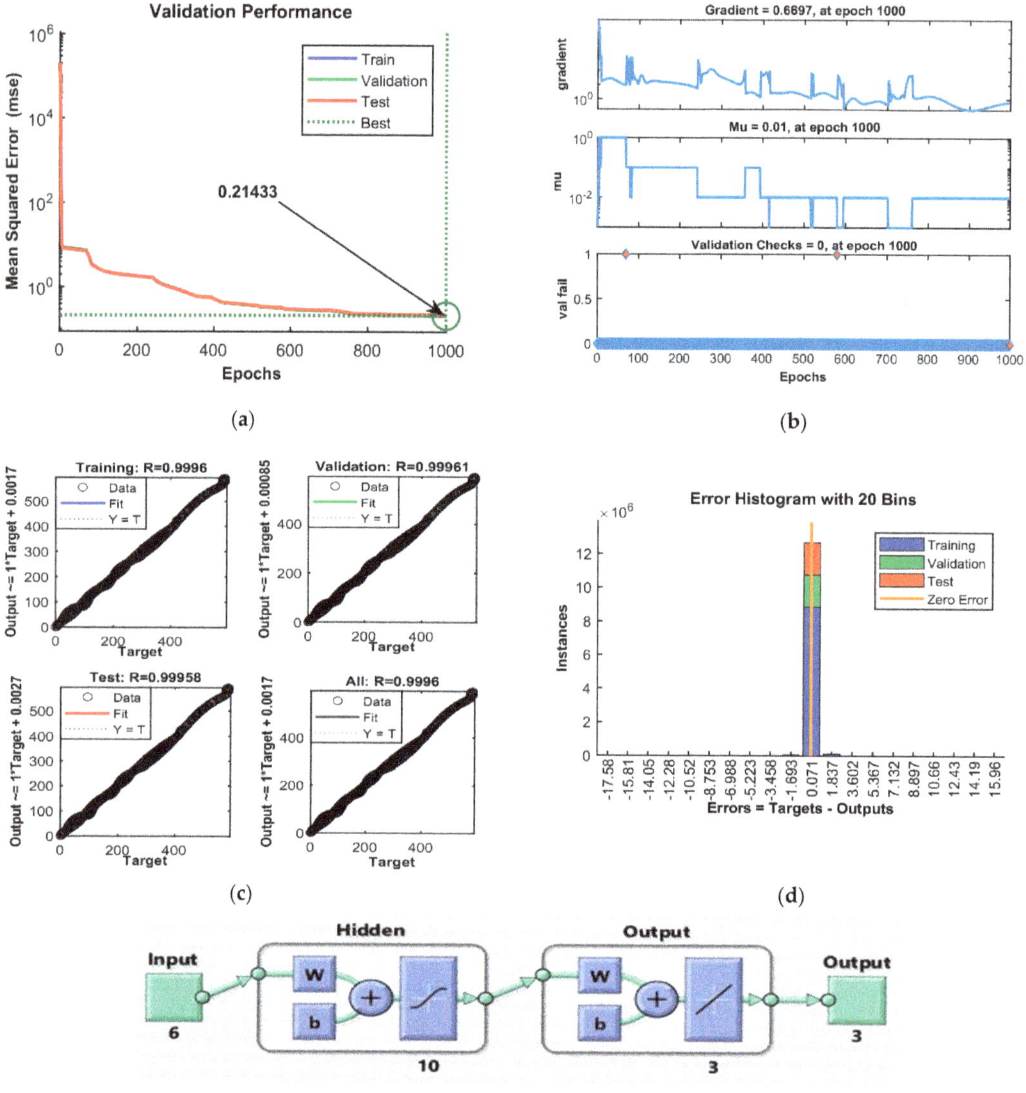

Figure 5. (**a**) Validation performance, (**b**) Training state, (**c**) Regression plot, (**d**) Error histogram, (**e**) Final trained neural network.

A phase-locked loop (PLL) is utilised to excerpt the important component of the point of common coupling (PCC) voltage, which is then used to generate the dq-0 reference axis. The PLL determines the phase and frequency of the PCC voltage, which is then used to generate a reference load voltage. By converting the input voltages from the PCC and the load voltages, a d-q-0 range is generated. The peak value of the load reference voltage is equal to the value of the d-axis component of the load reference voltage, as the load reference voltage must be phase-locked to the PCC voltage. On the q-axis, the components are kept constant at zero. The PCC voltage is subtracted from the load reference voltage to obtain the series compensator voltage. The variance between the PCC voltage and the load voltage is used to calculate the real voltages of the series compensators. To generate accurate reference signals, the voltage difference between the voltage of the reference and real series compensators must be made processed via PI controller; these signals are then utilised to produce control signals for the series compensators via an a-b-c converter for pulse width modulation (PWM).

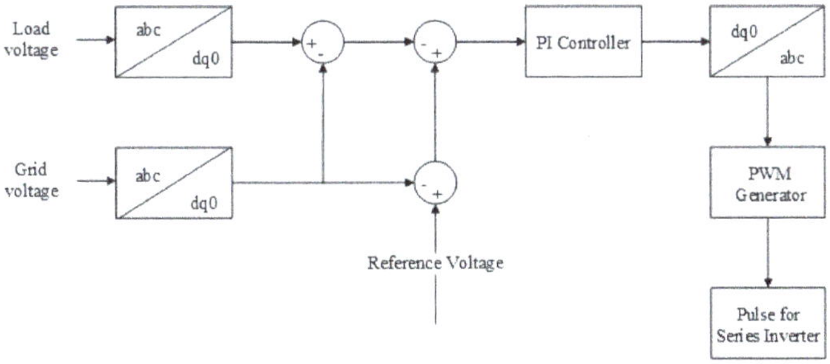

Figure 6. Series voltage source converter control logic.

4. Matlab/Simulink Model Details

In this segment, to validate the performance of the proposed control method, MATLAB/Simulink software is used to simulate the dynamic and steady-state performance of grid-connected solar PV systems and battery storage sourced UPQC model. In the simulation, a three-phase diode bridge rectifier with R-L load is used as a non-linear load (22% THD). The step size of the solver used in the simulation is 1×10^{-6} s. To handle these dynamic situations, the system has to cope with PCC voltage variations and solar irradiance changes. The MATLAB/Simulink simulation is shown in Figure 7.

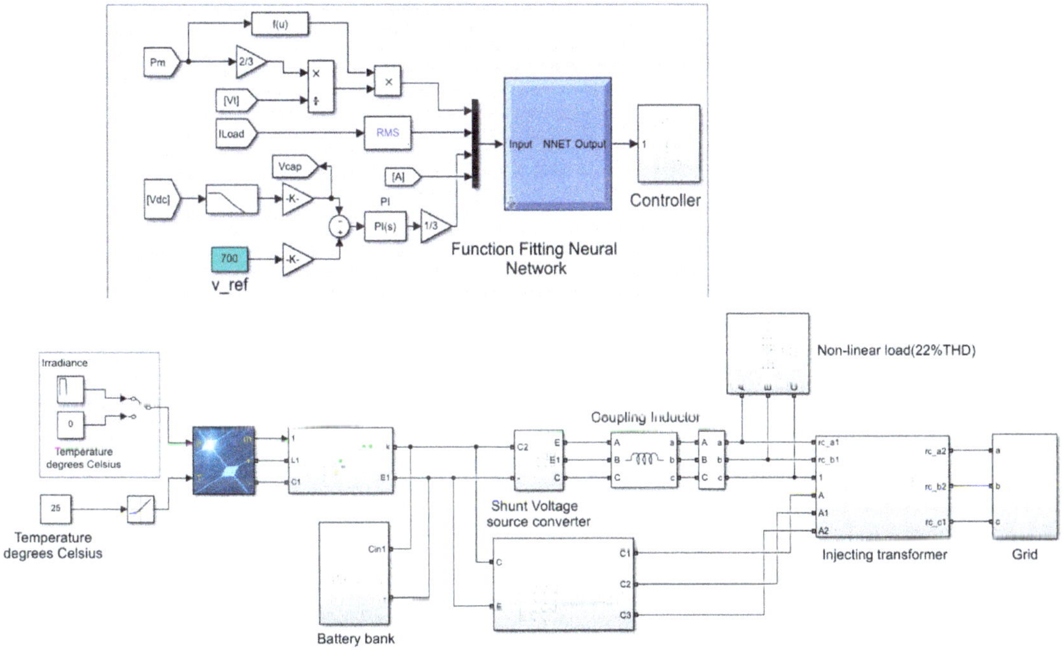

Figure 7. Matlab/Simulink simulation model of the proposed control technique.

5. Discussion

5.1. Results Analysis of the Proposed System under Voltage Sag and Voltage Swell Condition

This section discusses the voltage dips and swells analyses for the PV solar system and the battery-based UPQC. The temperature and photovoltaic irradiance are held at constant values of 25 °C and 1000 W/m^2, respectively. The response of the system parameters to the voltage drop is shown in Figure 8; these include photovoltaic module irradiance, PV current, PV voltage, photovoltaic power, battery voltage, battery current, grid voltage, load voltage, and voltage added into the system through the series voltage converter. From 1 s to 1.2 s, there is a voltage drop of 0.2 pu. The series voltage source converter injects a voltage in the contradictory phase of the disturbance to uphold the load voltage at its nominal value when the line voltage is disturbed.

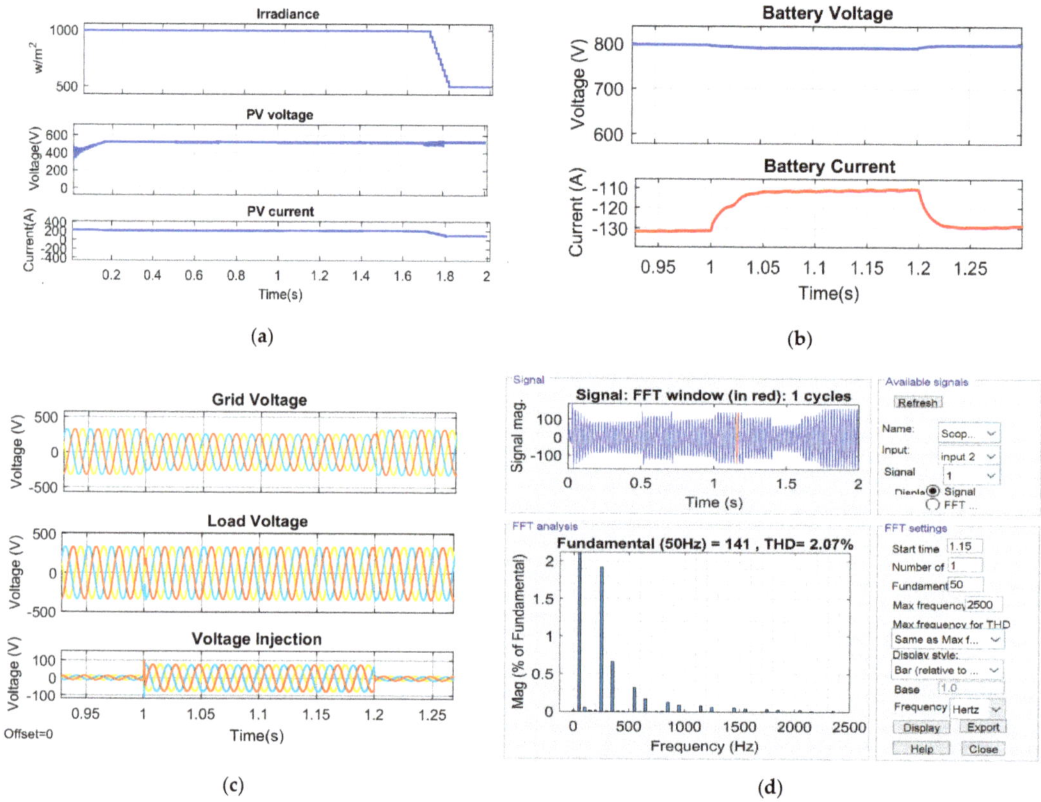

Figure 8. Response of the system parameters under voltage sag conditions. (**a**) Irradiance, PV voltage, power and current, (**b**) Battery voltage and current under voltage sag condition (**c**) Grid voltage, Load voltage and voltage injection under voltage sag condition (**d**) THD under voltage sag condition.

Figure 9 depicts the response of the system parameters under voltage swell conditions, such as irradiance of PV panel, PV voltage, PV current, PV power, battery voltage, battery current, grid voltage, load voltage, and the voltage added into the system through the series voltage source compensator. The voltage fluctuations of 0.2 pu are generated at 1.4 s to 1.6 s. The series voltage source converter injects a voltage in the contradictory phase of the disturbance to uphold the load voltage at its nominal value when the line voltage is disturbed.

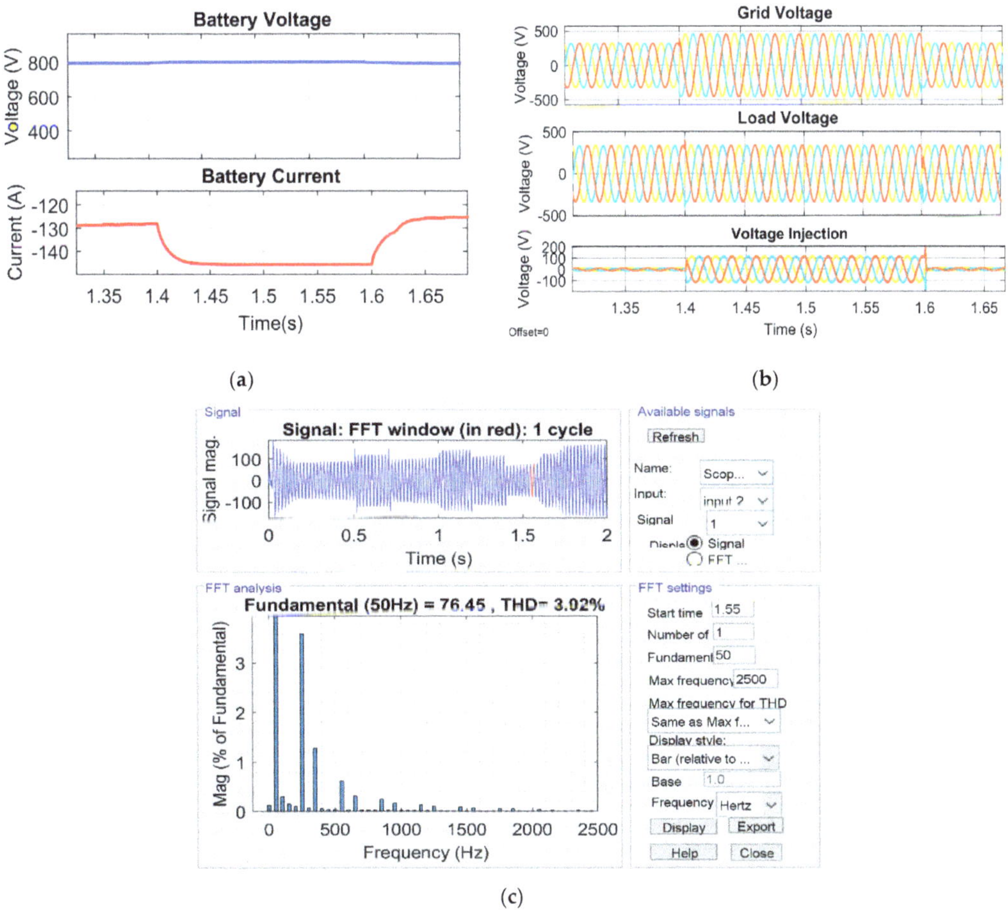

(c)

Figure 9. Response of the system parameters under voltage swell conditions. (**a**) Battery voltage and current under voltage sag condition (**b**) Grid voltage, Load voltage and voltage injection during voltage swell condition (**c**) THD during voltage swell condition.

5.2. Results Analysis of the Proposed System under Non-linear Load and Unbalanced Load

In this section, the analyses of the PV solar system and the battery-based UPQC under load unbalanced and non-linear conditions are presented. The PV solar temperature and radiation were kept at 25 °C and 1000 W/m^2, respectively. Figure 10 shows the response of system parameters under non-linear load conditions, such as PV panel irradiance, PV voltage, PV current, PV power, battery voltage, battery current, belt voltage and current, load voltage and current, and shunt inverter voltage and current. The non-linear load conditions are generated at 0.3 s to 0.5 s. Under these conditions, the system current remains sinusoidal thanks to the shunt voltage source converter, mitigating the harmonics caused by the non-linear load.

Figure 11 depicts the response of the system parameters under load unbalanced conditions. Parameters such as battery voltage, battery current, line current, line voltage load voltage and load current, and shunt inverter voltage and current are examined. The unbalanced load conditions are generated at 0.5 to 0.7 s by disconnecting one line of the load. In this condition, the line current remains sinusoidal due to the shunt voltage source converter and mitigates the harmonics caused by the unbalanced load.

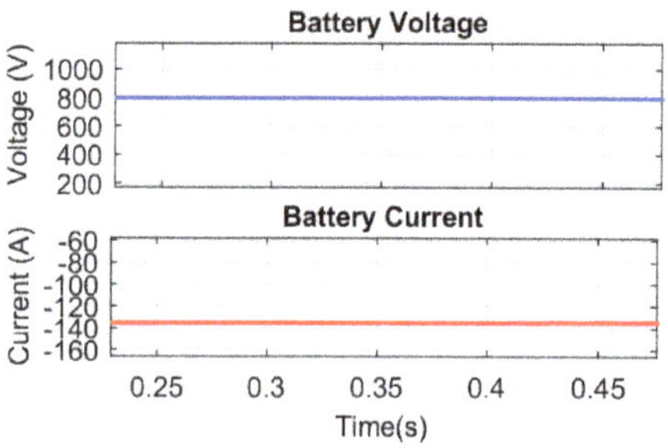

(a)

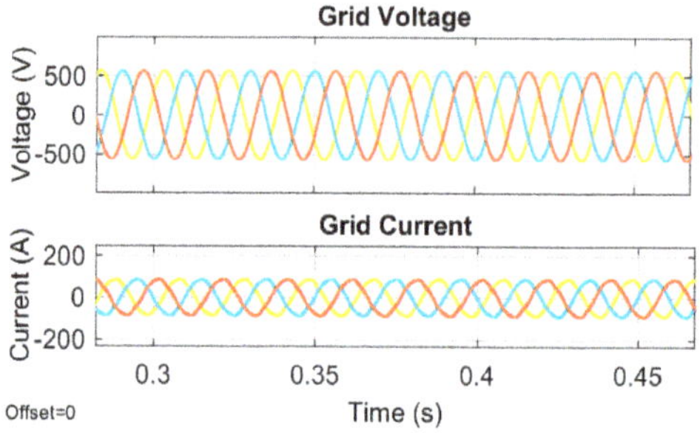

(b)

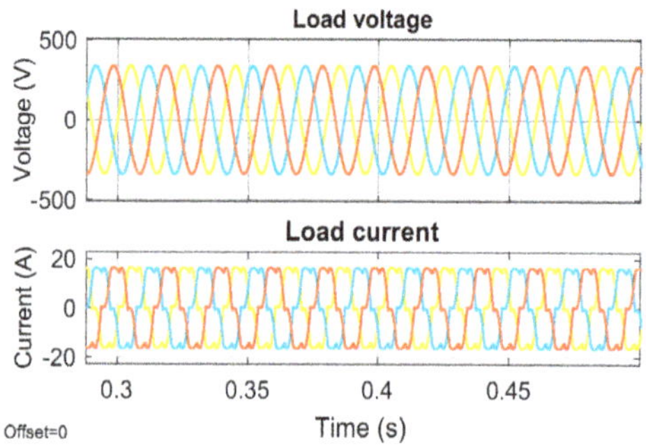

(c)

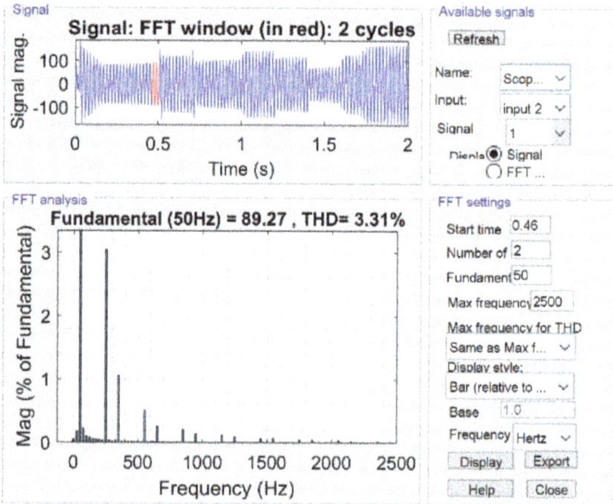

(d)

Figure 10. Response of the system parameters under non-linear load conditions. (**a**) Battery voltage and current under non-linear load condition, (**b**) Grid voltage and current under non-linear load condition, (**c**) Load voltage and current under non-linear load condition, and (**d**) THD under non-linear load condition.

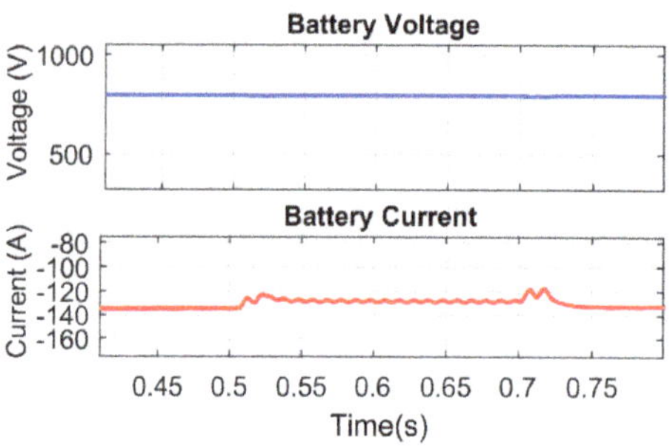

(a)

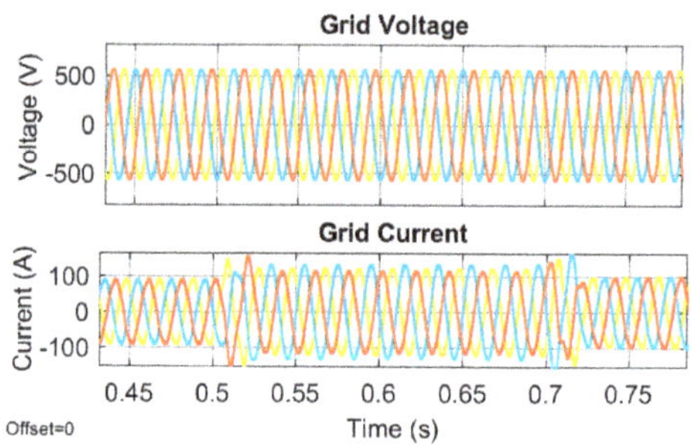

(b)

Figure 11. *Cont.*

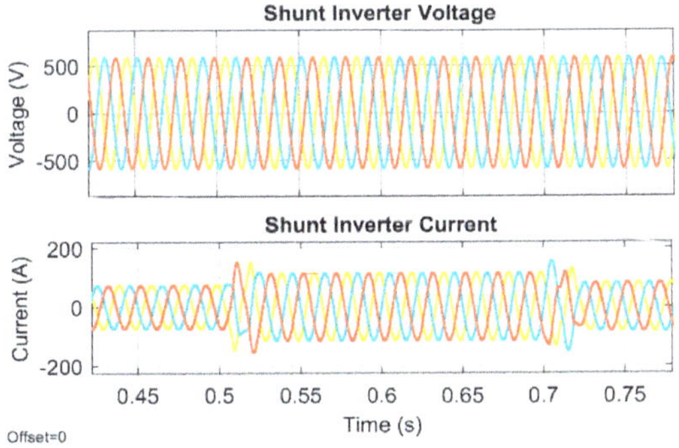

(c)

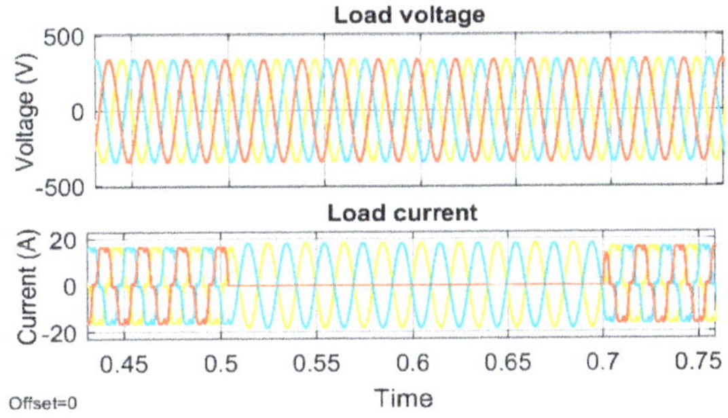

(d)

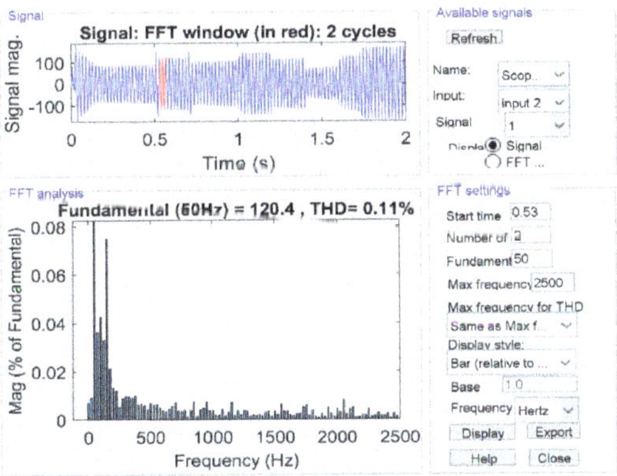

(e)

Figure 11. Response of the system parameters under unbalanced load conditions, (**a**) Battery voltage and current under unbalanced load conditions, (**b**) Grid voltage and current under unbalanced load conditions (**c**) Shunt inverter voltage and current under unbalanced load conditions, (**d**) Load voltage and current under unbalanced load conditions, (**e**) THD under unbalanced load conditions.

5.3. Comparison of the Proposed Neural Network Controller with PQ and SRF Control of Shunt Converter

In this section, the proposed neural network-based control of the shunt converter is compared with the PQ [10] and SRF [12] controls.

The total harmonic distortion of the grid current for different irradiation conditions is compared and presented in Table 1. From the analysis of the results, it is found that the proposed neural network controller meets the IEEE standards for total harmonic distortion at high and low irradiance conditions and that PQ and the SRF controller do not meet the IEEE standard under low irradiance conditions.

The total harmonic distortion of the grid current under different voltage drop conditions is compared and shown in Table 2. From the analysis of the results, it can be seen that the new proposed controller complies with the IEEE standard for total harmonic distortion and is always below 5%, but PQ and the SRF controller do not comply with the IEEE standard for high voltage drop.

The total harmonic distortion of the grid current under various voltage swell conditions is compared and shown in Table 3. From the examination of the results, the proposed neural network controller complies with the IEEE standard for total harmonic distortion and is always below 5%, while PQ and the SRF controller do not comply with the IEEE standard for high voltage swell conditions.

The total harmonic distortion of the line current under various unbalanced load conditions is compared and shown in Table 4. From the examination of the results, the proposed neural network controller meets the IEEE-519 standard for total harmonic distortion and is always below 5%, while PQ and the SRF controller do not meet the IEEE standard for some unbalanced load conditions.

Table 1. Comparisons Grid current THD for NN controller with PQ and SRF control for various irradiance conditions.

Irradiance (W/m^2)	THD of the Grid Current (%)		
	PQ Control [10]	SRF Control [12]	NN Control
1000	4.6	4.2	2.07
800	4.75	4.36	2.8
500	4.89	4.86	3.5
300	5.3	5.1	3.9
100	5.9	5.5	4.2

Table 2. Comparisons Grid current THD for NN controller with PQ and SRF control for various voltage sag conditions.

Voltage Sag (%)	THD of the Grid Current (%)		
	PQ Control [10]	SRF Control [12]	NN Control
5	4.32	4.21	2.75
10	4.51	4.42	2.9
15	4.58	4.71	3.1
20	5.45	5.19	3.4
30	5.95	5.64	3.7

Table 3. Comparisons Grid current THD for NN controller with PQ and SRF control for various voltage swell conditions.

Voltage Swell (%)	THD of the Grid Current (%)		
	PQ Control [10]	SRF Control [12]	NN Control
5	4.45	4.36	2.8
10	4.62	4.41	2.97
15	4.81	4.69	3.21
20	5.25	4.89	3.39
30	5.45	5.15	3.58

Table 4. Comparisons Grid current THD for NN controller with PQ and SRF control for various voltage unbalanced load conditions.

Unbalanced Load Conditions	THD of the Grid Current (%)		
	PQ Control [10]	SRF Control [12]	NN Control
Phase A open	5.05	4.89	3.31
Phase B open	5.12	5.01	3.45
Phase C open	4.95	5.06	3.35

6. Conclusions

Using variable irradiance, grid voltage fluctuations and non-linear and unbalanced load conditions of the three-phase solar PV and battery storage system based on UPQC architecture and dynamic performance were investigated. Non-linear load harmonics are reduced by the proposed Artificial Neural Network control. The obtained results show that the harmonic distortion of the grid current is within the standard limits of IEEE-519 in all test scenarios (less than 5%). When the irradiation, voltage and load unbalance are varied, the system remains stable. The introduction of neural network control has increased the performance of the d-q control, particularly under unbalanced circumstances; this model can be easily implemented with a simple and cheaper microcontroller, saving the cost of developing complex hardware to achieve similar efficiency of results. By combining distributed generation with power quality improvement, photovoltaics, and battery storage systems, UPQC is a promising option for the current distribution system. The idea proposed in this article is still expandable. Future work may attempt to implement the proposed model with a three- or five-level grid-connected photovoltaic battery UPQC system and also latest optimisation can be used for training the neural network for getting better power quality results.

Author Contributions: Conceptualization, O.E.O.; Formal analysis, K.P.; Methodology, O.E.O.; Software, O.E.O.; Supervision, Z.-H.L.; Writing—original draft, O.E.O.; Writing—review and editing, M.X. and A.J.R. All authors have read and agreed to the published version of the manuscript.

Funding: This work was sponsored by the Program of Foundation of Science and Technology Commission of Shanghai Municipality (22dz1206005, 22dz1204202), National Natural Science Foundation of China (12172228, 11572187), and Natural Science Foundation of Shanghai (22ZR1444400).

Conflicts of Interest: The authors declare no conflict of interest.

Nomenclature

UPQC	Unified Power Quality Conditioners
ANN	Artificial Neural Network
P&O	Perturb and Observe
THD	Total Harmonic Distortion
SRF	Synchronous reference frame theory
ANFIS	Adaptive Neuro-Fuzzy Inference system
FC	Fuel cell
SMES	Superconducting Magnetic Energy Storage System
DC	Direct Current
BESS	Battery Energy Storage System
PLL	Phased Locked Loop
Photovoltaics	PV
PCC	Point of common coupling
PWM	Pulse width modulation
I_v	Photovoltaics current
I_d	Diode Current
I_{sh}	Shunt Current
I	Output current
I_{sc}	Short circuit current
T	Temperature
G	Irradiance
K_i	Boltzmann Constant
V_{line_line}	Line to Line voltage of the grid
U_{dc_link}	Dc link Voltage of the Inverter
O_{factor}	Overload factor
V_{pp} or V_{pp}	Phase voltage of the grid
I_p	Phase current of the transformer
I_{l_ripple}	ripple current of the filter inductor
F_{ssw}	Inverter switching frequency
V_{svsc}	Voltage of the series voltage converter
I_{svsc_sag}	Current of the series voltage converter
T_{ratio}	Transformation ratio of the transformer

Appendix A

The seudo code briefly descrides the implementation of the Perturb and Observe maximum power point tracking algorithm, used in this work. In the algorithm, fastly, the string voltage and current are sampled then the power is calculated by multiplying the voltage and current. The algorithm compares the values of the current power and voltage with the previuos power and voltage then makes adjustments to the reference voltage according to the condition as shown in the seudo code below.

Algorithm A1: Perturb and Observe MPPT Algorithm

procedure $P\&O_{MPPT}\left(V_{st}, I_{st}, V_{ref}\right)$	
$V_{st,n} \leftarrow V_{st}$	
$I_{st,n} \leftarrow I_{st}$	Sample the string voltage and current
$P_{st,n} \leftarrow V_{st,n} \times I_{st,n}$	Calculate the string power
If $f_{init} \rightarrow$ False **then**	Check if the initialization has been performed or not,
$V_{ref} \leftarrow V_{st,n} - S_{\min}$	If not, reduce the string voltage reference then
$f_{init} \leftarrow True$	set the initialization flag to True
else	
If $P_{st,n} > P_{st,n-1}$ **then**	
If $V_{st,n} > V_{st,n-1}$ **then**	Check if the panel power and voltage is increased or not,
$V_{ref} \leftarrow V_{ref} + S_{\min}$	$if\ \frac{V_{st}}{dP_{st}} > 0$, increase V_{ref} by the step size $S_{\min}$
else	$if\ \frac{V_{st}}{dP_{st}} < 0$, decrease V_{ref} by the step size $S_{\min}$
$V_{ref} \leftarrow V_{ref} - S_{\min}$	
end if	
else	
If $V_{st,n} > V_{st,n-1}$ **then**	
$V_{ref} \leftarrow V_{ref} - S_{\min}$	Check if the panel power and voltage is decreased or not,
else	$if\ \frac{V_{st}}{dP_{st}} < 0$, decrease V_{ref} by the step size $S_{\min}$
$V_{ref} \leftarrow V_{ref} + S_{\min}$	$if\ \frac{V_{st}}{dP_{st}} > 0$, increase V_{ref} by the step size $S_{\min}$
end if	
end if	
end if	
$V_{st,n-1} \leftarrow V_{st,n}$	Update the previous voltage sample by the present voltage sample
$P_{st,n-1} \leftarrow P_{st,n}$	Update the previous power value by the present power value.
return V_{ref}	
end procedure	

References

1. *IEEE Std 1159*; IEEE Recommended Practice for Monitoring Electric Power Quality. Institute of Electrical and Electronics Engineers: Piscataway, NJ, USA, 2009; Volume 10, pp. 1–81. [CrossRef]
2. Javadi, A.; Woodward, L.; Al-Haddad, K. Real-Time Implementation of a Three-Phase THSeAF Based on a VSC and a P+R Controller to Improve the Power Quality of Weak Distribution Systems. *IEEE Trans. Power Electron.* **2018**, *33*, 2073–2082. [CrossRef]
3. Javadi, A.; Hamadi, A.; Woodward, L.; Al-Haddad, K. Experimental Investigation on a Hybrid Series Active Power Compensator to Improve Power Quality of Typical Households. *IEEE Trans. Ind. Electron.* **2016**, *63*, 4849–4859. [CrossRef]
4. Mansor, M.A.; Othman, M.M.; Musirin, I.; Noor, S.Z.M. Dynamic voltage restorer (DVR) in a complex voltage disturbance compensation. *Int. J. Power Electron. Drive Syst.* **2019**, *10*, 2222–2230. [CrossRef]
5. Ghosh, A.; Ledwich, G. Compensation of distribution system voltage using DVR. *IEEE Trans. Power Deliv.* **2002**, *17*, 1030–1036. [CrossRef]
6. Jothibasu, S.; Mishra, M.K. A Control Scheme for Storageless DVR Based on Characterization of Voltage Sags. *IEEE Trans. Power Deliv.* **2014**, *29*, 2261–2269. [CrossRef]
7. Farooqi, A.; Othman, M.M.; Abidin, A.F.; Sulaiman, S.I.; Radzi, M.A.M. Mitigation of power quality problems using series active filter in a microgrid system. *Int. J. Power Electron. Drive Syst.* **2019**, *10*, 2245–2253. [CrossRef]
8. Kumar, C.; Mishra, M.K. Operation and Control of an Improved Performance Interactive DSTATCOM. *IEEE Trans. Ind. Electron.* **2015**, *62*, 6024–6034. [CrossRef]
9. Hoon, Y.; Radzi, M.A.M.; Hassan, M.K.; Mailah, N.F. Control Algorithms of Shunt Active Power Filter for Harmonics Mitigation: A Review. *Energies* **2017**, *10*, 2038. [CrossRef]
10. Campanhol, L.B.G.; da Silva, S.A.O.; Goedtel, A. Application of shunt active power filter for harmonic reduction and reactive power compensation in three-phase four-wire systems. *IET Power Electron.* **2014**, *7*, 2825–2836. [CrossRef]
11. Sundarabalan, C.K.; Puttagunta, Y.; Vignesh, V. Fuel cell integrated unified power quality conditioner for voltage and current reparation in four-wire distribution grid. *IET Smart Grid* **2019**, *2*, 60–68. [CrossRef]

12. Devassy, S.; Singh, B. Design and performance analysis of three-phase solar PV integrated UPQC. In Proceedings of the 2016 IEEE 6th International Conference on Power Systems, New Delhi, India, 4–6 March 2016. [CrossRef]
13. Gee, A.M.; Robinson, F.; Yuan, W. A Superconducting Magnetic Energy Storage-Emulator/Battery Supported Dynamic Voltage Restorer. *IEEE Trans. Energy Convers.* **2017**, *32*, 55–64. [CrossRef]
14. Kollimalla, S.K.; Mishra, M.K.; Narasamma, N.L. Design and Analysis of Novel Control Strategy for Battery and Supercapacitor Storage System. *IEEE Trans. Sustain. Energy* **2014**, *5*, 1137–1144. [CrossRef]
15. Karanki, S.B.; Geddada, N.; Mishra, M.K.; Kumar, B.K. A Modified Three-Phase Four-Wire UPQC Topology with Reduced DC-Link Voltage Rating. *IEEE Trans. Ind. Electron.* **2013**, *60*, 3555–3566. [CrossRef]
16. Teke, A.; Saribulut, L.; Tumay, M. A Novel Reference Signal Generation Method for Power-Quality Improvement of Unified Power-Quality Conditioner. *IEEE Trans. Power Deliv.* **2011**, *26*, 2205–2214. [CrossRef]
17. Kinhal, V.G.; Agarwal, P.; Gupta, H.O. Performance Investigation of Neural-Network-Based Unified Power-Quality Conditioner. *IEEE Trans. Power Deliv.* **2011**, *26*, 431–437. [CrossRef]
18. Campanhol, L.B.G.; da Silva, S.A.O.; de Oliveira, A.A.; Bacon, V.D. Power Flow and Stability Analyses of a Multifunctional Distributed Generation System Integrating a Photovoltaic System with Unified Power Quality Conditioner. *IEEE Trans. Power Electron.* **2019**, *34*, 6241–6256. [CrossRef]
19. Hasan, K.; Othman, M.M.; Rahman, N.F.A.; Hannan, M.A.; Musirin, I. Significant implication of unified power quality conditioner in power quality problems mitigation. *Int. J. Power Electron. Drive Syst.* **2019**, *10*, 2231–2237. [CrossRef]
20. Han, B.; Bae, B.; Kim, H.; Baek, S. Combined Operation of Unified Power-Quality Conditioner with Distributed Generation. *IEEE Trans. Power Deliv.* **2006**, *21*, 330–338. [CrossRef]
21. Guerrero, J.M.; Loh, P.C.; Lee, T.-L.; Chandorkar, M. Advanced Control Architectures for Intelligent Microgrids—Part II: Power Quality, Energy Storage, and AC/DC Microgrids. *IEEE Trans. Ind. Electron.* **2013**, *60*, 1263–1270. [CrossRef]
22. Khadem, S.K.; Basu, M.; Conlon, M.F. Intelligent Islanding and Seamless Reconnection Technique for Microgrid with UPQC. *IEEE J. Emerg. Sel. Top. Power Electron.* **2015**, *3*, 483–492. [CrossRef]
23. Xiang, H.; Chen, B.; Yang, M.; Li, C. Altitude measurement based on characteristics reversal by deep neural network for VHF radar. *IET Radar Sonar Navig.* **2019**, *13*, 98–103. [CrossRef]
24. Agarwal, R.K.; Hussain, I.; Singh, B. Application of LMS-Based NN Structure for Power Quality Enhancement in a Distribution Network under Abnormal Conditions. *IEEE Trans. Neural Netw. Learn. Syst.* **2018**, *29*, 1598–1607. [CrossRef]
25. Kumar, A.; Mishra, V.; Ranjan, R. Low Voltage Ride through Capability and Power Quality Improvement in Grid Connected PV System Using ANN Tuned UPQC. In Proceedings of the 2021 IEEE 4th International Conference on Computing, Power and Communication Technologies (GUCON), Kuala Lumpur, Malaysia, 24–26 September 2021. [CrossRef]
26. Gupta, N.; Institute of Engineering & Technology. Artificial neural network and synchrosqueezing wavelet transform based control of power quality events in distributed system integrated with distributed generation sources. *Int. Trans. Electr. Energy Syst.* **2021**, *31*, e12824. [CrossRef]
27. Kumar, A.S.; Rajasekar, S.; Raj, P.A.-D.-V. Power Quality Profile Enhancement of Utility Connected Microgrid System Using ANFIS-UPQC. *Procedia Technol.* **2015**, *21*, 112–119. [CrossRef]
28. Dheeban, S.; Selvan, N.M. ANFIS-based Power Quality Improvement by Photovoltaic Integrated UPQC at Distribution System. *IETE J. Res.* **2021**, 1–19. [CrossRef]

energies

MDPI

Article

Studying the Optimal Frequency Control Condition for Electric Vehicle Fast Charging Stations as a Dynamic Load Using Reinforcement Learning Algorithms in Different Photovoltaic Penetration Levels

Ibrahim Altarjami [1] and Yassir Alhazmi [2,*]

[1] Electrical Engineering Department, Taibah University, Madinah 44256, Saudi Arabia; itarjmi@taibahu.edu.sa
[2] Electrical Engineering Department, Umm Al-Qura University, Makkah 24382, Saudi Arabia
* Correspondence: yahazmi@uqu.edu.sa; Tel.: +966-555514490

Abstract: This study investigates the impact of renewable energy penetration on system stability and validates the performance of the (Proportional-Integral-Derivative) PID-(reinforcement learning) RL control technique. Three scenarios were examined: no photovoltaic (PV), 25% PV, and 50% PV, to evaluate the impact of PV penetration on system stability. The results demonstrate that while the absence of renewable energy yields a more stable frequency response, a higher PV penetration (50%) enhances stability in tie-line active power flow between interconnected systems. This shows that an increased PV penetration improves frequency balance and active power flow stability. Additionally, the study evaluates three control scenarios: no control input, PID-(Particle Swarm Optimization) PSO, and PID-RL, to validate the performance of the PID-RL control technique. The findings show that the EV system with PID-RL outperforms the other scenarios in terms of frequency response, tie-line active power response, and frequency difference response. The PID-RL controller significantly enhances the damping of the dominant oscillation mode and restores the stability within the first 4 s—after the disturbance in first second. This leads to an improved stability compared to the EV system with PID-PSO (within 21 s) and without any control input (oscillating more than 30 s). Overall, this research provides the improvement in terms of frequency response, tie-line active power response, and frequency difference response with high renewable energy penetration levels and the research validates the effectiveness of the PID-RL control technique in stabilizing the EV system. These findings can contribute to the development of strategies for integrating renewable energy sources and optimizing control systems, ensuring a more stable and sustainable power grid.

Keywords: optimal condition; dynamic load; renewable energy penetration; reinforcement algorithm; electric vehicle; charging station

Citation: Altarjami, I.; Alhazmi, Y. Studying the Optimal Frequency Control Condition for Electric Vehicle Fast Charging Stations as a Dynamic Load Using Reinforcement Learning Algorithms in Different Photovoltaic Penetration Levels. *Energies* **2024**, *17*, 2593. https://doi.org/10.3390/en17112593

Academic Editors: Veerapandiyan Veerasamy, Shailendra Singh and Sunil Kumar Singh

Received: 24 April 2024
Revised: 21 May 2024
Accepted: 24 May 2024
Published: 28 May 2024

1. Introduction

The power system is facing major changes and challenges as a result of the growing integration of renewable energy sources and the rising popularity of electric vehicles (EVs). It becomes increasingly important to ensure system stability as the penetration of renewable energy increases. Furthermore, the rapid expansion of EVs calls for a dependable and effective infrastructure for charging them, which calls for efficient frequency control. Thus, the goal of this research is to determine the best scenario for a dynamic load in frequency regulation for electric vehicle fast charging stations by utilizing a reinforcement learning (RL) algorithm and renewable energy penetration [1–3]. There are several environmental and financial advantages to using renewable energy sources, such as wind and solar energy [2,4,5]. Figure 1 shows the percentage of CO_2 emissions attributable to different sectors: industry, transportation, residential, electricity and heat, and other sectors worldwide [6].

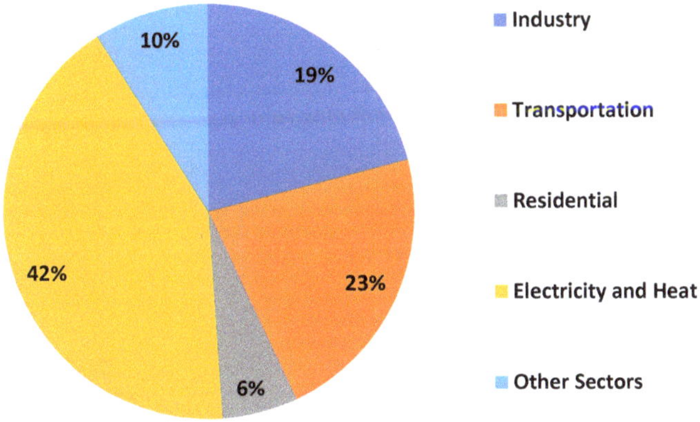

Figure 1. The percentage of world CO_2 emissions attributable to each sector (IEA, 2016) [6] Source: IEA/OECD CO_2 emissions from fuel combustion, 2016.

The renewable sources' intermittency makes it difficult to keep the system stable because variations in generation might cause frequency variances. Therefore, it is crucial to investigate how the penetration of renewable energy affects the system's stability and to create plans to address any possible problems [5,7]. Furthermore, the widespread use of EVs has increased the demand for infrastructure related to charging. In order to meet the changing requirements of electric vehicles (EVs), provide user convenience, and encourage the wider use of electric mobility, fast charging stations are essential. When there is a significant demand for EV charging, efficient frequency control is essential for controlling the charging load and preserving grid stability [8–13].

The increasing integration of renewable energy sources and the massive growth of EVs are transforming the global energy landscape as shown in Figure 2. Renewable energy sources, like wind and solar energy, provide sustainable and environmentally friendly substitutes for conventional fossil fuel-based power. The broad deployment of EVs intends to lessen the transportation sector's reliance on fossil fuels and greenhouse gas emissions at the same time. These advancements, nonetheless, provide new difficulties for the power system and call for creative fixes to guarantee grid stability and effective infrastructure for charging [14–16].

Although it is much desired, intermittent generation is produced by renewable energy sources because they are weather-dependent. The power grid may experience frequency fluctuations as a result of the fluctuating output of renewable energy, which could jeopardize system stability. For power systems to function dependably, a steady frequency must be maintained because fluctuations might affect linked equipment's performance and upset the supply and demand balance. Thus, it is crucial to comprehend how the penetration of renewable energy affects system stability in order to guarantee the dependable integration of clean energy sources. Furthermore, a reliable and effective charging infrastructure is now required due to the rising popularity of EVs. Long-distance driving and reduced charging times for EV users are made possible by EV fast charging stations, which are essential in satisfying the growing demand for quick charging. Nevertheless, these fast charging stations' high power requirements can put pressure on the grid, causing variations in frequency and voltage. During times of heavy demand for EV charging, efficient frequency control is crucial to managing the charging load and maintaining grid stability [17–20].

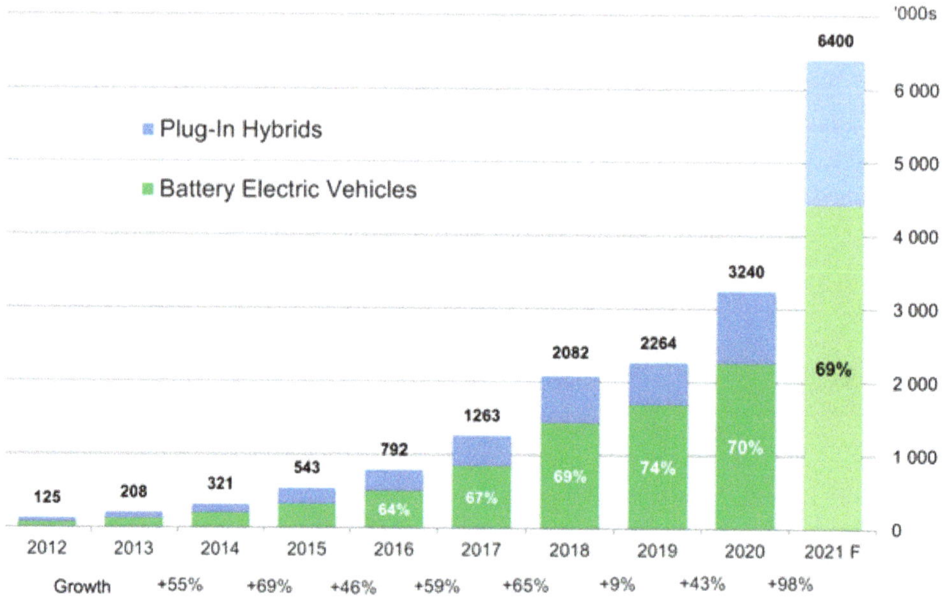

Figure 2. The global electric vehicle growth [13].

2. Related Works

A secondary controller was created and used in [21] to control the system frequency in a networked microgrid system. Energy-storing units (ESUs), synchronous generators, and renewable energy resources (RESs) made up the suggested power system. Photovoltaic (PV) and wind turbine generator (WTG) units are two examples of renewable energy sources. A flywheel and a battery made up the ESU. Despite various load demand variations and renewable energy sources, the deployed (Mayfly Algorithm) MA-PID controller proved its ability to deliver and control the system frequency effectively. The study's validation comparisons provided verification of this [21].

A novel form of power system surfaced as a means of fulfilling China's carbon emission reduction objectives for the years 2030 and 2060. One of the most significant challenges facing this field of technical innovation is coordinating the energy storage systems (ESSs), load, network infrastructures, and generation. It is necessary to estimate the associated static and dynamic model parameters accurately and effectively in order to manage and run these grid resources in a secure, reliable, and cost-effective manner. Power systems have made use of deep reinforcement learning (DRL), a popular artificial intelligence technique in recent years, especially in control-related domains. DRL is a potent optimization tool that produces promising results when used to solve parameter estimate issues. As a result, the DRL techniques used to estimate the parameters of power system models, including load and generator models, were thoroughly addressed in this study [22].

When compared to synchronous generation power grids, inverter-based networks are not physically able to produce a substantial amount of inertia. This calls for greater in-depth research on stability issues related to these kinds of systems. Consequently, in power systems with significant penetration rates of renewable energy, when many inverters and dispersed energy sources are involved, it is imperative to utilize suitable analytical techniques for voltage stability analysis. This study reviewed the literature in-depth on voltage stability analyses in power systems that include a significant amount of renewable energy. There was a discussion of several upgrade techniques and assessment plans for voltage stability in power networks with various distributed energy sources. The study also conducted a thorough examination of the simulation verification models and voltage

stability analysis techniques currently in use for microgrids. The many distributed generator types, microgrid designs, and modes of operation were taken into consideration when reviewing the conventional and enhanced voltage stability analysis methods. Furthermore, a detailed evaluation of the voltage stability indices was conducted, taking into account the applicable conditions. These indices are crucial in voltage stability assessments [23].

The goal of the project was to model a fleet of EVs in order to offer Frequency Containment Reserve for Disturbance (FCR-D) upwards and Firm Frequency Response (FFR) services while adhering to current laws and regulations. The Nordic 45 test model (N45) is an enhanced version of the Nordic 44 test model (N44) that was created to perform more accurate simulations of the Nordic power system. To evaluate various aspects of electric vehicle (EV) performance as frequency reserve providers in Kundur's two-area power system model (K2A) and the new N45 test model, ten simulation cases were developed. After presenting and analyzing the simulation findings, a conclusion about the viability of EVs as frequency reserve suppliers in the current Nordic power system was reached. Because EV charging equipment and measurement devices have an inherent time delay, it was found that deploying EVs for FCR-D and FFR services was not currently possible. However, it was found that the usefulness of using EVs as frequency reserve providers might be improved if the time delay of these components was much lowered, enabling nearly instantaneous activation of EVs' frequency response following a fault occurrence [24].

Reference [25] presented a two-stage transfer and deep learning-based method to predict dynamic frequency in power systems after disturbances and find the best event-based load-shedding plan to keep the frequency of the system stable. Convolutional Neural Networks (CNNs) and Long Short-Term Memory (LSTM) networks were coupled in the suggested method to efficiently use the input data's spatial and temporal measurements. The CNN-LSTM technique greatly improved the timeliness of online frequency control while maintaining good accuracy and efficacy, according to simulation findings performed on the IEEE 118-bus system. Furthermore, test cases conducted on the South Carolina 500-bus system and the New England 39-bus system showed that the transfer learning procedure produced accurate results even in the presence of a shortage of training data [25].

Another study proposed a system to enable a thorough examination of grid transient stability, based on several machine-learning algorithms. Furthermore, the framework made it easier to thoroughly identify new patterns that appeared in the dynamic response of individual synchronous generators in systems with reduced inertia. Given that crucial operating points fluctuate in systems with high renewable penetration, the framework's capacity to evaluate all hourly operating points over the course of a study year was a significant advantage. An IEEE-39 system was used as a case study, integrating inverter-based resources at different penetration levels and locations, to illustrate the advantages of the suggested platform [26].

The literature assessment indicates that further research and exploration of the optimal condition as a reinforcement learning algorithm for electric vehicle fast charging stations and a dynamic load for frequency control using renewable energy integration is still possible. To overcome these obstacles and achieve optimal frequency regulation and effective operation of fast charging stations, this research suggests using a reinforcement learning (RL) algorithm in conjunction with renewable energy penetration. Because RL algorithms may learn optimal control policies through interactions with the environment, they are especially well suited for dynamic and unpredictable systems, such as electric grids and networks for electric vehicle charging. There are two goals that this study aims to achieve. First, in order to evaluate the effect of renewable energy penetration on system stability, it looks at the frequency response and active power flow under various photovoltaic (PV) penetration scenarios. Second, it endeavors to assess how well the suggested RL algorithm performs in terms of attaining optimal frequency control and guaranteeing the efficient operation of EV fast charging stations.

After the introduction and the related work in Sections 1 and 2, respectively, the rest of this article has been designed as follows: Section 3 describes the proposed methodology and it covers system identifications and simulation tools as well as the mathematical model of EV charging stations. In addition, Section 3 describes the development of merging traditional PID control with RL using the adaptive PID controller and Q-learning. In Section 4, the experimental setup and Kundur's model details have been explained. The analysis of case studies as well as the results have been highlighted in Section 5. Finally, Sections 6 and 7 cover the discussion and conclusion of this research including the main findings and future work.

3. Methodology

This section introduces the methodology of this research. It involves a Proportional-Integral-Derivative (PID) controller, reinforcement learning (RL), design-adaptive PID controller using Quality (Q)-learning, and experimental setup.

The increasing integration of renewable energy sources in today's modern power systems creates an urgent need to address the decreasing system inertia [1]. A potential solution to this problem is to combine RL techniques with the conventional Proportional-Integral-Derivative (PID) control as seen in Figure 3. Power system operators face a distinct set of issues when integrating renewable energy sources. Variations in power generation are caused by the intrinsic variability and uncertainty of renewable sources, which distinguishes them from conventional power plants. As a result, the overall inertia of the system tends to diminish, which is important for maintaining stability. It is crucial to use reinforcement learning approaches to improve the performance of current control mechanisms, like PID control, in order to overcome this deterioration.

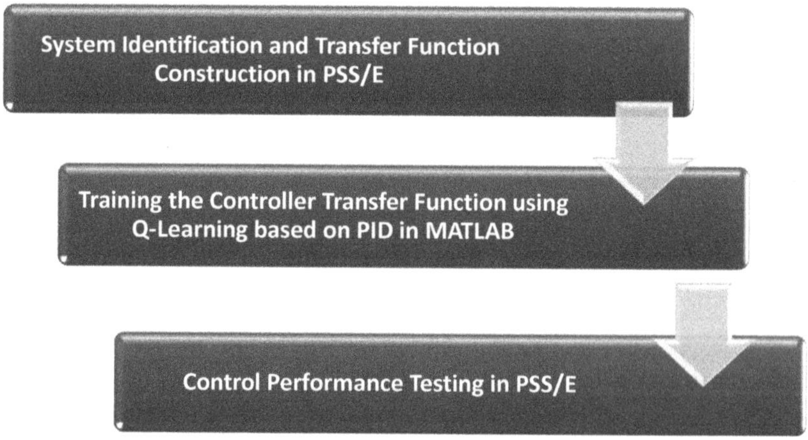

Figure 3. Stages of controller design.

Therefore, in order to preserve power system stability during the integration of renewable energy sources, creative control measures are required. Through the integration of PID control and reinforcement learning, the control system can be made more capable of responding to changing circumstances and improving overall system performance. The evaluation results show this approach's efficacy and highlight its potential for a range of power system applications. Subsequent investigations may concentrate on enhancing the control approach and investigating supplementary prospects for its integration into practical power systems.

The Power System Simulation for Engineering (PSS/E) is used to model the power system with great care from the beginning of the design process. Dynamic equations relevant to the control actuator (a governor or a power electronic converter) are generated within this framework. The transfer function is then trained using the model-free RL algorithm known as Q-learning. The first step in this process is to define the actions, the state space, and a reward function that are all carefully customized for the power system.

The final phase is to validate the developed controller in multiple PSS/E scenarios. After being fitted with PID parameters that are taken out of the Q-learning procedure, the controller is tested to determine how reliable and effective it is. The primary goal is to make sure that the controller improves the power system's general stability and reliability while simultaneously mitigating the difficulties brought on by significant renewable integration.

3.1. System Identification and Transfer Function Construction in PSS/E

A popular software program for power system analysis is called PSS/E. Techniques for system identification are used to collect information and examine how the system reacts to various inputs or disruptions. The transfer function of the system, which depicts the connection between its input and output, is then estimated using this data.

In order to construct a mathematical model, system identification involves creating a dynamic system representation using measurement data. Three crucial processes are included in the model development process, which is based on the following measurements [2]:

Stage 1: input–output selection: this refers to choosing the best actuation and observation signals.

Stage 2: system identification: building the functional model is the task of this step.

Stage 3: controller design: this entails creating a controller using the system that has been identified.

Initially, a set of differential equations called the subspace state-space model or difference equations are used to generate a transfer function model [3]. Then, using system identification techniques, a model is created based on real measurements [4].

The measurement-guided model structure makes use of the Output Error (OE) polynomial model. Within the category of polynomial models, this model is a particular configuration that is used, in this case, for model identification using probing measurements. Conventional transfer functions are described by means of OE models, in which the system is defined by the combined effects of the measured outputs and the inputs (represented by injected noise) [5].

Figure 3 shows the simulated response comparison between validation data (y_1) and model_OE 94.02 %. Amplitude per time can be obtained according to Figure 4.

The system structure can be described as follows, assuming that the injected noise signal has enough information to identify a system model with the chosen order of n as Equation (1) [7]:

$$y(k) = \frac{B(z,\theta)}{F(z,\theta)}u(k) + e(k) = \frac{b_1 z^{-1} + \cdots + b_n z^{z} \cdot n_b}{1 + f_1 z^{-1} + \cdots + f_n z^{-n}}u(k) + e(k) \tag{1}$$

The OE mathematical model is therefore represented by Equation (2):

$$\overline{y}(t) + \sum_{k=1}^{n} u_k \overline{y}(t-k) = \sum_{j=1}^{n} b_j u(t-j) + e(t) \tag{2}$$

where t denotes the sampled data index, e represents the injected noise process, u and $\overline{y}$ are the input and output of the model, a_k is the model coefficient denominator, and b_j is the model coefficient numerator.

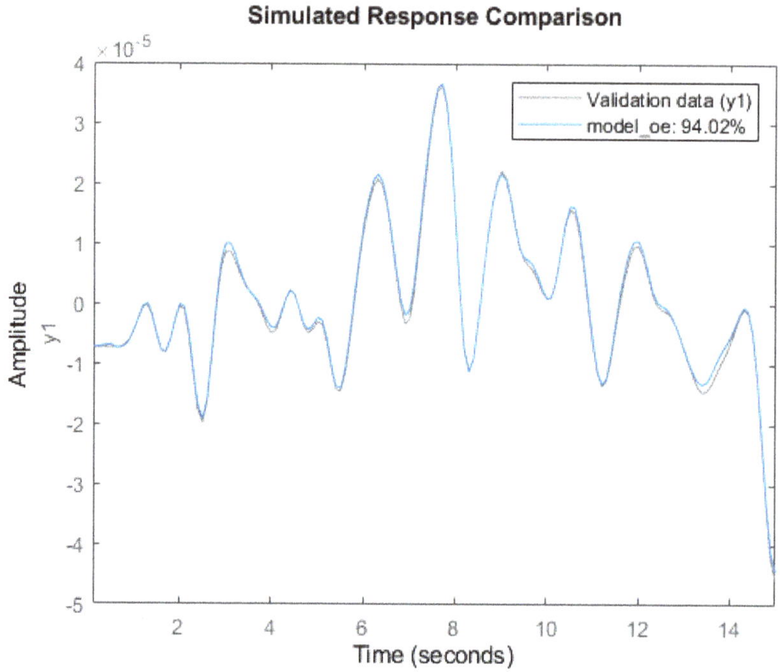

Figure 4. Transfer function validation.

For time-domain validation, the created model and the real model responses are compared. To achieve fit accuracy between the created system and the real model response, the original data that was used to build the model must be available. This is known as the actual system response. Equation (3) is used to express how accurate the system that was built is [8] and is as follows:

$$\text{Accuracy Index} = \left(1 - \frac{\| Y_i - \hat{Y}_i \|}{\| Y_i - \overline{Y}_i \|}\right) \times 100 \quad (3)$$

where Y_i and $\hat{Y}_i$ denote the constructed model and actual system responses, respectively, and $\hat{Y}_i$ is the mean of the actual system response over several periods. The index should approach 1, indicating that the constructed model response aptly represents the actual system.

3.2. Mathematical Computations and Analysis of Aggregated Electric Vehicles for Frequency Regulation

To use EVs in frequency regulation, multiple EVs must be connected to the power grid. An EV aggregator is an EV control center that manages the battery charge and discharge behavior of all EVs in the aggregator. The dynamic model of the *i*th EV in the EV aggregator is presented by the first-order transfer function as in Equation (4):

$$G_{EV} = \frac{K_{EV}}{ST_{EV} + 1} \quad (4)$$

where K_{EV} and T_{EV} are the gain and time constants of the *i*th EV battery system, respectively.

The transfer function of $e^{-s\rho_i}$ is used to describe the communication delay from an EV aggregator to the ith EV and the scheduling delay in the EV aggregator. ρ_i is the delay time taken to receive control signals from the EV aggregator. All EVs' delays ρ_i and time constants T_{EV} (i = 1, 2...) are considered to be equal on average, denoted by ρ_i and T_{EV}, respectively. This assumption yields an aggregated model of numerous EVs consisting of a single delay function and one EV dynamic.

The usage of an aggregation model of electric vehicles appears to be suitable since a cluster of several EVs and traditional generators are regulated jointly to modify their power injection to follow load disturbances.

To calculate the stability region and delay the margin, the characteristics equations of the single region LFC-EV system was determined as in Equation (5).

$$\Delta(s) = P\ (S) + Q\ (S)e^{-S\rho} \tag{5}$$

$\Delta(s)$ is the characteristic of the equation, while $P(S)$ and $Q(S)$ represent two polynomial with actual coefficients based on systems characteristics. The polynomial $P(S)$ and $Q(S)$ are given by Equation (6):

$$\begin{cases} Q(S) = Q_0 + SQ_1 + S^2Q_2 + S^3Q_3 + S^4Q_4 \\ P\ (S) = P_0 + SP_1 + S^2P_2 + S^3P_3 + S^4P_4 + S^3P_3 + S^4P_4 \end{cases} \tag{6}$$

3.3. Linear Proportional-Integral-Derivative (PID) Controller

The proportional, integral, and derivative components make up the three basic portions of the linear Proportional-Integral-Derivative (PID) controller, a commonly used control mechanism. Equation (7) [9] determines the PID controller's output in its discrete-time version.

$$u(k) = Kp\ *\ e(k) + Ki\ *\ \Sigma e(n) + Kd\ *\ (e(k) - e(k-1)) \tag{7}$$

The output of the controller at step k is represented by $u(k)$ in this equation, the error at step k is indicated by (k), and the proportional, integral, and derivative gains are denoted by Kp, Ki and Kd, respectively.

In proportion to the present mistake, the proportional term $Kp * e(k)$ adds to the controller's output. It assists in lowering steady-state mistakes and offers prompt remedial action in response to the error. The cumulative sum of previous errors is taken into consideration by the integral term, $Ki * \Sigma e(n)$. By gradually changing the controller's output, it guarantees that any persistent fault will finally be fixed. The system's ability to respond to persistent disturbances is enhanced and steady-state errors are reduced thanks to the integral term.

The rate of change of the error is taken into account in the derivative term, $Kd * (e(k) - e(k-1))$. It exerts a dampening effect on the system's response and predicts future changes in the error. The derivative term contributes to the reduction of overshoot and the improvement of controller stability. To attain the intended control performance, the PID controller can be tuned by adjusting the values of Kp, Ki, and Kd. Effective regulation and control of the system are made possible by the proportional, integral, and derivative gains, which establish the relative contributions of each component to the output of the controller

3.4. Reinforcement Learning (RL)

Through the use of reinforcement learning, a machine learning technique, an agent is trained to make choices and behave in a way that maximizes a cumulative reward. The way that both people and animals learn by making mistakes served as inspiration. Through interaction with the environment and feedback in the form of incentives or penalties for its activities, the agent in reinforcement learning learns to make the best decisions by experimenting with different tactics and taking advantage of the most lucrative ones. The agent iteratively refines its decision-making skills and creates a policy that maximizes its long-term payoff by mapping states to actions. The applications of reinforcement learning have been effectively implemented in a number of fields, such as autonomous systems, gaming, and robotics.

A kind of machine learning called reinforcement learning (RL) places a strong emphasis on solving problems repeatedly in order to achieve predetermined goals. It is essentially based on how creatures learn by adjusting their tactics or control laws in response to their surroundings, all without a prior understanding of the underlying system model [10].

Think of this as an agent interacting with its surroundings, as shown in Figure 5. Throughout this conversation [11]:

- The agent observes the current state, denoted as S_t from a set of possible states S, and chooses an action A_t from a set of possible actions A.
- Upon receiving the chosen action, the environment transitions to a new state S_{t+1}, also from S and provides a scalar reward R_{t+1} from a subset of real numbers R.
- In response, the agent receives this reward, finds itself in the new state S_{t+1}, and determines its next action A_{t+1} from A.

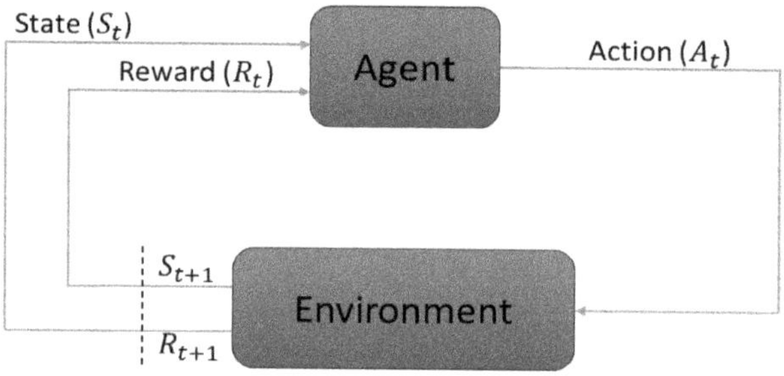

Figure 5. Reinforcement learning (Q-learning) process.

This cycle perpetuates until the agent reaches the terminal state S_t from S. The agent's primary objective is to formulate an optimal control policy that maximizes its cumulative discounted rewards over time, referred to as the expected discounted return G_T as shown in Equation (8) [10], where γ is the discounting factor, $0 < \gamma < 1$.

$$G_t \doteq R_{t+1} + \gamma R_{t+2} + \gamma^2 R_{t+3} + \ldots = \sum_{k=0}^{\infty} \gamma^k R_{t+k+1} \tag{8}$$

The fundamentals of Markov Decision Processes (MDPs) are closely related to the reinforcement learning (RL) problem [11]. A crucial feature of an MDP is its ability to guarantee that, regardless of an agent's past experiences in different states, the current state contains all the necessary information to make decisions. Generally speaking, RL algorithms fall into two main categories: policy-based and value-based [11]. Within the value-based paradigm of reinforcement learning, a unique value function is utilized to determine the value or importance of every state. Under policy π, the state-value function is expressed as follows in Equation (9) when evaluating from a certain state:

$$v_\pi(s) \doteq \mathbb{E}_\pi[G_t \mid S_t = s] = \mathbb{E}_\pi\left[\sum_{k=0}^{\infty} \gamma^k R_{t+k+1} \mid S_t = s\right] \tag{9}$$

where $\mathbb{E}_\pi$ denotes the expectation under policy π. Similarly, under policy π, the action-value function, denoted as q_π, which represents the value of taking action a in state s, can be described in Equation (10) [10]:

$$q_\pi(s,a) \doteq \mathbb{E}_\pi[G_t \mid S_t = s, A_t = a] = \mathbb{E}_\pi\left[\sum_{k=0}^{\infty} \gamma^k R_{t+k+1} \mid S_t = s, A_t = a\right] \tag{10}$$

Of all the action-value functions, the one that stands paramount is the optimal action-value function, articulated as Equation (11):

$$q_*(s,a) \doteq \max_{\pi} q_\pi(s,a), \tag{11}$$

The Q-learning algorithm is an off-policy, value-based learning methodology within reinforcement learning (RL). Within this framework, the action-value function, denoted as Q, aims to emulate the optimal action-value function, $\hat{q}^*$ by bypassing the adherence to the present policy. Its update mechanism is articulated as in Equation (12) [12,27–30]:

$$Q(S_t, A_t) \leftarrow Q(S_t, A_t) + \alpha\left[R_{t+1} + \gamma \max_a Q(S_{t+1}, a) - Q(S_t, A_t)\right] \tag{12}$$

where α denotes the learning rate.

3.5. *Design Adaptive PID Controller Using Q-Learning*

To create an adaptive PID controller that learns using Q-learning, the fundamental ideas of the traditional PID controller must be seamlessly integrated with the reinforcement learning paradigms that are inherent in Q-learning. The following is a structured framework for conceiving such an integrated system (see Figure 6):

Figure 6. The structured framework for conceiving an integrated system.

- State: This indicates the current position of the system. In control challenges, the state usually contains the instantaneous error, denoted as e(t), and its temporal derivative, denoted as (de(t))/dt.
- Action: This refers to changes made to the PID parameters. In real-world scenarios, actions could include increasing, decreasing, or maintaining the current PID gains.
- Reward: Known as scalar feedback, the reward offers an assessment of the agent's effectiveness. Here, possible metrics could be anything from the opposite of the absolute error to more complex assessment instruments that capture the system's operational intelligence.
- Q-table initialization: Start with a tabular framework in which states characterize rows and the actions that correspond define the columns. The internal values of this matrix,

known as Q-values, are then gradually improved based on the iterative feedback from the overall system.

The steps of an algorithm can be illustrated as follows:

- Action selection: using an exploration approach like ε-greedy, choose an action (change PID parameters) at each time step based on the current state.
- Take action: apply the PID controller to the system and adjust the PID settings.
- Observe reward: Calculate the reward and assess how the system responded. A PID controller's objective could be to arrive at a predetermined point with the least amount of oscillation and settling time. Either the negative absolute error or a more complex function, such as the integral of time-weighted absolute error (ITAE), can be used to calculate the reward.
- Update Q-values: adjust the state-action pair's Q-value by applying the Q-learning update rule.
- Loop: repeat this process for a specified number of episodes or until the Q-values converge.

The Q-values will direct the PID controller's actions once the Q-table has been sufficiently trained. The course of action with the highest Q-value is the best one in a particular state. Since Q-learning has historically operated on discrete sets of states and actions, both the states and the actions must be discretized. For more complicated situations, neural networks (deep Q-learning) are one continuous version and approximator that can be used. To make sure the system is stable and adaptive, it should be tested in a variety of circumstances after it has been taught. A framework for combining Q-learning with PID controllers was explained in this overview.

4. Experimental Setup

The system model used in this study is depicted in Figure 7. It consists of two interconnected areas: Area 1 (generation) and Area 2 (loading). In Area 1, the generation units include Generator G1 with a capacity of 700 MW, and Generator G2, also with a capacity of 700 MW. In Area 2, the generation units include Generator G3 with a capacity of 719 MW, and Generator G4 with a capacity of 700 MW. The total generation capacity of the system is 2819 MW.

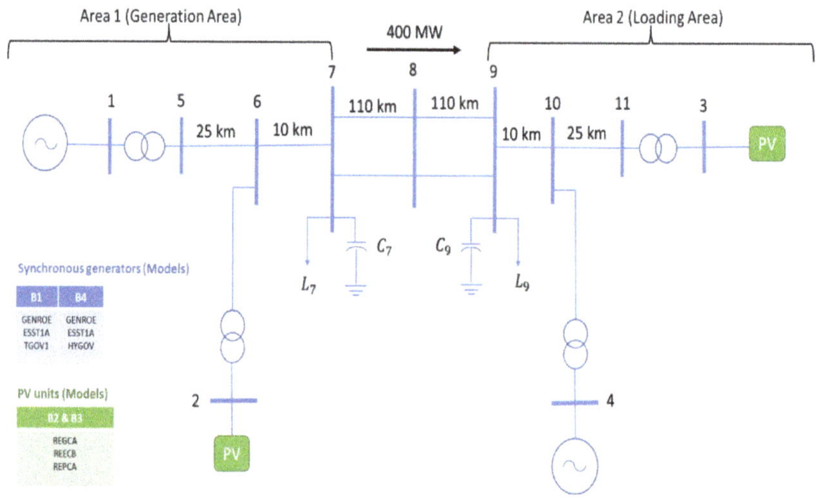

Figure 7. Kundur system model in PSSE.

The network is structured with several buses and transmission lines connecting the generators and loads. In Area 1, Bus 1 is connected to Generator G1 and Bus 2 to Generator

G2. In Area 2, Buses 3 and 4 are connected to Generators G3 and G4, respectively. The two areas are interconnected through a series of transmission lines with specific distances between buses: 25 km between Bus 1 and Bus 5, 10 km between Bus 5 and Bus 6, 110 km between Bus 6 and Bus 7, another 110 km between Bus 7 and Bus 8, and a further 110 km between Bus 8 and Bus 9. From Bus 9 to Bus 10, the distance is 10 km, followed by 25 km from Bus 10 to Bus 11.

In addition, the network includes reactive power elements at specific buses to manage voltage levels and stability. Capacitor C7 with a capacity of 200 MVAr and Load L7 with a capacity of 967 MW are present at Bus 7, while Capacitor C9 with a capacity of 350 MVAr, and Load L9 with a capacity of 1767 MW are at Bus 9. The total active power of the investigated system is 2734 M. There is a power flow of 400 MW from Area 1 to Area 2, indicating the amount of transfer of power through the interconnection.

This model represents a typical high-voltage transmission network, which is used to analyze the power flow, stability, and reliability of the interconnected power system. The configuration ensures a robust system that can manage significant loads and maintain stable operations across the interconnected areas.

Kundur's model is shown in Figure 7. In Kundur's paradigm, power normally moves from Area 1 to Area 2, with two machines (G and PV) where G2 and G3 are replaced by PV in each area. The PSSE program was used to run simulations in order to assess the efficacy of the methods presented in this study. The PSSE model used in the simulations was obtained from the University of Illinois at Urbana-Champaign's Illinois Center for a Smarter Electric Grid (ICSEG) [PSSE Models]. In order to assess the effectiveness of adopting PID control and examine the simulation findings, a comparison was carried out with a focus on four particular buses. These buses (Bus 1, Bus 3, Bus 7, and Bus 9) were selected based on an observability analysis which means two buses in the generation area and two buses in the loading area are required to observe the dynamic response of the system stability. In each area, two buses selected should observe the generation bus (Bus 1 and Bus 3) and the other two buses should be selected to observe the load buses (Bus 7 and Bus 9).

5. Results and Sensitivity Analysis

This section presents and highlights the results of studying the impact of renewable energy penetration on system stability, as well as the performance validation of the PID-RL control.

5.1. Impact of Renewable Energy Penetration on System Stability

The impact of high renewable energy penetration on system stability has been presented. Three scenarios have been studied and compared as follows: no photovoltaic (PV), 25% photovoltaic (PV)—700 MW, and 50% photovoltaic (PV)—1400 MW. The purpose of the study is to examine the performance of utilizing renewable energy penetration on the system stability of stations. The comparison of frequency difference responses for the three scenarios is shown in Figure 8. It refers to an electrical power system's capability to keep a steady frequency in the face of fluctuations in the ratio of generation to demand. Variations in renewable energy production can result in demand as well as supply imbalances that negatively influence system frequency. Therefore, it is crucial to investigate the frequency response in order to guarantee system stability in these scenarios with a significant penetration of renewable energy. Three scenarios—none, 25%, and 50% photovoltaic (PV)—have been evaluated in the study that is being presented during a disturbance of generation reduction of 50 MW. As seen in Figure 8, the findings suggest that, in contrast to the scenarios with 25% and 50% PV penetration, the frequency response is more stable in the absence of renewable energy. On the other hand, tie-line active power, which represents the active power flow between interconnected systems, has high active power oscillation with 25% and 50% PV penetration. Conversely, when it comes to the flow of active power between interconnected systems, the study compared three scenarios: no photovoltaic (PV), 25%

photovoltaic (PV), and 50% photovoltaic (PV). The findings revealed that the third scenario, with 50% PV penetration, exhibited lower stability in tie-line active power compared to the scenarios with no PV and 25% PV. This suggests that higher levels of PV penetration can contribute to improved stability in the flow of active power between interconnected systems as shown in Figure 9.

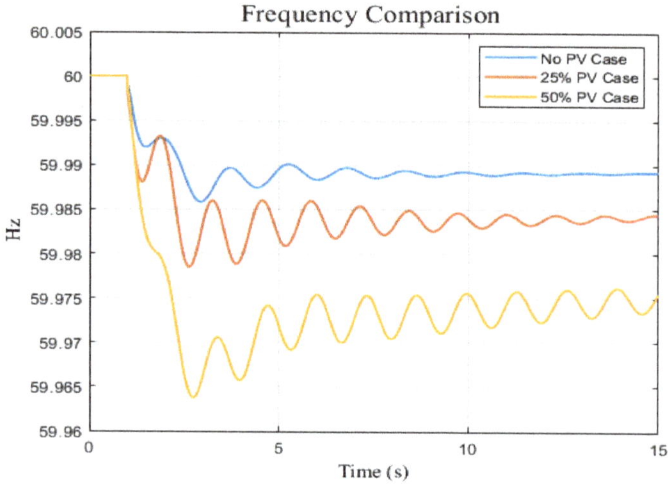

Figure 8. Kundur system model in PSSE frequency responses for three penetration scenarios: no PV, 25% of PV, and 50% of PV.

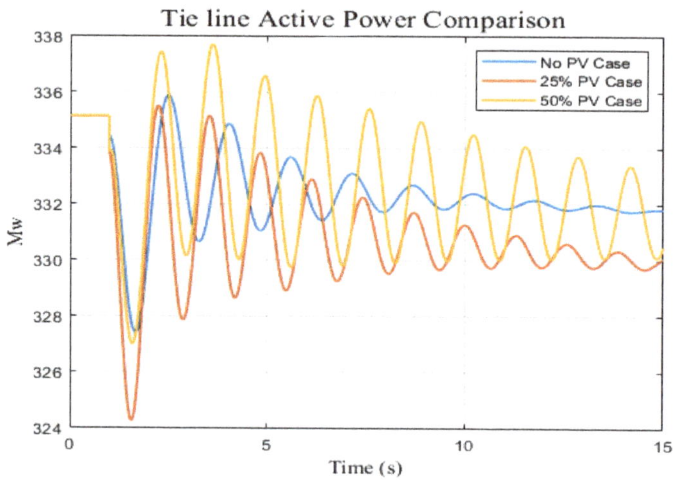

Figure 9. Tie-line active power response comparison for three penetration scenarios: no PV, 25% of PV, and 50% of PV.

Intriguing findings were obtained from the frequency differential response between the loading area and the generation area. Three different scenarios were investigated in the study: none, 25%, and 50% photovoltaic (PV). The results showed that in comparison to the scenario with no PV, the third scenario, with 50% PV penetration, showed less stability in frequency response between the loading area and the generation area. Additionally, it was observed that the 50% PV scenario exhibited lower stability than the 25% PV scenario.

These results suggest that higher levels of PV penetration cannot contribute to enhanced stability in maintaining frequency balance between the generations and loading areas without adding controllers as shown in Figure 10.

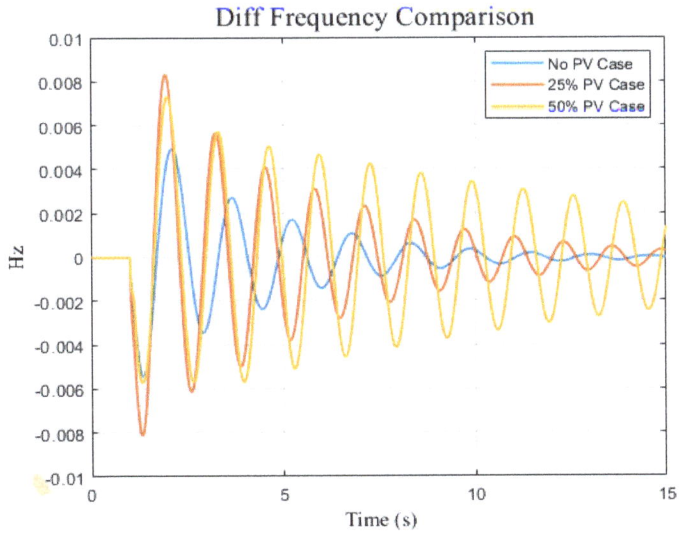

Figure 10. Frequency difference response between generation area and loading area for three penetration scenarios: no PV, 25% of PV, and 50% of PV.

5.2. PID-RL Control Performance Validation and Testing

Furthermore, the study conducted a performance validation of the PID-RL (Proportional-Integral-Derivative reinforcement learning) control technique, which combines traditional control theory with reinforcement learning algorithms. The validation process involved analyzing the performance of PID controllers and a base case scenario where no control input was applied to each EV station placement bus. This comparison allowed for an assessment of the effectiveness by utilizing a PID controller with a Particle Swarm Optimization (PSO) model for EV stations [19] and comparing it with a PID controller using RL and with no control input. By evaluating these different control approaches, the study aimed to determine their respective performances and identify the most suitable control strategy for regulating and stabilizing the power system in varying conditions.

The study conducted a comparative analysis of the stability of the electric vehicle (EV) system with 1% capacity (30 MW) under three control scenarios: no control input, PID-PSO [19], and PID-RL (reinforcement learning). The highest PV penetration level was selected (50%) and a disturbance of 50 MW generation reduction occurred in the first second. The results, as depicted in Figures 11–13, highlighted that adding the EV system with PID-RL exhibited superior performance in terms of frequency response, tie-line active power response, and frequency difference response between the generation area and loading area, respectively. These findings indicated that the PID controller with RL optimization significantly enhanced the damping of the dominant oscillation mode within the EV system, resulting in improved stability when compared to both the EV system with PID-PSO and the EV system without any control input.

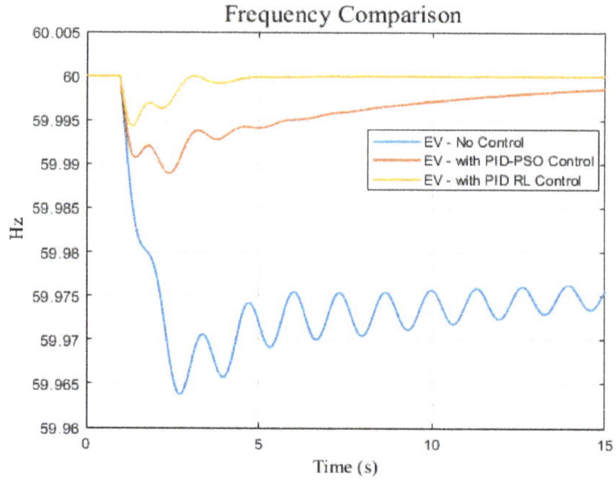

Figure 11. Frequency responses for three penetration scenarios for EV system with no control, PID-PSO control [19], and PID-RL control.

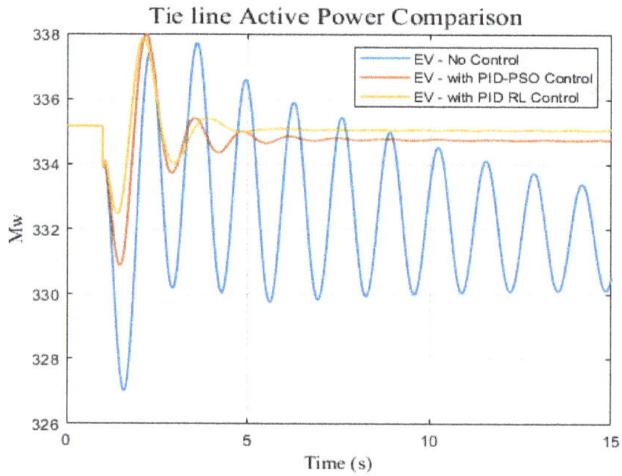

Figure 12. Tie-line active power response for three penetration scenarios for EV system with no control, PID-PSO control [19], and PID control.

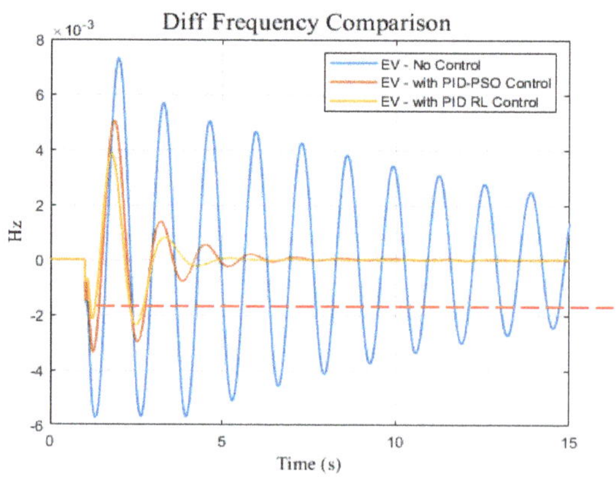

Figure 13. Frequency difference responses between generation area and loading area for three penetration scenarios for EV system with no control, PID-PSO control [19], and PID-RL control.

In order to study the sensitivity analysis of the results, two scenarios have been investigated. The first one is when the different magnitudes of generation reduction are considered—50 MW and 100 MW—whereas the PID-RL is used with an EV fast charging station of 1%. Figure 14 shows that the scenario of 100 MW generation reduction exceeds ±0.2 Hz operational limits for a primary response (10 s) while the statutory limit is ±0.5 Hz [31]. However, the integration of PID-RL enhances the frequency response for the same reduction and meets the operational frequency limits (Figure 14). The second sensitivity analysis scenario is applied to investigate different sizes of EV charging stations (0.75%, 1%, and 2%) regarding 21 MW, 30 MW, and 60 MW, respectively. The large size of the EV charging station as dynamic load positively enhances the frequency response can be seen in Figure 15 for the same PV penetration level and the same generation reduction disturbance.

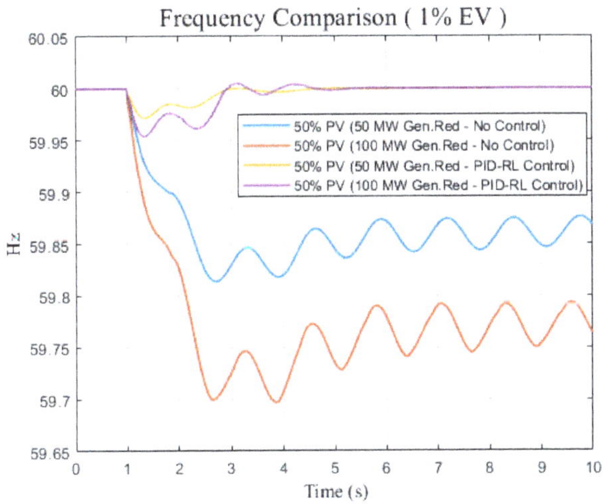

Figure 14. Frequency difference responses for different generation reduction disturbances for 50% PV and 1% EV systems with no control and PID-RL control.

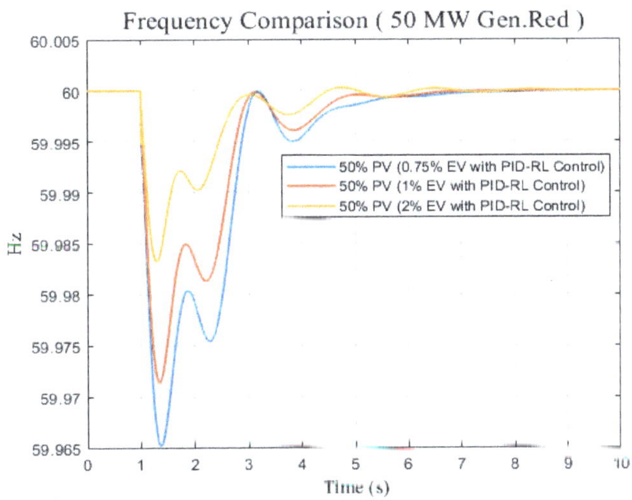

Figure 15. Frequency difference responses for different sizes of EV charging stations (0.75%, 1%, 2%) and a 50 MW generation reduction disturbance for 50% PV and PID-RL control.

6. Discussion

As discussed in the literature [17–24], using EVs to enhance system stability requires a fast response without any delay. As shown in Figures 11–13, the system restored its stability within 4 s after the disturbance of 50 MW generation reduction occurred when PID-RL was used. The allowable time to restore the system stability is within 60 s after a disturbance has occurred. Moreover, the highest power flow between the interconnected areas occurred when PID-RL was used, whereas the active power oscillation was the lowest when PID-RL was used.

Comparing our proposed PID-RL controller technique with the PID-PSO proposed in [19], Figure 11 shows that our proposed technique has a faster frequency response compared to PID-PSO. Therefore, when an EV fast charging station is modeled as a dynamic load for frequency control, the PID-RL is the best controller option in terms of a fast frequency response.

According to [32], a comparison was conducted between the Genetic Algorithm (GA)—PID, Grey Wolf Optimizer (GWO)—PID, PSO-PID, and JAYA-PID in terms of the following: Peak Undershoot frequency (ΔHz), Settling Time (s) for both Area 1 and Area 2 as well as Peak Undershoot for power deviation in tie-line (MW) and Settling Time (s). To validate the fast response of our proposed PID-RL, a similar comparison to [32] is prepared in Table 1.

Table 1. A comparison of the dynamic response of the PID-RL and various controllers of optimized PID parameters with different approaches [32].

Methods	Performance Indices					
	Area 1		Area 2		Tie-Line Power Deviation	
	Peak Undershoot (ΔHz)	Settling Time (s)	Peak Undershoot (ΔHz)	Settling Time (s)	Peak Undershoot (ΔMW)	Settling Time (s)
GA-PID [32]	0.059	44	0.105	52	55	45
PSO-PID [19]	0.031	24	0.013	21.5	9	19.5
GWO-PID [32]	0.024	21	0.011	18	7	17
JAYA-PID [32]	0.02	18	0.0085	15	4	11
PID-RL	0.005	4	0.004	4	3	5

The comparison shows how fast the PID-RL controller is compared to others thanks to the training stage of the controller using Q-learning. All optimized parameters of PID controller techniques—in the literature—have taken certain optimum values based on the stability problem; however, the power of reinforcement learning is that PID parameters are dynamic and change their values based on the best rewards of Q-learning as explained in Section 3. This fast response makes the proposed PID-RL suitable for a dynamic EV fast charging for load control with high performance, and it overcomes the problem of the delay discussed in the literature [17–24].

7. Conclusions

In summary, this research looked at two crucial areas: how the use of renewable energy affects system stability and how well the PID-RL control method performs. Three scenarios were assessed regarding the effects of the penetration of renewable energy: 25% PV, 50% PV, and no PV. The results showed that a higher degree of PV penetration (50%) contributed to the ability to improve the stability in tie-line active power flow between interconnected systems, even while the absence of renewable energy produced a more stable frequency response. This implies that higher PV penetration can improve stability in preserving active power flow and frequency balance. The study investigated three control scenarios: no control input, PID-PSO [19], and proposed PID-RL before moving on to the PID-RL control performance validation. The outcomes showed that in terms of frequency response, tie-line active power response, and frequency difference response, the EV system with PID-RL performed better than the other scenarios. In comparison to the EV system with PID-PSO and the EV system without control input, the PID controller with RL optimization greatly increased the damping of the dominant oscillation mode, resulting in increased stability. Overall, this study confirms the efficacy of the PID-RL control technique in stabilizing the EV system and offers insightful information about how the penetration of renewable energy affects system stability. These results can aid in the formulation of plans for incorporating renewable energy sources and improving control mechanisms in order to guarantee a more reliable and sustainable electrical grid. This research can be extended to different renewable energy resources such as wind turbines, and different electrical grids and models.

Author Contributions: Conceptualization, Y.A. and I.A.; methodology, Y.A. and I.A.; software, I.A.; validation, Y.A. and I.A.; investigation, Y.A. and I.A.; resources, I.A.; data curation, I.A.; writing—original draft preparation, Y.A.; writing—review and editing, Y.A. and I.A.; visualization, I.A.; supervision, Y.A.; project administration, Y.A. All authors have read and agreed to the published version of the manuscript.

Funding: This research received no external funding.

Data Availability Statement: Data are contained within the article.

Conflicts of Interest: The authors declare no conflict of interest.

References

1. Kroposki, B.; Johnson, B.; Zhang, Y.; Gevorgian, V.; Denholm, P.; Hodge, B.M.; Hannegan, B. Achieving a 100% renewable grid: Operating electric power systems with extremely high levels of variable renewable energy. *IEEE Power Energy Mag.* **2017**, *15*, 61–73. [CrossRef]
2. Liu, H.; Zhu, L.; Pan, Z.; Guo, J.; Chai, J.; Yu, W.; Liu, Y. Comparison of mimo system identification methods for electromechanical oscillation damping estimation. In Proceedings of the 2016 IEEE Power and Energy Society General Meeting (PESGM), Boston, MA, USA, 17–21 July 2016; pp. 1–5.
3. Ljung, L. System identification toolbox. In *The Matlab User's Guide*; MathWorks Incorporated: Natick, MA, USA, 2011.
4. Nan, J.; Yao, W.; Wen, J.; Peng, Y.; Fang, J.; Ai, X.; Wen, J. Wide-area power oscillation damper for DFIG-based wind farm with communication delay and packet dropout compensation. *Int. J. Electr. Power Energy Syst.* **2021**, *124*, 106306. [CrossRef]
5. Ogata, K.; Yang, Y. *Modern Control Engineering*; Prentice Hall: Indianapolis, IN, USA, 2002; Volume 5.
6. Kumar, M.; Panda, K.P.; Naayagi, R.T.; Thakur, R.; Panda, G. Comprehensive Review of Electric Vehicle Technology and Its Impacts: Detailed Investigation of Charging Infrastructure, Power Management, and Control Techniques. *Appl. Sci.* **2023**, *13*, 8919. [CrossRef]
7. Xu, Z.; Cheng, C.; Li, Y. Static source error correction model based on MATLAB and simulink. In Proceedings of the IEEE 2019 Prognostics and System Health Management Conference (PHM-Qingdao), Qingdao, China, 25–27 October 2019; pp. 1–5.
8. Eriksson, R.; Söder, L. Wide-area measurement system-based subspace identification for obtaining linear models to centrally coordinate controllable devices. *IEEE Trans. Power Deliv.* **2011**, *26*, 988–997. [CrossRef]
9. Kumawat, A.K.; Kumawat, R.; Rawat, M.; Rout, R. Real time position control of electrohydraulic system using PID controller. *Mater. Today Proc.* **2021**, *47*, 2966–2969. [CrossRef]
10. Pongfai, J.; Su, X.; Zhang, H.; Assawinchaichote, W. PID controller autotuning design by a deterministic Q-SLP algorithm. *IEEE Access* **2020**, *8*, 50010–50021. [CrossRef]
11. Naeem, M.; Rizvi, S.T.H.; Coronato, A. A gentle introduction to reinforcement learning and its application in different fields. *IEEE Access* **2020**, *8*, 209320–209344. [CrossRef]
12. Yao, Y.; Ma, N.; Wang, C.; Wu, Z.; Xu, C.; Zhang, J. Research and implementation of variable-domain fuzzy PID intelligent control method based on Q-Learning for self-driving in complex scenarios. *Math. Biosci. Eng.* **2023**, *20*, 6016–6029. [CrossRef] [PubMed]
13. Ghosh, A. Possibilities and challenges for the inclusion of the electric vehicle (EV) to reduce the carbon footprint in the transport sector: A review. *Energies* **2020**, *13*, 2602. [CrossRef]
14. Khalid, J.; Ramli, M.A.; Khan, M.S.; Hidayat, T. Efficient load frequency control of renewable integrated power system: A twin delayed DDPG-based deep reinforcement learning approach. *IEEE Access* **2022**, *10*, 51561–51574. [CrossRef]
15. Dong, C.; Sun, J.; Wu, F.; Jia, H. Probability-based energy reinforced management of electric vehicle aggregation in the electrical grid frequency regulation. *IEEE Access* **2020**, *8*, 110598–110610. [CrossRef]
16. Xu, P.; Zhang, J.; Gao, T.; Chen, S.; Wang, X.; Jiang, H.; Gao, W. Real-time fast charging station recommendation for electric vehicles in coupled power-transportation networks: A graph reinforcement learning method. *Int. J. Electr. Power Energy Syst.* **2022**, *141*, 108030. [CrossRef]

17. Hussain, A.; Bui, V.H.; Kim, H.M. Deep reinforcement learning-based operation of fast charging stations coupled with energy storage system. *Electr. Power Syst. Res.* **2022**, *210*, 108087. [CrossRef]
18. Amir, M.; Zaheeruddin Haque, A.; Kurukuru, V.B.; Bakhsh, F.I.; Ahmad, A. Agent based online learning approach for power flow control of electric vehicle fast charging station integrated with smart microgrid. *IET Renew. Power Gener.* 2022, *Early View.*
19. Albert, J.R.; Selvan, P.; Sivakumar, P.; Rajalakshmi, R. An advanced electrical vehicle charging station using adaptive hybrid particle swarm optimization intended for renewable energy system for simultaneous distributions. *J. Intell. Fuzzy Syst.* **2022**, *43*, 4395–4407. [CrossRef]
20. Wu, Y.; Wang, Z.; Huangfu, Y.; Ravey, A.; Chrenko, D.; Gao, F. Hierarchical operation of electric vehicle charging station in smart grid integration applications—An overview. *Int. J. Electr. Power Energy Syst.* **2022**, *139*, 108005. [CrossRef]
21. Boopathi, D.; Jagatheesan, K.; Anand, B.; Samanta, S.; Dey, N. Frequency regulation of interlinked microgrid system using mayfly algorithm-based PID controller. *Sustainability* **2023**, *15*, 8829. [CrossRef]
22. Wang, Y.; Chai, B.; Lu, W.; Zheng, X. A review of deep reinforcement learning applications in power system parameter estimation. In Proceedings of the IEEE 2021 International Conference on Power System Technology (POWERCON), Haikou, China, 8–9 December 2021; pp. 2015–2021.
23. Liang, X.; Chai, H.; Ravishankar, J. Analytical methods of voltage stability in renewable dominated power systems: A review. *Electricity* **2022**, *3*, 75–107. [CrossRef]
24. Teigenes, M. Provision of Primary Frequency Control from Electric Vehicles in the Nordic Power System. Master's Thesis, NTNU, Trondheim, Norway, 2023.
25. Xie, J.; Sun, W. A transfer and deep learning-based method for online frequency stability assessment and control. *IEEE Access* **2021**, *9*, 75712–75721. [CrossRef]
26. Meridji, T.; Restrepo, J. Machine Learning-Based Platform for the Identification of Critical Generators-Context of High Renewable Integration. Available online: https://papers.ssrn.com/sol3/papers.cfm?abstract_id=4728247 (accessed on 24 April 2024).
27. Cao, J.; Zhang, M.; Li, Y. A review of data-driven short-term voltage stability assessment of power systems: Concept, principle, and challenges. *Math. Probl. Eng.* **2021**, *2021*, 1–12. [CrossRef]
28. Tejeswini, M.V.; Raglend, I.J. Modelling and sizing techniques to mitigate the impacts of wind fluctuations on power networks: A review. *Int. J. Ambient Energy* **2022**, *43*, 3600–3616. [CrossRef]
29. Nguyen-Hoang, N.D.; Shin, W.; Lee, C.; Chung, I.Y.; Kim, D.; Hwang, Y.H.; Youn, J.; Maeng, J.; Yoon, M.; Hur, K.; et al. Operation Method of Energy Storage System Replacing Governor for Frequency Regulation of Synchronous Generator without Reserve. *Energies* **2022**, *15*, 798. [CrossRef]
30. Schneider, K.P.; Sun, X.; Tuffner, F.K. Adaptive Load Shedding as Part of Primary Frequency Response To Support Networked Microgrid Operations. *IEEE Trans. Power Syst.* **2023**, *39*, 287–298. [CrossRef]
31. Nedd, M.; Browell, J.; Bell, K.; Booth, C. Containing a Credible Loss to Within Frequency Stability Limits in a Low-Inertia GB Power System. *IEEE Trans. Ind. Appl.* **2020**, *56*, 1031–1039. [CrossRef]
32. Annamraju, A.; Nandiraju, S. Coordinated control of conventional power sources and PHEVs using jaya algorithm optimized PID controller for frequency control of a renewable penetrated power system. *Prot. Control Mod. Power Syst.* **2019**, *4*, 1–13. [CrossRef]

MDPI AG
Grosspeteranlage 5
4052 Basel
Switzerland
Tel.: +41 61 683 77 34

Energies Editorial Office
E-mail: energies@mdpi.com
www.mdpi.com/journal/energies

www.ingramcontent.com/pod-product-compliance
Lightning Source LLC
Chambersburg PA
CBHW040749250926
51850CB00007BB/1444
9783725823260